HOW HIGH IS YOUR *TQ*?
(Tube Quotient)

Try answering these tantalizing teasers
on TV's all-time all-stars and see how you score.

• Who played Beaver in LEAVE IT TO BEAVER?
• Name the four Monkees.
• Who was the star of RAMAR OF THE JUNGLE?
• From what city did AMERICAN BANDSTAND originate?
• What was the name of the TV movie that
served as the pilot of KOJAK?
• What was Dinah Shore's theme song?
…Who loved Lucille Ball?
…Was James Bond ever a TV series?
…Who used the slogan "Have Gun, Will Travel"?
…What was the original name
of the ED SULLIVAN SHOW?
…Where is the home planet of Mr. Spock?

These are just a few of the
fun-crammed questions and amazing facts
you'll find when you play TV's top quiz game.

**THE TV GUIDE
QUIZ BOOK**

Plus: 192 original program pages.

THE

TV GUIDE

QUIZ BOOK

**Edited by Stan Goldstein
and Fred Goldstein**

THE TV GUIDE QUIZ BOOK
A Bantam Book/May 1978

ISBN 0-553-11811-0

Published simultaneously in the United States and Canada

PRINTED IN THE UNITED STATES OF AMERICA

0 9 8 7 6 5 4 3 2 1

INTRODUCTION

Almost everyone knows Mary Tyler Moore, but who remembers Mona McCluskey? Or Camp Runamuck? Or the Chicago Teddy Bears? Not many people. That's what makes the television/nostalgia game fun, and this book makes it easy for everyone to play. *The TV Guide Quiz Book* is divided into three parts. In Part One the questions are asked, and in Part Two they are answered. No doubt you'll find many of the questions easy but others are harder, and some of them are tough enough to be a trivia buff's delight. And the range of the questions is as wide as television itself, covering the medium's most memorable moments and those that most everyone would just as soon forget.

Part Three places the programs in their original settings, as they were originally printed in *TV Guide's* program pages from 1953 through 1974. This section can be read merely for reference, or it can be read for the nostalgic pleasure of seeing what America has been watching since 25 years ago. These pages will nudge your memory, helping you to recall where you were, what you were doing, and whom you were doing it with when these shows were on.

In all, *The TV Guide Quiz Book* contains nearly a quarter century of television history. Sitting down with this book is like having all these years of TV flash before your eyes—with one big difference. This time around, you won't see any commercials.

CONTENTS

Part One
QUESTIONS

COMEDY

QUESTIONS

COMEDY

		An-swer Page	Pro-gram Page
31.	For what company did Ralph Kramden work?	88	—
32.	What did Ed Norton do for a living?	88	—
33.	Who played Ed Norton on THE HONEYMOONERS?	88	—
34.	To what fraternal organization did Ralph Kramden belong?	88	—
35.	Who was the star of the early-Fifties comedy BONINO?	88	123
36.	Barry Nelson and Joan Caulfield played man and wife in what early-Fifties situation comedy?	88	—
37.	WHERE'S RAYMOND? was the name of a weekly situation comedy that starred what song-and-dance man?	88	124
38.	Who starred in the role of Cosmo Topper?	88	126
39.	What was the name of Cosmo Topper's wife?	88	—
40.	Name the husband-and-wife ghosts on TOPPER. Who played them?	88	126
41.	What was the name of the St. Bernard on TOPPER? What did he drink?	88	—
42.	Name the president of the bank where Cosmo Topper worked.	88	—
43.	On the TOPPER series, what was the name of Topper's maid?	88	—
44.	What was the full name of the character played by William Bendix on THE LIFE OF RILEY?	88	—
45.	What was the name of Riley's next-door neighbors? Who played them?	89	159
46.	What was the name of Riley's daughter? Who played her?	89	—
47.	Name Riley's wife. Who played her?	89	159

4

QUESTIONS

COMEDY

QUESTIONS

QUESTIONS

An-
swer
Page

Pro-
gram
Page

117. What was the name of the Captain of the ship? Who played him?   90 167

118. What was the name of Susanna's best friend on OH! SUSANNA? Who played her?   90 167

119. On OH! SUSANNA, what was the name of the English steward? Who played him?   90 167

120. On THOSE WHITING GIRLS, who played the part of Penny?   90 168

121. Name the character played by Eve Arden on THE EVE ARDEN SHOW.   90 172

122. What did she do for a living?   90 172

123. What were the names of Eve Arden's two daughters on THE EVE ARDEN SHOW?   90 —

124. Who said "What a revoltin' development this is"?   90 —

125. On THE REAL McCOYS, what was Grampa's first name?   90 —

126. Who played Grampa McCoy on THE REAL McCOYS?   90 175

127. Who played Luke on THE REAL McCOYS?   90 175

128. What was the name of Luke's wife? Who played her?   90 175

129. What was the name of the Mexican farm hand on THE REAL McCOYS? Who played him?   90 175

130. What was the name of Grampa McCoy's best friend?   90 175

131. What state did the McCoys come from?   90 —

132. Who played Hassie on THE REAL McCOYS?   90 175

COMEDY

QUESTIONS

QUESTIONS

COMEDY

		An- swer Page	Pro- gram Page
212.	In CAR 54, WHERE ARE YOU? what were the first names of Toody and Muldoon? Who played them?	92	201
213.	Name Toody's wife. Who played her?	92	201
214.	What was the name of the Captain on CAR 54, WHERE ARE YOU?	92	201
215.	Who was the creator of this series? What other comedy series did he create?	92	201
216.	Who played Patrolman Schnauzer on CAR 54, WHERE ARE YOU?	92	—
217.	What black comedian played Officer Anderson on this series?	92	201
218.	What famous head of state was mentioned in the theme song of CAR 54, WHERE ARE YOU?	92	—
219.	What did Dick Van Dyke do for a living on THE DICK VAN DYKE SHOW? What character did he play?	92	205
220.	What was the name of Dick's wife? Who played her?	92	205
221.	What was the name of their son? Who played him?	92	205
222.	What was Buddy's last name on THE DICK VAN DYKE SHOW? Who played him?	92	205
223.	Who played Sally?	92	205
224.	Name the producer and chief hatchet man on The Alan Brady Show. Who played him?	92	205
225.	What famous comedian produced THE DICK VAN DYKE SHOW?	92	205
226.	Name the next-door neighbors on THE DICK VAN DYKE SHOW. Who played them?	92	—

QUESTIONS

17

COMEDY

QUESTIONS

COMEDY

QUESTIONS

		Answer Page	Program Page
303.	On GOMER PYLE, USMC, what was the name of Gomer's drill sergeant? Who played him?	94	234
304.	Who played Gomer?	94	234
305.	Who played Duke on GOMER PYLE, USMC?	94	234
306.	Who played Gilligan on GILLIGAN'S ISLAND?	94	236
307.	Who played the Skipper?	94	236
308.	Name the shipwrecked actress on GILLIGAN'S ISLAND. Who played her?	94	236
309.	What was Mr. Howell's first name on GILLIGAN'S ISLAND? Who played him?	94	236
310.	What was the name of the professor on GILLIGAN'S ISLAND?	94	236
311.	Who was the star of I DREAM OF JEANNIE?	94	245
312.	What was the name of Jeannie's master? Who played him?	94	245
313.	What was Tony's highest rank on I DREAM OF JEANNIE?	94	—
314.	What was the name of the base psychiatrist on I DREAM OF JEANNIE? Who played him?	94	245
315.	Who was Tony's best friend on I DREAM OF JEANNIE? Who played him?	94	240
316.	Who was the star of GET SMART?	94	240
317.	In GET SMART, what was Smart's first name?	94	240
318.	What was the name of the organization for which Smart worked?	94	240

COMEDY

QUESTIONS

COMEDY

QUESTIONS

25

	An- swer Page	Pro- gram Page

385. What character was eventually adopted and made part of the family on MY THREE SONS? Who played him? — 95 252

386. What were Dave and Larry's names on GOOD MORNING WORLD? Who played them? — 95 258

387. What did they do for a living? — 95 258

388. What was the name of Dave's wife on GOOD MORNING WORLD? Who played her? — 95 258

389. Name Dave and Larry's boss on GOOD MORNING WORLD. Who played him? — 95 258

390. What comedy series starred Paula Prentiss and Richard Benjamin? Name the characters they played. — 95 259

391. Name the character known as THE FLYING NUN. Who played her? — 96 260

392. Who played Carlos on THE FLYING NUN? — 96 260

393. Eve Arden and Kaye Ballard starred in what late-Sixties comedy series? — 96 —

394. Who was the star of JULIA? — 96 266

395. What was Julia's occupation? Where did she work? — 96 266

396. Name the doctor for whom Julia worked. Who played him? — 96 266

397. What was Julia's last name? — 96 —

398. What was the name of Julia's son? — 96 266

399. Who played Oscar on the series HE AND SHE? — 96 —

400. Name the character played by Doris Day in her series debut. — 96 262

	An-swer Page	Pro-gram Page
401. Who played Myrna on THE DORIS DAY SHOW?	96	—
402. Name the sheepdog on THE DORIS DAY SHOW	96	262
403. What were the names of Doris's two sons on THE DORIS DAY SHOW?	96	262
404. Who played Nicholson on THE DORIS DAY SHOW?	96	—
405. What weekly comedy series starred Robert Morse?	96	263
406. Who played Gloria on THAT'S LIFE?	96	263
407. In what comedy series did Lucille Ball star after I LOVE LUCY and THE LUCY SHOW? What character did she play?	96	265
408. What was the name of Lucy's brother-in-law? Who played him?	96	265
409. Name Lucy's two children. Who played them?	96	—
410. Who played Blondie on the late-Sixties revival of the series of the same name?	96	264
411. Who played Dagwood?	96	264
412. Name the actor who played Dithers on BLONDIE.	96	264
413. THE UGLIEST GIRL IN TOWN was set in what city?	96	—
414. What was the name of the character known as THE UGLIEST GIRL IN TOWN? What was her secret?	96	264
415. Who played THE UGLIEST GIRL IN TOWN?	96	264
416. There have been two comedy series titled DOC. Name the doctor on the first one. Who played him?	96	272

		An- swer Page	Pro- gram Page
417.	Who played Sam on MAYBERRY R.F.D.?	96	278
418.	What was the name of Sam's son on MAYBERRY R.F.D.?	96	278
419.	Who was the star of THE GOVERNOR AND J.J.?	96	279
420.	What was the name of the Governor on THE GOVERNOR AND J.J.? Who was J.J.?	96	—
421.	What now famous TV star played Orrin Hacker on the first episode of THE GOVERNOR AND J.J.?	96	—
422.	Name the father on THE COURTSHIP OF EDDIE'S FATHER. Who played him?	96	280
423.	Who played Eddie on THE COURT-SHIP OF EDDIE'S FATHER?	96	280
424.	Name the housekeeper on THE COURTSHIP OF EDDIE'S FATHER.	96	280
425.	Name the main characters on THE GOOD GUYS. Who played them?	96	276
426.	What kind of business did THE GOOD GUYS run?	96	276
427.	Who played Mike and Carol Brady on THE BRADY BUNCH?	96	276
428.	Name the housekeeper on THE BRADY BUNCH. Who played her?	96	276
429.	How many sons did Mike have on THE BRADY BUNCH before he married Carol? How many daughters did Carol have?	96	—
430.	Name Mike's sons on THE BRADY BUNCH.	96	276
431.	Name Carol's daughters.	96	276

QUESTIONS

QUESTIONS

COMEDY

DRAMA/ ADVENTURE

		An- swer Page	Pro- gram Page
15.	What was the name of Lassie's first television family?	98	—
16.	Who was the star of WATERFRONT? What character did he play?	98	139
17.	On what night of the week did CBS broadcast ALFRED HITCHCOCK PRESENTS?	98	146
18.	What original television drama was later made into a movie that won an Academy Award for best picture? Who starred in the television version?	98	—
19.	Who was the star of THE ADVEN- TURES OF ROBIN HOOD?	98	149
20.	What actor played Prince John on THE ADVENTURES OF ROBIN HOOD?	98	—
21.	Name the anonymous philanthropist on THE MILLIONAIRE.	98	—
22.	What was the name of the emissary on THE MILLIONAIRE? Who played him?	98	148
23.	Who was the main character of the dramatic series THE HUNTER? Who played him?	98	151
24.	Name the two lieutenants who were featured on THE 77th BENGAL LANCERS. Who played them?	98	158
25.	What was the name of the fort on THE 77th BENGAL LANCERS? Name the commandant.	98	—
26.	Who played CAPTAIN GALLANT OF THE FOREIGN LEGION?	98	—
27.	What was the name of Captain Gallant's son? Who played him?	98	—

QUESTIONS

QUESTIONS

		An- swer Page	Pro- gram Page
	character on ADVENTURES IN PARADISE? Who played him?	99	195
61.	Name the schooner on ADVENTURES IN PARADISE.	99	195
62.	Who played Clay on ADVENTURES IN PARADISE?	99	195
63.	What were the names of the two wanderers on ROUTE 66? Who played them?	99	197
64.	What make car did they drive?	99	—
65.	Who was the creator and host of THE TWILIGHT ZONE?	99	200
66.	Who starred as Jungle Jim in the series of that name?	99	—
67.	Name the host of G.E. THEATER.	99	201
68.	Russ Andrews and Steve Banks were insurance investigators on what show? Who played them?	99	206
69.	Who was the host of the mystery series THRILLER?	99	204
70.	Who was the star of WINDOW ON MAIN STREET? What character did he play?	99	203
71.	Who was the star of RESCUE 8?	99	212
72.	Who played Larry and Mike on THE AQUANAUTS?	99	212
73.	Who played Fathers O'Malley and Fitzgibbons in the TV series GOING MY WAY?	99	213
74.	Name the director of the community center on GOING MY WAY. Who played him?	99	213
75.	Who played Nick Alexander on SAINTS AND SINNERS?	99	209

DRAMA/ADVENTURE

	Answer Page	Program Page

92. What was the name of the high school where Mr. Novak taught? — 100 — —

93. Name the principal on MR. NOVAK. Who played him? — 100 — 221

94. Can you complete the titles of the following soap operas? — 100 — 142

   A. THE INNER _____
   B. THE ROAD OF _____
   C. THE GUIDING _____
   D. THE GREATEST _____
   E. _____WINDOWS
   F. ONE MAN'S _____
   G. THE BRIGHTER _____
   H. THE _____STORM
   I. _____ LOVE

95. Frank Sinatra, Eva Marie Saint, and Paul Newman starred in a PRODUCERS' SHOWCASE presentation of what famous play? — 100 — 145

96. What was the name of the agent on THE MAN FROM U.N.C.L.E.? Who played him? — 100 — 232

97. What did the acronym U.N.C.L.E. stand for on THE MAN FROM U.N.C.L.E.? — 100 — 232

98. Who was the head of U.N.C.L.E.? Who played him? — 100 — 232

99. Name the enemy organization on THE MAN FROM U.N.C.L.E. — 100 — 232

100. What was the name of Solo's assistant? Who played him? — 100 — 232

101. What was used as a front for U.N.C.L.E. headquarters? — 100 — —

102. On what two nights of the week was PEYTON PLACE seen on the ABC network? — 100 — —

	An- swer Page	Pro- gram Page
103. Who played Allison MacKenzie on PEYTON PLACE?	100	235
104. What was the name of Allison's mother?	100	235
105. Who played Rodney Harrington on PEYTON PLACE?	100	235
106. Michael was the first name of what PEYTON PLACE M.D.?	100	235
107. What was the last name of PEYTON PLACE's George and Julie?	100	235
108. Who was Allison's rival for Rodney's affection on PEYTON PLACE? Who played her?	100	235
109. What was the name of the submarine on VOYAGE TO THE BOTTOM OF THE SEA?	100	229
110. Name the sub's commander. Who played him?	100	229
111. On VOYAGE TO THE BOTTOM OF THE SEA, what was Admiral Nelson's first name. Who played him?	100	229
112. What was Admiral Nelson the director of?	100	229
113. Who was the sinister adversary on VOYAGE TO THE BOTTOM OF THE SEA?	100	229
114. Who played Dr. Fred Wilson on VOYAGE TO THE BOTTOM OF THE SEA?	100	229
115. What was the name of the bomber group commander on 12 O'CLOCK HIGH?	100	234
116. On 12 O'CLOCK HIGH, what was Wiley Crowe's rank? Who played him?	100	234

	An- swer Page	Pro- gram Page
117. In what series did Dennis Weaver star after leaving GUNSMOKE?	100	236
118. Who were the three stars of THE ROGUES?	100	—
119. What was the name of the character who replaced Jeff Miller on LASSIE? Who played him?	100	177
120. Who played Uncle Petrie on LASSIE?	100	177
121. What was the name of the original undercover agent who headed the strike force on MISSION:IMPOSSIBLE? Who played him?	100	248
122. What was the name of the original master of disguise on MISSION: IMPOSSIBLE? Who played him?	100	248
123. On MISSION:IMPOSSIBLE, what was Barney's specialty?	100	248
124. What was the name of the weight-lifter on MISSION:IMPOSSIBLE? Who played him?	100	248
125. Who was the original female member of the MISSION:IMPOSSIBLE strike force? Who played her?	100	248
126. What self-destructed every week on MISSION:IMPOSSIBLE?	100	—
127. What was the first name of Lieutenant Hanley on the series COMBAT? Who played him?	100	215
128. What was the name of the squad leader on COMBAT? Who played him?	100	215
129. Who played Private Braddock on COMBAT?	100	215
130. Who was the star of the TV series TARZAN?	100	253
131. What was the name of THE GREEN		

QUESTIONS

DRAMA/ADVENTURE

160. What was the ship's source of power on STAR TREK? — 101 / —

161. On THE FUGITIVE, what was the first name of Kimble's wife? Who played her in guest appearances? — 101 / 256

162. What actor played the part of the one-armed man? — 101 / 256

163. Who finally was revealed as the killer of Kimball's wife on the final episode of THE FUGITIVE? — 101 / —

164. Name the main character on RUN FOR YOUR LIFE. Who played him? — 101 / 259

165. What character replaced Dan Briggs as the leader of the Impossible Missions Force? Who played him? — 101 / —

166. Who played Garrison in the series GARRISON'S GORILLAS? — 101 / 261

167. Name the main character on IT TAKES A THIEF. For what organization did he work? — 101 / 266

168. Who was the star of IT TAKES A THIEF? — 101 / 266

169. Complete the titles of the following soap operas: — 101 / —
    A. AS THE WORLD _____
    B. ANOTHER _____
    C. SEARCH FOR _____
    D. DARK _____
    E. ALL MY _____ — 102
    F. THE YOUNG AND THE _____
    G. _____OF OUR LIVES
    H. THE _____OF NIGHT

170. On LAND OF THE GIANTS, what did the giants call the earthlings? — 102 / —

QUESTIONS

CRIME

QUESTIONS

		An- swer Page	Pro- gram Page
216.	Name the detective on M SQUAD. Who played him?	103	173
217.	Who played Nick and Nora Charles on THE THIN MAN?	103	173
218.	What was the name of Nick and Nora's dog?	103	—
219.	Who was the star of the police drama HARBOR COMMAND?	103	—
220.	Who played McGraw in the series MEET McGRAW?	103	172
221.	Halloran and Muldoon were detectives on what police drama? Who played them?	103	179
222.	Who played PETER GUNN?	103	181
223.	Who played Edie on PETER GUNN?	103	181
224.	Who played Lieutenant Jacoby on PETER GUNN?	103	—
225.	Name the two private eyes on 77 SUNSET STRIP. Who played them?	103	182
226.	On 77 SUNSET STRIP, what character was known for combing his hair? Who played him?	103	182
227.	What was the name of the detective Lieutenant on THE LINEUP? Who played him?	103	182
228.	What was the name of the Inspector? Who played him?	103	182
229.	Who was the narrator on THE UN-TOUCHABLES?	103	186
230.	Who played Eliot Ness on THE UN-TOUCHABLES?	103	186
231.	What actor periodically appeared as Al Capone on THE UNTOUCHA-BLES?	103	—

QUESTIONS

DRAMA/ADVENTURE

QUESTIONS

DOCTORS/LAWYERS

DRAMA/ADVENTURE

QUESTIONS

		An-swer Page	Pro-gram Page
	secretary on THE TRIALS OF O'BRIEN? Who played her?	106	245
349.	Who played the Great McGonigle on THE TRIALS OF O'BRIEN?	106	245
350.	On "THE LAWYERS" segment of THE BOLD ONES, what was the name of the law firm's senior partner? Who played him?	106	275
351.	Who played Brian Darrell on "THE LAWYERS" segment of THE BOLD ONES?	106	275
352.	What was Neil's relation to Brian on "THE LAWYERS" segment of THE BOLD ONES? Who played him?	106	275
353.	Dr. Paul Lochner was the hospital Chief of Staff on what show? Who played him?	106	—
354.	Who starred as MARCUS WELBY, M.D.?	106	279
355.	What was the name of Marcus Welby's young motorcycle riding associate? Who played him?	106	279
356.	Name the doctor who was the principal character on MEDICAL CEN-TER Who played him?	106	—
357.	What football star appeared in the first episode of MEDICAL CENTER?	106	—
358.	On "THE DOCTORS" segment of THE BOLD ONES, three actors played the doctors. Name them.	106	273
359.	Who played Matt Lincoln on the series of the same name?	106	—
360.	What kind of doctor was MATT LINCOLN?	106	—

QUESTIONS

GAME/QUIZ

QUESTIONS

		Answer Page	Program Page
28.	Identifying a rebus puzzle was the gimmick on what A.M. game show?	107	—
29.	Who was the emcee of TO TELL THE TRUTH?	107	165
30.	What child prodigy won almost $200,000 on THE $64,000 QUESTION?	107	—
31.	When Joyce Brothers appeared on THE $64,000 QUESTION, what was her category?	107	—
32.	On YOU BET YOUR LIFE, what animal delivered the secret word?	107	—
33.	Who was the host of THE $64,000 CHALLENGE?	107	—
34.	Kitty Carlisle was a regular panelist on what television game show?	107	165
35.	What game show did Allen Ludden host?	107	217
36.	Who was the host of QUEEN FOR A DAY?	107	—
37.	The original NAME THAT TUNE had three different hosts. Name them.	107	—
38.	Name the host of YOU DON'T SAY!	107	217
39.	What daytime game show was hosted by Johnny Carson?	107	—
40.	Who was the original host of CONCENTRATION?	107	—
41.	Comedian Jan Murray was the emcee of what Fifties A.M. game show?	107	—
42.	What quiz show used tic-tac-toe as its format? Who was the emcee?	107	—
43.	Who was the host of PEOPLE ARE FUNNY?	107	138
44.	Who succeeded Garry Moore as the host of I'VE GOT A SECRET?	107	—

QUESTIONS

CHILDREN'S SHOWS

QUESTIONS

CHILDREN'S SHOWS

QUESTIONS

CHILDREN'S SHOWS

QUESTIONS

78. In what year was "The Wizard of Oz" first shown on TV?

79. Who replaced Ed McConnell as the host of SMILIN' ED'S GANG?

NEWS/ DOCUMENTARIES/ SPORTS

QUESTIONS

	Answer Page	Program Page
28. Name the original three commentators on NFL MONDAY NIGHT FOOT-BALL.	110	—
29. Who replaced Hugh Downs as host of TODAY in 1971?	110	—
30. Name the family shown on the documentary AN AMERICAN FAMILY.	110	300
31. Who was the anchorman on CHRONOLOG?	110	—
32. Who replaced Edward R. Murrow as the host of PERSON TO PERSON?	110	—

VARIETY

	An-swer Page	Pro-gram Page

14. Marion Marlowe was a regular on what variety show? — 111 122

15. What television star was said to "own" Tuesday nights? — 111 134

16. Who were the stars of YOUR SHOW OF SHOWS? — 111 —

17. What famous sister act appeared frequently with Arthur Godfrey? — 111 122

18. What Hawaiian personality appeared with Arthur Godfrey? — 111 122

19. Julius La Rosa began his career as a regular on what variety show? — 111 122

20. Mel Brooks used to be a staff writer on what early-Fifties variety show? — 111 —

21. Who was the original host of TONIGHT? — 111 162

22. Name the announcer on THE ARTHUR GODFREY SHOW. — 111 —

23. What were the last names of Tex and Jinx? — 111 —

24. Carl Reiner was a regular on what early-Fifties variety show? — 111 —

25. Who was Garry Moore's announcer on THE GARRY MOORE SHOW? — 111 179

26. THE TEXACO STAR THEATRE was later known as_____? — 111 134

27. Who did Jimmy Durante say goodnight to at the end of each show? — 111 —

28. Frank Parker and LuAnn Simms were singers on what show? — 111 122

29. Name the singing group that backed up Dinah on THE DINAH SHORE SHOW. — 111 139

30. On what network did Johnny Carson

	An-swer Page	Pro-gram Page
first appear as host of his own variety show?	111	143
31. What well-known newspaperman hosted his own variety show in the early Fifties?	111	159
32. What was the name of Lawrence Welk's orchestra?	111	—
33. What Hollywood star hosted HOLLYWOOD AND THE STARS?	111	223
34. Who was Steve Allen's announcer when he hosted TONIGHT? Who was his musical director?	111	162
35. By what nickname is Jackie Gleason known?	111	236
36. Who immediately succeeded Steve Allen as host of TONIGHT?	111	162
37. What character on THE STEVE ALLEN SHOW said "Heigh-ho, Steverino"? Who played him?	111	174
38. Name the regular on THE STEVE ALLEN SHOW who became famous for always being nervous.	112	174
39. On THE STEVE ALLEN SHOW, who played the "Man on the Street" who could never remember his name?	112	174
40. Eddie Fisher was a regular on what mid-Fifties television variety show?	112	172
41. What was the name of Red Skelton's Western buffoon?	112	172
42. Who was the orchestra leader on THE RED SKELTON SHOW?	112	172
43. In what year did SATURDAY NIGHT AT THE MOVIES debut?	112	202

VARIETY

	An-swer Page	Pro-gram Page
44. What feature film was the first movie shown on NBC's SATURDAY NIGHT AT THE MOVIES?	112	202
45. What weekly series replayed silent films?	112	213
46. What was the name of the talk show hosted by David Susskind?	112	—
47. On what variety show did Pat Boone make his first appearance?	112	—
48. Who succeeded Jack Lescoulie as the host of TONIGHT?	112	—
49. Red Foley was the host of what country and Western variety show?	112	176
50. Who was the star of the early-Fifties spectacular "Peter Pan"?	112	141
51. Who played Captain Hook?	112	141
52. What was the name of the mentalist who had his own network show in the mid-Fifties?	112	144
53. Who was Jack Paar's announcer when he hosted TONIGHT?	112	—
54. What was Dinah Shore's theme song?	112	194
55. In what year did Johnny Carson take over TONIGHT?	112	211
56. Who was Johnny Carson's announcer when he first took over TONIGHT?	112	211
57. Who was the orchestra leader of TONIGHT when Johnny Carson first became host?	112	211
58. Who was Dick Clark's long-time announcer on AMERICAN BANDSTAND?	112	—
59. From what city did AMERICAN BAND-STAND originate?	112	—

QUESTIONS

QUESTIONS

WESTERNS

QUESTIONS

QUESTIONS

QUESTIONS

Part Two
ANSWERS

COMEDY

1. Pinky Lee
2. Mr. Thornberry/ Don DeFore
3. Ricky
4. Hilliard
5. Connie Brooks/ Eve Arden
6. Mr. Boynton/ Robert Rockwell
7. The school principal/ Gale Gordon
8. Mrs. Davis
9. June Collyer
10. Marie Wilson
11. Susie MacNamara/ Ann Sothern
12. Don Porter
13. Wally Cox
14. Tony Randall
15. Peg Lynch and Alan Bunce
16. Joan Davis
17. Brad Stevens/ Jim Backus
18. Gale Storm
19. Freddie/Bob Hayden
20. Vern Albright/ Charles Farrell
21. Honeywell and Todd
22. Mrs. Odettes
23. Willie
24. Vern's girlfriend
25. Elliott and Goulding
26. Jackie Gleason
27. THE PEOPLE'S CHOICE
28. THE HONEY-MOONERS
29. Audrey Meadows
30. Mrs. Manicotti
31. The Gotham Bus Company
32. Sewer worker
33. Art Carney
34. The Royal Order of Raccoons
35. Ezio Pinza
36. MY FAVORITE HUSBAND
37. Ray Bolger
38. Leo G. Carroll
39. Henrietta
40. George and Marion Kerby/Robert Sterling and Anne Jeffreys
41. Neil/Martinis
42. Mr. Schuyler
43. Katie
44. Chester A. Riley

45. Jim and Honeybee Gillis/Tom D'Andrea and Gloria Blondell
46. Babs/Lugene Sanders
47. Peg/Marjorie Reynolds
48. Junior/Wesley Morgan
49. Danny Williams
50. Jean Hagen
51. Sherry Jackson/ Rusty Hamer
52. Leon Ames and Lurene Tuttle
53. STANLEY
54. Elena Verdugo
55. Arbuckle
56. Harry Von Zell
57. Larry Keating/ Bea Benaderet
58. Ricardo
59. William Frawley
60. Tropicanna
61. MacGillicutty
62. Betty Ramsey/ Mary Jane Croft
63. THOSE WHITING GIRLS
64. Molly Goldberg
65. Ray McNulty
66. Comstock College
67. Peggy/Phyllis Avery
68. Eddie Anderson
69. Don Wilson
70. A Maxwell

71. Mary Livingstone
72. IT'S ALWAYS JAN/ Jan Stewart
73. Gertrude Berg
74. YOU'LL NEVER GET RICH
75. Phil Silvers
76. Ernie
77. The motor pool
78. Colonel Hall/ Paul Ford
79. Camel Cigarettes
80. Fort Baxter/ Roseville, Kansas
81. Rocco/Harvey Lembeck
82. Corporal
83. Herbie Faye
84. Joe E. Ross
85. The mess hall
86. The Andersons
87. Betty, Bud, Kathy/ Elinor Donahue, Billy Gray, Lauren Chapin
88. Margaret/Jane Wyatt
89. Jim Anderson
90. Spring Byington
91. Ruskin
92. Henshaw
93. Pete/Harry Morgan
94. PETE AND GLADYS
95. Frances Rafferty
96. Chuck/Dwayne Hickman
97. Photographer

98. Schultzy/
    Ann B. Davis
99. Margaret/
    Rosemary DeCamp
100. Tim Moore
101. Sapphire/
     Ernestine Wade
102. George Stevens
103. The Mystic Knights of
     the Sea Lodge
104. Cab driver
105. Algonquin J.
     Calhoun
106. Alvin Childress and
     Spencer Williams
107. Brown
108. Lightnin'
109. Celia/Carol Burnett
110. Willard Waterman
111. Throckmorton
112. Birdie
113. Ann Baker
114. Dexter/Bobby Ellis
115. Susanna Pomeroy/
     Gale Storm
116. S.S. Ocean Queen
117. Captain Huxley/
     Roy Roberts
118. Nugey/ZaSu Pitts
119. Cedric/
     Jimmy Fairfax
120. Kathy Nolan
121. Liza Hammond
122. Novelist
123. Mary and Jenny
124. Chester Riley

125. Amos
126. Walter Brennan
127. Richard Crenna
128. Kate/Kathy Nolan
129. Pepino/
     Tony Martinez
130. George MacMichael
131. West Virginia
132. Lydia Reed
133. Sock Miller
134. Rollo/Dick Wesson
135. Mandy/Pat Breslin
136. J. B. Barker
137. Cleo
138. THE BROTHERS
139. Marjorie Lord
140. Tonoose/
     Hans Conried
141. Angela Cartwright
142. THE ANN
     SOTHERN SHOW
143. Goldie/manicurist
144. Stone
145. Alex/Carl Betz
146. Doctor
147. Mary and Jeff/
     Shelley Fabares and
     Paul Petersen
148. Bartley House
149. Jay North
150. Henry and Alice/
     Herbert Anderson
     and Gloria Henry
151. Mitchell
152. The Wilsons
153. Cleaver

154. Jerry Mathers
155. Ward and June/
    Hugh Beaumont and
    Barbara Billingsley
156. Wally/Tony Dow
157. Wally's best friend
158. Haskell/
    Ken Osmond
159. Theodore
160. Julie Foster
161. Hal Towne
162. FIBBER McGEE
    AND MOLLY
163. THE MANY LOVES
    OF DOBIE GILLIS
164. Dwayne Hickman
165. Thalia Menninger/
    Tuesday Weld
166. Herbert and
    Winifred/Frank
    Faylen and Florida
    Friebus
167. Maynard G. Krebs/
    Bob Denver
168. A grocery
169. Zelda Gilroy
170. Chatsworth Osborne
    Jr.
171. Ruth/Elaine Stritch
172. Shirley Boone
173. Sherwood
174. Greenwich Village
175. To be an actress
176. Appopolous/
    Leon Belasco
177. Peter Lind Hayes
    and Mary Healy

178. Lindsey
179. Leslie/Merry Martin
180. Wilma/Bea
    Benaderet
181. SALLY
182. Truesdale
183. Mrs. Banford/
    Marion Lorne
184. John Forsythe/
    Bentley Gregg
185. Attorney
186. Peter/Sammee Tong
187. Hollywood,
    California
188. Kelly
189. Jasper
190. Auntie Bee/
    Frances Bavier
191. Andy's deputy
192. Fife/Don Knotts
193. Mayberry
194. Opie/Ronny Howard
195. Taylor
196. THE DANNY
    THOMAS SHOW
197. Gomer Pyle/Jim
    Nabors
198. Ellie/
    Elinor Donahue
199. Floyd
200. Chick
201. Martha/Abby Dalton
202. Captain/
    Roscoe Karns
203. Henry Kulky
204. Porter

205. Harry Morgan and Cara Williams
206. Frank Aletter
207. Violet and Iris
208. Bedrock
209. Pebbles
210. Fred and Wilma/ Alan Reed and Jean Vander Pyl
211. Barney and Betty Rubble/Mel Blanc and Bea Benaderet
212. Gunther and Francis/Joe E. Ross and Fred Gwynne
213. Lucille/Beatrice Pons
214. Captain Block
215. Nat Hiken/ SERGEANT BILKO
216. Al Lewis
217. Nipsey Russell
218. Nikita Khrushchev
219. Comedy writer/ Rob Petrie
220. Laura/Mary Tyler Moore
221. Ritchie/Larry Mathews
222. Sorrell/Morey Amsterdam
223. Rose Marie
224. Melvin Cooley/ Richard Deacon
225. Carl Reiner
226. Jerry and Millie Helper/Jerry Paris

and Ann Morgan Guilbert
227. Mike, Robbie, and Chip/Tim Considine, Don Grady, and Stanley Livingston
228. Steve Douglas/Fred MacMurray
229. Tramp
230. Bub/William Frawley
231. Shirley Booth
232. George and Dorothy Baxter/Don DeFore and Whitney Blake
233. Bill Dana/José Jiminez
234. Dean Jones
235. U.S.S. *Appleby*
236. Lieutenant (j.g.)
237. Virgil Stoner/ Jack Albertson
238. Harvey Lembeck and Beau Bridges
239. The Clampetts
240. Buddy Ebsen
241. Irene Ryan
242. Jed's daughter/ Donna Douglas
243. Jethro/Max Baer
244. Milburn Drysdale/ Raymond Bailey
245. Jane Hathaway/ Nancy Kulp
246. Jethro's mother/ Bea Benaderet
247. Captain Bing-hamton/Joe Flynn

ANSWERS

248. Lieutenant Commander
249. Quinton/ Ernest Borgnine
250. Parker/Tim Conway
251. Carpenter/ Bob Hastings
252. Lucy Carmichael
253. Jerry/Jimmy Garrett
254. Westchester County
255. Vivian Bagley/Vivian Vance
256. Widow
257. Charley Halper/ Sid Melton
258. Bunny/Pat Carroll
259. HOW TO MARRY A MILLIONAIRE/ Barbara Eden, Merry Anders, and Lori Nelson
260. Harry and Arch/ John Astin and Marty Ingels
261. Carpenters
262. Patty Lane and Cathy Lane/cousins
263. Richard/ Eddie Applegate
264. Martin and Natalie/ William Schallert and Jean Byron
265. Glynis Johns
266. Keith/Keith Andes
267. Larry Blyden
268. Lois, Rusty, and Terry

269. Washington, D.C.
270. Kathy/Inger Stevens
271. Imogene Coca
272. A talking horse
273. Wilbur Post
274. Connie Hines
275. Walter Andrews/ Walter Brennan
276. Shady Rest Hotel
277. Kate/Bea Benaderat
278. Edgar Buchanan
279. Billie Joe, Bobbie Joe, and Betty Joe
280. The Cannonball Express
281. Sammy Jackson
282. Ben Whitledge
283. Connie Stevens
284. Jeff Conway/ Ron Harper
285. George Burns
286. Walter Burnley/ John McGiver
287. Elizabeth Montgomery
288. Darrin/Dick York
289. Samantha's mother/Agnes Moorehead
290. McMann and Tate
291. Larry Tate/ David White
292. Gladys Kravitz/ Alice Pearce
293. Tabitha

294. Aunt Clara/Marion Lorne
295. 43 Mockingbird Lane
296. Fred Gwynne
297. Frankenstein
298. A vampire/Al Lewis
299. Lily/ Yvonne DeCarlo
300. Eddie
301. Their niece
302. VALENTINE'S DAY/Tony Franciosa
303. Sergeant Carter/ Frank Sutton
304. Jim Nabors
305. Ronnie Schell
306. Bob Denver
307. Alan Hale
308. Ginger Grant/ Tina Louise
309. Thurston/Jim Backus
310. Professor Hinkley
311. Barbara Eden
312. Tony Nelson/ Larry Hagman
313. Major
314. Dr. Bellows/ Hayden Rorke
315. Roger Healey/ Bill Daily
316. Don Adams
317. Maxwell
318. CONTROL
319. 86
320. Agent 99/Barbara Feldon
321. Edward Platt
322. Fang
323. KAOS
324. Mel Brooks and Buck Henry
325. The Cone of Silence
326. A telephone
327. Ray Walston
328. Martin
329. Bill Bixby/ newspaper reporter
330. O'Hara
331. Mrs. Brown/ Pamela Britton
332. Ann Sothern
333. Dave Crabtree/ Jerry Van Dyke
334. Captain Manzini/ Avery Schreiber
335. A 1928 Porter
336. Barbara/Maggie Pierce
337. Patricia Crowley
338. Jim
339. Fort Courage
340. Captain Parmenter/ Ken Berry
341. Sergeant O'Rourke/ Forrest Tucker
342. Larry Storch
343. Dobbs/ James Hampton
344. Wrangler Jane/ Melody Patterson

ANSWERS

345. Edward Everett Horton
346. Frank de Kova
347. Sally Field
348. John/Peter Deuel
349. Jeff
350. Don Porter
351. Johnny Tillotson
352. PETTICOAT JUNCTION
353. Oliver Wendell Douglas
354. Lisa/Eva Gabor
355. MONA McCLUSKEY
356. CAMP RUNAMUCK
357. Debbie Watson
358. Uncle Lucius/ Frank McGrath
359. Stalag 13
360. Bob Crane/Colonel
361. Colonel Wilhelm Klink/Werner Klemperer
362. Sergeant Schultz/ John Banner
363. LeBeau/ Robert Clary
364. Newkirk/ Richard Dawson
365. MAYBERRY R.F.D.
366. Bill Davis/ Brian Keith
367. French/ Sebastian Cabot
368. Jody, Buffy, and Cissy

369. Peter Tork, David Jones, Mickey Dolenz, and Mike Nesmith
370. Peter Christopher/ Michael Callan
371. Greta/Patricia Harty
372. Ned and Phyllis/ Phyllis Diller
373. Dave and Julie/ Peter Deuel and Judy Carne
374. San Francisco
375. Stan/Rich Little
376. Richard Mulligan/ TV actor
377. Marie/Marlo Thomas
378. Actress
379. Don Hollinger/ Ted Bessell
380. Lew/Lew Parker
381. Rosemary De Camp
382. Restaurant
383. Roger Smith
384. Uncle Charley/ William Demarest
385. Ernie/Barry Livingston
386. Lewis and Clark/ Joby Baker and Ronnie Schell
387. Disc jockeys
388. Linda/Julie Parrish
389. Hutton/Billy De Wolfe
390. HE AND SHE/Dick and Paula Hollister

391. Sister Bertrille/
Sally Field
392. Alejandro Rey
393. THE MOTHERS-
IN-LAW
394. Diahann Carroll
395. Nurse/an aerospace
center
396. Dr. Chegley/
Lloyd Nolan
397. Baker
398. Corey
399. Jack Cassidy
400. Doris Martin
401. Rose Marie
402. Nelson
403. Billy and Toby
404. McLean Stevenson
405. THAT'S LIFE
406. E. J. Peaker
407. HERE'S LUCY/
Lucy Carter
408. Harry/Gale Gordon
409. Kim and Craig/Lucy
and Desi Arnaz
410. Patricia Harty
411. Will Hutchins
412. Jim Backus
413. London
414. Timmie/"she" was
really a man
415. Peter Kastner
416. Doc Fillmore/
Forrest Tucker
417. Ken Berry
418. Mike

419. Dan Dailey
420. Governor
Drinkwater/
his daughter
421. Carroll O'Connor
422. Tom/Bill Bixby
423. Brandon Cruz
424. Mrs. Livingston
425. Rufus and Bert/
Bob Denver and
Herb Edelman
426. A diner
427. Robert Reed and
Florence Henderson
428. Alice/Ann B. Davis
429. Three apiece
430. Peter, Greg, and
Bobby
431. Cindy, Marcia, and
Jan
432. Monte Markham
433. Pat Harrington
434. Chet Kincaid/
gym teacher
435. Brian and Verna
436. 1970
437. Richards
438. Gavin MacLeod
439. Herschel Bernardi
440. Lil/Sue Ann
Langdon
441. NANCY
442. BAREFOOT IN THE
PARK
443. Jack Klugman and
Tony Randall

444. Sportswriter
445. Photographer
446. Juliet Mills
447. Everett/Richard Long
448. Andy Griffith
449. Jerry Van Dyke
450. Keith
451. Shirley Jones
452. 1971
453. TO ROME WITH LOVE
454. TILL DEATH DO US PART
455. Albert and Jane Miller/Larry Hagman and Donna Mills
456. David Wayne
457. Sandy Duncan
458. Jenny/Hope Lange
459. Preston
460. Fannie Flagg
461. Jim Howard
462. SHIRLEY'S WORLD/Shirley Logan
463. The Roaring Twenties

464. Dean Jones
465. Big Nick
466. Joe/Richard S. Castellano
467. Harry/Grant's Tomb
468. Steinberg/David Birney
469. Meredith Baxter
470. Fitzgerald
471. Walt
472. Wayne Rogers
473. Lieutenant Colonel Henry Blake/McLean Stevenson
474. OZZIE'S GIRLS
475. James Coco/a state unemployment office
476. THE TEXAS WHEELERS/Jack Elam
477. Robert Dreyfuss/Paul Sand
478. The bass fiddle
479. Fred/Steve Landesberg
480. Jodie Foster

DRAMA/ ADVENTURE

1. I REMEMBER MAMA
2. BIG STORY
3. Jon Hall
4. TERRY AND THE PIRATES
5. Herbert Philbrick/ Richard Carlson
6. THE HALLS OF IVY
7. William Todhunter Hall/Ronald Colman
8. CLIMAX
9. Barry Nelson
10. Alistair Cooke
11. George Cleveland
12. Tommy Rettig
13. Jan Clayton/ Ellen Miller
14. Porky/Don Keeler
15. Miller
16. Preston Foster/ John Herrick
17. Sunday
18. "Marty"/Rod Steiger
19. Richard Greene
20. Donald Pleasence
21. John Beresford Tipton

22. Michael Anthony/ Marvin Miller
23. Bart Adams/ Barry Nelson
24. Lieutenants Rhodes and Storm/ Phil Carey and Warren Stevens
25. Fort Oghora/ Colonel Standish
26. Buster Crabbe
27. Cuffy/Cullen Crabbe
28. Matt Anders/ Brian Keith
29. Sheena/ Irish McCalla
30. Robert Newton
31. Dan Tempest/ Robert Shaw
32. KRAFT THEATRE
33. Ruth Martin/ Cloris Leachman
34. SCHLITZ PLAYHOUSE OF STARS
35. Guy Williams
36. Los Angeles
37. Tornado

38. Juan Ortega/
    Anthony Caruso
39. Sergeant Garcia
40. Joe Kirkwood, Jr.
41. Thursday
42. YOU ARE THERE
43. Veterinarians
44. NORTHWEST
    PASSAGE
45. SEA HUNT/
    Lloyd Bridges
46. Dean Fredericks
47. Air Force
48. Charles Bronson/
    Mike Kovac
49. Glenn Barton/
    George Nader
50. Darren McGavin/
    Captain Grey
    Holden
51. The *Enterprise*
52. Ben Frazer/
    Burt Reynolds
53. THE HALLMARK
    HALL OF FAME
54. Julie Harris
55. *Fortuna*
56. Gambling
57. John Vivyan/
    Ross Martin
58. The Spice Islands
59. William Reynolds
60. Adam Troy/
    Gardner McKay
61. The *Tiki*
62. James Holden

63. Tod Stiles and
    Buz Murdock/
    Martin Milner and
    George Maharis
64. Corvette
65. Rod Serling
66. Johnny Weissmuller
67. Ronald Reagan
68. THE INVESTI-
    GATORS/James
    Franciscus and
    James Philbrook
69. Boris Karloff
70. Robert Young/
    Cameron Brooks
71. Jim Davis
72. Jeremy Slate and
    Ron Ely
73. Gene Kelly and
    Leo G. Carroll
74. Tom Colwell/
    Dick York
75. Nick Adams
76. Mark Grainger/
    John Larkin
77. Liz Hogan/
    Barbara Rush
78. Ed Begley
79. Mark Richman
80. Redigo/
    Richard Egan
81. Tal/Ryan O'Neal
82. George C. Scott/
    Neil Brock
83. Cicely Tyson
84. David Janssen
85. Doctor

86. Philip Gerard/
    Barry Morse
87. Murdering his wife
88. The one-armed man
89. Derailment of a train
90. John/
    James Franciscus
91. English
92. Jefferson High
93. Albert Vane/
    Dean Jagger
94. A) Flame B) Life
    C) Light D) Gift
    E) Golden F) Family
    G) Day H) Secret
    I) First
95. OUR TOWN
96. Napoleon Solo/
    Robert Vaughn
97. United Network
    Command for Law
    Enforcement
98. Alexander
    Waverly/
    Leo G. Carroll
99. THRUSH
100. Illya Kuryakin/
     David McCallum
101. A dry-cleaning shop
102. Tuesday and
     Thursday
103. Mia Farrow
104. Constance/
     Dorothy Malone
105. Ryan O'Neal
106. Dr. Rossi
107. Anderson

108. Betty Anderson/
     Barbara Parkins
109. The *Seaview*
110. Lee Crane/
     David Hedison
111. Harriman/
     Richard Basehart
112. The Nelson
     Oceanographic
     Research Institute
113. Dr. Gamma
114. Eddie Albert
115. General Savage/
     Robert Lansing
116. Major General/
     John Larkin
117. KENTUCKY JONES
118. Gig Young,
     David Niven, and
     Charles Boyer
119. Timmy/Jon Provost
120. George Chandler
121. Dan Briggs/
     Steven Hill
122. Rollin Hand/
     Martin Landau
123. Electronics
124. Willy/Peter Lupus
125. Cinnamon Carter/
     Barbara Bain
126. The tape recording
127. Gil/Rick Jason
128. Sergeant
     Saunders/
     Vic Morrow
129. Shecky Greene
130. Ron Ely

ANSWERS

131. Kato/Bruce Lee
132. Van Williams
133. Black Beauty
134. April Dancer/
     Stefanie Powers
135. Mark Slate/
     Noel Harrison
136. Irwin Allen
137. Tony Newman and
     Doug Phillips/
     James Darren and
     Robert Colbert
138. Dr. Raymond Swain
139. Lee Meriwether
140. North Africa
141. Sam/Christopher
     George
142. Moffitt, Hitchcock,
     and Pettigrew
143. The Robinsons
144. Dr. Smith/
     Jonathan Harris
145. Guy Williams and
     June Lockhart/
     Angela Cartwright
146. Wednesday and
     Thursday
147. Adam West/
     Burt Ward
148. Neil Hamilton
149. Aunt Harriet/
     Madge Blake
150. Alan Napier
151. A) Cesar Romero
     B) Burgess Meredith
     C) Julie Newmar
     D) David Wayne

E) Otto Preminger
F) Victor Buono
G) Frank Gorshin
H) Vincent Price
I) Edward Everett
   Horton
152. 1966
153. The U.S.S.
     Enterprise
154. United Starship
155. Captain James T.
     Kirk/William Shatner
156. Vulcan/
     Leonard Nimoy
157. Human and Vulcan
158. Captain Pike/
     Jeffrey Hunter
159. Gene Roddenberry
160. Integrated
     matter and
     antimatter
161. Helen/
     Diane Brewster
162. Bill Raisch
163. The one-armed man
164. Paul Bryan/
     Ben Gazzara
165. Jim Phelps/
     Peter Graves
166. Ron Harper
167. Alexander Mundy/
     S.I.A.
168. Robert Wagner
169. A) Turns
     B) World
     C) Tomorrow
     D) Shadows

E) Children
F) Restless
G) Days
H) Edge
170. The little people
171. Gary Conway
172. London
173. A black bear
174. Dennis Weaver/Mark
175. THE SURVIVORS/ Harold Robbins
176. Walt Whitman High
177. Seymour Kaufman/Michael Constantine
178. Karen Valentine
179. The Carlyles
180. Lana Turner and George Hamilton
181. Robert Stack
182. Mrs. Peel
183. Ben Richards/ Christopher George
184. Fletcher
185. Kelly Robinson and Alexander Scott/ Robert Culp and Bill Cosby
186. Tennis professional and trainer
187. Rod Serling
188. THE MAN AND THE CITY/ Thomas Jefferson Alcala
189. Samantha Eggar/ Yul Brynner
190. Alistair·Cooke
191. Dr. Simon Locke/ Sam Groom
192. Tony Blake/ Bill Bixby
193. Kolchak/ Darren McGavin
194. EMERGENCY!
195. Dave Barrett

CRIME

196. Reed Hadley/ Captain Braddock
197. TREASURY MEN IN ACTION
198. Jack Webb
199. Officer Frank Smith/ Ben Alexander
200. Joe
201. Kent Taylor
202. Richard Denning and Barbara Britton
203. Ralph Bellamy
204. Hugh Marlowe
205. Florenz Ames
206. Rod Cameron and Lynn Bari

207. HIGHWAY PATROL/ Dan Mathews
208. David Brian
209. Mike Lanyard/ Louis Hayward
210. "Ten-four"
211. Mark Stevens/ Steve Wilson
212. Bruce Seton
213. David Janssen
214. Mary Tyler Moore
215. Chicago
216. Lieutenant Ballinger/Lee Marvin
217. Peter Lawford and Phyllis Kirk
218. Asta
219. Wendell Corey
220. Frank Lovejoy
221. NAKED CITY/ John McIntire and James Franciscus
222. Craig Stevens
223. Lola Albright
224. Herschel Bernardi
225. Stuart Bailey and Jeff Spencer/ Efrem Zimbalist, Jr., and Roger Smith
226. Kookie/Edd Byrnes
227. Lieutenant Guthrie/ Warner Anderson
228. Inspector Greb/ Tom Tully
229. Walter Winchell
230. Robert Stack

231. Neville Brand
232. Greasy Thumb
233. Frank Nitti/ Bruce Gordon
234. THE LAWLESS YEARS
235. Johnny/ John Cassavetes
236. The Crescendo
237. Dorothy Provine
238. The Charleston Club
239. Scott Norris and Pat Garrison/Rex Reason and Donald May
240. Chris Higbee/ Gary Vinson
241. Dixie
242. Checkmate, Incorporated
243. Eric Ambler
244. Don Corey and Jed Sills/ Anthony George and Doug McClure
245. Professor Carl Hyatt/ Sebastian Cabot
246. Tom Lopaka and Tracy Steele/ Robert Conrad and Anthony Eisley
247. Cricket
248. Poncie Ponce
249. Miami
250. Van Williams
251. Troy Donahue
252. Richard Denning

253. Holbrook and Ballard/
     Robert Taylor and Mark Goddard
254. Tige Andrews
255. Victor Jory
256. Robert Lansing
257. Teddy/
     Gena Rowlands
258. Gene Barry
259. Amos
260. Regis Toomey
261. Henry
262. Gary Conway
263. Nick Anderson/
     Ben Gazzara
264. John Egan/
     Chuck Connors
265. Efrem Zimbalist, Jr.
266. Inspector
267. Lewis
268. Arthur Ward/
     Philip Abbott
269. Jim Rhodes/
     Stephen Brooks
270. Barbara/Lynn Loring
271. Quinn Martin
272. Dennis Cole
273. Burt Reynolds
274. Iroquois
275. Mike Connors
276. Intertect
277. Joe
278. Mike Haines/
     Jack Warden
279. Jeff and Johnny/
     Robert Hooks and Frank Converse
280. Raymond Burr
281. Robert T. Ironside
282. Barbara Anderson
283. Adam Greer/
     Tige Andrews
284. Pete Cochran and Linc Hayes/
     Michael Cole and Clarence Williams III
285. Julie Barnes/
     Peggy Lipton
286. Mark Sanger
287. Gannon/
     Harry Morgan
288. McCLOUD
289. George Maharis/
     Ralph Bellamy
290. Vanessa/
     Yvette Mimieux
291. Paul Ryan/
     Robert Conrad
292. Harry Morgan
293. David Janssen
294. James
295. Malloy and Reed/
     Martin Milner and Kent McCord
296. Bumper Morgan/
     William Holden
297. Cassie Walters
298. James Franciscus
299. Michael
300. He was blind
301. Insurance investigator

302. New Orleans
303. Marlyn Mason
304. "Prescription: Murder"/Peter Falk
305. "See How They Run"/John Forsyth, Senta Berger, Franchot Tone, Jane Wyatt, George Kennedy
306. "The Marcus-Nelson Murders"
307. Richard Roundtree
308. POLICE STORY
309. Teresa Graves

DOCTORS/LAWYERS

310. Reed Hadley/ Bart Matthews
311. Dr. Konrad Styner/ Richard Boone
312. Raymond Burr
313. Barbara Hale
314. Paul Drake/ William Hopper
315. Arthur Tragg/ Ray Collins
316. Hamilton Burger/ William Talman
317. Pat O'Brien/ Roger Perry
318. Jim McKay
319. Abraham Lincoln Jones/ James Whitmore
320. Marsha
321. C. E. Carruthers
322. E. G. Marshall/ Lawrence Preston
323. Robert Reed
324. Joan Hackett
325. Helen Donaldson
326. Richard Chamberlain
327. Blair General Hospital
328. Dr. Gillespie/ Raymond Massey
329. Vincent Edwards
330. Neurosurgery
331. Chief Resident
332. Dr. David Zorba/ Sam Jaffe
333. Maggie Graham/ Bettye Ackerman
334. Ted/Harry Landers
335. Nick Kanavaras
336. Dr. Harold Jensen
337. Nurse Wills/ Jeanne Bates
338. Liz Thorpe/ Shirl Conway
339. Gail Lucas/Zina Bethune
340. Psychiatry

341. Paul Richards
342. York Hospital
343. Dr. Edward Raymer/
     Eduard Franz
344. Joseph Campanella
345. Peter Falk
346. Daniel
347. Katie
348. Miss G./Elaine Stritch
349. David Burns
350. Walter Nichols/
     Burl Ives
351. Joseph Campanella
352. Neil was Brian's
     younger brother/
     James Farentino
353. MEDICAL
     CENTER/
     James Daly
354. Robert Young
355. Dr. Kiley/
     James Brolin
356. Dr. Joe Gannon/
     Chad Everett
357. O. J. Simpson
358. E. G. Marshall,
     John Saxon,
     David Hartman
359. Vince Edwards
360. Psychiatrist
361. Arthur Hill
362. Lee Majors
363. Monte Markham
364. Harry Guardino and
     Dane Clark
365. James Stewart
366. Richard Crenna
367. Edward Asner
368. Barry Newman
369. Maggie/
     Susan Howard
370. San Remo

GAME/QUIZ

1. Mike Wallace
2. LIFE BEGINS AT 80
3. Herb Shriner
4. WHAT'S MY LINE
5. STRIKE IT RICH/ Warren Hull
6. Bud Collyer/ Roxanne
7. Sylvania
8. Groucho Marx
9. Hatcheck girl
10. MASQUERADE PARTY
11. George Fenneman
12. Garry Moore/ Henry Morgan and Bill Cullen
13. Bert Parks
14. Dorothy Kilgallen, Arlene Francis, and Bennett Cerf/ John Daly
15. Jack Barry
16. Ralph Edwards
17. 20 QUESTIONS
18. Art Baker
19. THE BIG PAYOFF
20. Hal March
21. Edgar Bergen
22. Jack Barry
23. Herb Stempel
24. $143,000/Mrs. Vivienne Nearing
25. The Terry twins
26. DOTTO/Jack Narz
27. Eddie Hilgemeier, Jr.
28. CONCENTRATION
29. Bud Collyer
30. Robert Strom
31. Boxing
32. A duck
33. Ralph Story
34. TO TELL THE TRUTH
35. PASSWORD
36. Jack Bailey
37. George de Witt, Bill Cullen, and Red Benson
38. Tom Kennedy
39. WHO DO YOU TRUST?
40. Hugh Downs
41. TREASURE HUNT
42. TIC TAC DOUGH/ Jack Barry
43. Art Linkletter
44. Steve Allen
45. Bob Eubanks
46. Jim Lange
47. Joe Garagiola

CHILDREN'S SHOWS

1. Princess Summerfall Winterspring/ The Tinka Tonkas
2. Chief Thunderthud
3. Doodyville
4. The Peanut Gallery
5. Polka Dottie
6. DING DONG SCHOOL
7. Ray Forrest
8. Richard Simmons
9. Yukon King
10. The Royal Canadian Mounted Police
11. Jack Sterling
12. THE MERRY MAILMAN
13. Al Hodge
14. Mary Hartline/ Claude Kirchner
15. Paul Winchell
16. Knucklehead Smith
17. Rocky Jones
18. KUKLA, FRAN and OLLIE
19. George and Jane
20. Judy
21. Chief Thunderthud
22. Ed Herlihy
23. Don Herbert
24. DISNEYLAND
25. Paul Tripp
26. TIME FOR BEANY
27. WINKY DINK AND YOU/Jack Barry
28. SMILIN' ED'S GANG
29. Richard Webb
30. Fort Apache
31. Lee Aaker
32. The 101st Cavalry
33. James Brown
34. Biff
35. Buzz
36. Pat Miekle
37. Video Rangers
38. Vena Ray
39. Beulah
40. Burr Tillstrom
41. Fran Allison
42. Fess Parker
43. George Russell/ Buddy Ebsen
44. Phineas T. Bluster
45. Inspector Henderson/ Robert Shayne

46. Corky/
    Mickey Braddock
47. He was a clown/
    Noah Beery, Jr.
48. Robert Lowery
49. Bimbo
50. CAPTAIN
    KANGAROO
51. TINKER'S
    WORKSHOP
52. MY FRIEND FLICKA
53. Johnny Washbrook/
    Gene Evans
54. Bob Keeshan
55. SHIRLEY
    TEMPLE'S
    STORYBOOK
56. Richard Webb
57. Ikky/Sid Melton
58. Alvin
59. THE SHARI LEWIS
    SHOW
60. The Broken Wheel
    Ranch
61. Bobby Diamond

62. Jim/Peter Graves
63. Jellystone National
    Park
64. Porter Ricks/
    Brian Kelly
65. Jim Backus
66. Fat Albert
67. SESAME STREET
68. The Muppets/
    Jim Henson
69. Clarabell
70. THE MICKEY
    MOUSE CLUB/
    Tim Considine
71. Mouseketeers
72. George Reeves
73. Jack Larson
74. Noel Neill and
    Phyllis Coates
75. Jimmy Dodd
76. Cubby
77. Annette Funicello
78. 1956
79. Andy Devine

NEWS/
DOCUMENTARIES/
SPORTS

1. SEE IT NOW/
   Edward R. Murrow
2. Dave Garroway
3. John Cameron
   Swayze
4. Friday
5. Edward R. Murrow
6. Drew Pearson
7. Charles Collingwood
8. OUTLOOK
9. Jim Leaming
10. Walter Cronkite
11. John Gunther
12. Howard K. Smith
13. ABC
14. Walter Cronkite
15. John Chancellor
16. Charles Collingwood
17. CBS
18. Walter Cronkite
19. AS THE WORLD
    TURNS
20. Tom Harmon
21. Douglas Edwards
22. The Green Bay
    Packers and the
    Kansas City Chiefs
23. 1968/Mike Wallace
    and Harry Reasoner
24. Frank Reynolds and
    Jules Bergman
    (ABC); Walter Cron-
    kite (CBS); Chet
    Huntley, David
    Brinkley, and Frank
    McGee (NBC)
25. Walter Cronkite
26. John Huston
27. The New York Jets
    and the Cleveland
    Browns
28. Keith Jackson,
    Howard Cosell,
    and Don Meredith
29. Frank McGee
30. The Loud family
31. Garrick Utley
32. Charles Collingwood

VARIETY

1. THE ED SULLIVAN SHOW and STEVE ALLEN
2. THE PERRY COMO SHOW
3. Dennis Day
4. THE TOAST OF THE TOWN
5. THE COLGATE COMEDY HOUR
6. Eddie Cantor
7. The June Taylor Dancers
8. The Pennsylvanians
9. KOVACS UNLIMITED
10. Jackie Gleason
11. Milton De Lugg
12. Ed Wynn
13. ARTHUR GODFREY AND FRIENDS, ARTHUR GODFREY'S TALENT SCOUTS, and ARTHUR GODFREY TIME
14. ARTHUR GODFREY AND FRIENDS
15. Milton Berle
16. Sid Caesar and Imogene Coca
17. The McGuire Sisters
18. Haleloke
19. ARTHUR GODFREY AND FRIENDS
20. YOUR SHOW OF SHOWS
21. Steve Allen
22. Tony Marvin
23. McCrary and Falkenburg
24. YOUR SHOW OF SHOWS
25. Durward Kirby
26. THE MILTON BERLE SHOW
27. "Mrs. Calabash, wherever you are"
28. ARTHUR GODREY AND FRIENDS
29. The Skylarks
30. CBS
31. Walter Winchell
32. The Champagne Music Makers
33. Joseph Cotten
34. Gene Rayburn/ Skitch Henderson
35. The Great One
36. Jack Lescoulie
37. Gordon Hathaway/ Louis Nye

38. Don Knotts
39. Tom Poston
40. THE GEORGE GOBEL SHOW
41. Deadeye
42. David Rose
43. 1961
44. "How to Marry a Millionaire"
45. SILENTS PLEASE
46. OPEN END
47. ARTHUR GODFREY'S TALENT SCOUTS
48. Jack Paar
49. JUBILEE, U.S.A.
50. Mary Martin
51. Cyril Ritchard
52. Dunninger
53. Hugh Downs
54. "See the U.S.A."
55. 1962
56. Ed McMahon
57. Skitch Henderson
58. Charlie O'Donnell
59. Philadelphia
60. SING ALONG WITH MITCH
61. Jack Linkletter
62. The Nick Castle Dancers
63. Charlie Weaver
64. THAT WAS THE WEEK THAT WAS/David Frost
65. THE ED SULLIVAN SHOW
66. "I Want to Hold Your Hand," "She Loves You," "All My Loving," "Till There Was You," and "I Saw Her Standing There"
67. THE RED SKELTON SHOW
68. THE JACKIE GLEASON SHOW
69. Miami Beach
70. Ray Bloch
71. Frank Fontaine
72. SHINDIG
73. The King Family
74. Jimmy O'Neill
75. Arthur Godfrey
76. Allen Funt
77. Dorothy Collins
78. Dick Clark
79. Saturday
80. Arthur Treacher
81. 1968/NBC
82. THE GOLDDIGGERS
83. 1969/ Candice Bergen
84. 1967
85. Leslie Uggams
86. Paul Lynde and Stanley Myron Handelman
87. Woody Allen
88. Ernestine

89. Geraldine Jones/ "The devil made me do it"
90. Uncle Miltie
91. HALF THE GEORGE KIRBY COMEDY HOUR
92. Peggy Cass
93. Topo Gigio
94. Arte Johnson
95. Henry Gibson
96. Gladys Ormphby/ Ruth Buzzi
97. "Say good night, Dick"
98. "Good night, Dick"
99. The Flying Fickle Finger of Fate Award
100. "Sock it to me"

WESTERN

1. Champion
2. Melody Ranch
3. Jingles/Andy Devine
4. Buttermilk
5. The Queen of the West
6. Pat Brady/Nellybelle
7. Bullet/Trigger
8. "Happy Trails to You"
9. Mineral City
10. Duncan Renaldo
11. Pancho/Leo Carillo
12. Sky King/Kirby Grant
13. THE LONE RANGER
14. DEATH VALLEY DAYS
15. Wild Bill Hickok
16. Gail Davis
17. Tagg/Jimmy Hawkins
18. Lofty Craig
19. Bill Williams
20. William Boyd
21. Edgar Buchanan
22. Topper
23. 1955
24. Milburn Stone
25. Dodge City
26. Amanda Blake/The Long Branch
27. Dennis Weaver/Goode
28. Sam
29. WYATT EARP/Hugh O'Brian
30. John Hart and Clayton Moore
31. Jay Silverheels
32. Jock Mahoney
33. ZANE GREY THEATRE
34. Dick Jones
35. Edgar Buchanan
36. THE TALES OF WELLS FARGO/Dale Robertson
37. Scott Forbes
38. Doc Holliday/Douglas Fowley
39. Tom Jeffords/John Lupton
40. Michael Ansara
41. Bret
42. Bart
43. Seth Adams/Ward Bond
44. Robert Horton
45. Flint

46. Charlie
47. Josh Randall/
    Steve McQueen
48. Clint Walker
49. Ty Hardin
50. Bodie
51. Chuck Connors/
    Lucas McCain
52. Mark/
    Johnny Crawford
53. THE
    CALIFORNIANS/
    Richard Coogan
54. Paladin/
    Richard Boone
55. A chess knight
56. Matt Rockford/
    George Montgomery
57. Audrey Totter
58. Rex Reason
59. The Cartwrights
60. The Ponderosa
61. Ben/Lorne Greene
62. Adam/Pernell
    Roberts
63. Little Joe/
    Michael Landon
64. Hoss
65. THE VIRGINIAN
66. John Hart
67. TRACKDOWN
68. Gene Barry
69. Dodge City
70. Johnny Ringo/
    Don Durant
71. Christopher Colt/
    Wayde Preston

72. BOOTS AND
    SADDLES
73. Simon Fry/
    Henry Fonda
74. Clay McCord/
    Allen Case
75. Fran
76. LARAMIE
77. John Smith
78. Jess Harper/
    Robert Fuller
79. Robert Crawford, Jr.
80. RAWHIDE/
    Rowdy Yates
81. Bronco Layne/
    Ty Hardin
82. An insurance
    company
83. Sam Logan/
    Robert Rockwell
84. WICHITA TOWN/
    Mike Dunbar and
    Ben Matheson
85. Lane Temple/
    John Smith
86. Johnny Yuma/
    Nick Adams
87. Clay Hollister/
    Pat Conway
88. THE TALL MAN/
    Barry Sullivan and
    Clu Gulager
89. James Drury
90. Henry Garth/
    Lee J. Cobb
91. Doug McClure
92. Betsy/Roberta Shore

93. John McIntire
94. Jack Lord
95. Rodeo rider
96. The Gold Buckle
97. Bruce Dern and Warren Oates
98. Jock Mahoney
99. Pahoo/X Brands
100. Burt Reynolds/ Quint Asper
101. Inger
102. Fess Parker
103. Yadkin
104. Mingo/Ed Ames
105. Rebecca/ Israel and Jemima
106. Chuck Connors/ McCord
107. A coward
108. The Barkleys
109. Lee Majors
110. Victoria/Barbara Stanwyck
111. Dr. Miguelito Loveless/ Michael Dunn
112. James/ Robert Conrad
113. Artemus Gordon/ Ross Martin
114. Barry Sullivan
115. Dale Robertson
116. John Mills
117. Jim Crown/ Stuart Whitman
118. MacGregor
119. Sam/Glenn Ford
120. Hannibal Heyes and Jed "Kid" Curry
121. Richard Boone
122. Caine/ David Carradine

Part Three
PROGRAM PAGES

FRIDAY, APRIL 10, 1953

2 WCBS-TV • 4 WNBT • 5 WABD • 6 WNHC-TV CONN. • 7 WABC-TV • 9 WOR-TV
11 WPIX • 13 WATV • 43 WICC-TV CONN.

Friday

APRIL 10

Day
8:00 p.m. (

❹ ❻ These Two—Comedy
Pinky's friends throw a surprise birthday party for him. Pinky Lee, Martha Stewart.

❺ Madison Sq. Garden—Films
Basketball: Harlem Globetrotters, N.Y.C. High School All-Stars. Wrestling: Vic Holbrook vs. Chuck Morgan; the Canadian Angel vs. Clyde Steeves. Boxing: Sylvester Jones vs. Shirley Pembleton; Dan Bucceroni vs. Dave Davey.

❼ The Stu Erwin Show—Series
A hostess and her guest discover they are wearing the same outfit.

❾ Broadway TV Theater
Meg Mundy and William Prince in "Wuthering Heights," Emily Brontë's classic.

⓫ MOVIE—"Montana Mike"
(Orig. title: "Heaven Only Knows")—Comedy. An angel makes an error in charting a saloon-keeper's destiny, and is sent to earth to rectify the error. Robert Cummings.

7:45 ❷ Perry Como Show—Music
Helen O'Connell, Bob Eberle, and Russ Case and his orchestra take over the show while Perry, the Fontanes, and Mitchell Ayres are in Durham, North Carolina, for the annual Chesterfield show.

❹ News—J. C. Swayze

8:00 ❷ ❻ MAMA—Peggy Wood
When one of Katrin's classmates becomes engaged, Katrin decides she, too, should be betrothed.

❹ RCA VICTOR SHOW—Day
Continuing hits efforts to discourage the beauty-contest winner, Dennis brings up his contract with Jack Benny, which contains a "marriage clause." Sponsored by RCA Victor.

❺ CITY ASSIGNMENT—Drama
"Father and Son." While covering a criminal line-up, Steve and Lorelei recognize the son of a police lieutenant.

❼ OZZIE & HARRIET—Sketch
A "double your money back" guarantee on a package of pancake mix gives Ricky food for thought. The Nelson family.

❽ WEEK-END BUILDER

8:30 ❷ MY FRIEND IRMA—Comedy
Irma and Jane try to help a violinist get into the Philharmonic Orchestra.

❹ ❻ LIFE OF RILEY—Comedy
Vacation hardships ruin Riley's plan to "get away from it all." William Bendix.

❺ DARK OF NIGHT—Drama
An on-location drama from A.B.C. Freight Forwarding Service, New York City.

❼ THE LITTLE SHOW—Music
Pagliacci with Lucille Norman.

⓭ SENATOR KENNETH HAND
A political talk by one of the candidates for the Republican nomination for Governor of New Jersey.

8:45 ❼ RUDOLPH HALLEY—Report
⓫ FILM HIGHLIGHTS

8:55 ⓫ HEADLINE NEWS

9:00 ❷ PLAYHOUSE OF STARS
Victor Jory in "The Mirror." A successful man's wall of smug self-assurance crumbles.

❹ ❻ BIG STORY—Crime Drama
C. D. Johnston of the St. Paul Dispatch-Pioneer Press uncovers a confidence game.

❺ ⓭ LIFE BEGINS AT 80—Panel
Oldtimers discuss problems of living.

❼ APPOINTMENT WITH LOVE

❾ NEWS BULLETINS

⓫ BASKETBALL—Bud Palmer
N.Y. Knickerbockers vs. Minneapolis Lakers, World Series playoffs.

9:15 ❾ SPORTS—Harry Wismer

9:30 ❷ OUR MISS BROOKS—Comedy
Connie Brooks almost loses a golden opportunity with Mr. Boynton. Eve Arden. 9th in ARB's N.Y.C. ratings.

❹ ❻ ALDRICH FAMILY—Series
Henry enters a baby-picture contest to

Sullivan and Lamour — 8:00 p.m. (2) ⑤   Cantor — 8:00 p.m. (4)   Randall — 9:00 p.m. (4) ⑥   Menjou — 10:30 p.m. (4)

6:45 ⑥ **Charles Laughton**—Readings
⑦ **Other Lands, Other Places**
⑪ **Sports**—Hy Turkin
⑬ **Picture News**

7:00 ② **GENE AUTRY**—Western
Outlaws try to steal Champion.

④ ⑥ **RED SKELTON**—Comedy
(1) Red portrays a pickle salesman. (2) Red gives an elephant a bath. (3) Red tries to get a loan from a finance company. Red's guests: the Bell Sisters.

⑤ **GEORGETOWN U. FORUM**
"Social Security For the Aged." Guests: Mr. Leonard J. Calhoun, attorney, and Mr. Allen D. Marshall; secretary of the pension board for General Electric.

⑦ **YOU ASKED FOR IT**—Baker
(1) Log-rolling dog. (2) Badminton's greatest exhibition player. (3) A dance on a balloon 3,000 feet in the air. (4) Description of an unseen person. (5) Newspapers torn into picture masterpieces.

⑪ **IT HAPPENED THIS WEEK**
⑬ **MOVIE**—"A Lawman Is Born" Western with Johnny Mack Brown.
㊸ **WHAT'S YOUR TROUBLE?**

7:15 ㊸ **FILM SHORT**

7:30 ② ⑥ **PVT. SECRETARY**—Comedy
Susie's bad case of spring fever receives an abrupt cure. Ann Sothern.

④ **MR. PEEPERS**—Wally Cox
Mr. Peepers revamps his lectures to interest a Chinese student. Marion Lorne.

⑤ **MOVIE**—"Candles at Nine"
(English) Melodrama, set in an old estate, about the violent death of an old miser. Jessie Matthews, John Stuart.

⑦ ㊸ **PLYMOUTH PLAYHOUSE**
Chapters from the ABC Album. Today: Part II of "A Tale of Two Cities," with Wendell Corey, Wanda Hendrix, and Judith Evelyn.

Charles Dickens' classic story of the French Revolution. Original music by Dimitri Tiomkin, composer of "High Noon."

⑨ **BROADWAY TV THEATER**
Conrad Nagel and Claire Luce in "Interference," with Scott McKay and Anne Burr. An eminent physician discovers that his wife is being blackmailed.

⑪ **MOVIE**—"Night Train"
(English) The celebrated spy film directed by Carol (Third Man) Reed. Margaret Lockwood, Rex Harrison, Paul Henreid.

8:00 ② ⑥ **TOAST OF THE TOWN**
Ed Sullivan presents: Dorothy Lamour with a Hawaiian instrumental group; the Goofers, novelty act; Byron Nelson, golf star.

④ **COLGATE COMEDY HOUR**
Tonight's star is Eddie Cantor. Guests: tenor Jan Peerce, singer Connie Russell and dancer Billy Daniel. ARB rates the Bob Hope Comedy Hour 8th in N.Y.C.

⑦ ㊸ **ALL-STAR NEWS**
⑬ **MOVIE**—"Men of San Quentin" Chronicles the achievements of a humane warden. J. Anthony Hughes.

8:55 ⑪ **NEWS BULLETINS**

9:00 ② **FRED WARING**—Music
The maestro and his Pennsylvanians present songs for Mother's Day.

④ ⑥ **TV PLAYHOUSE**—Drama
Tony (Mr. Peepers' friend Wes) Randall in "A Little Something in Reserve," with Geoffrey Lumb, Kendall Clark and Barbara Bolton. A lighthearted young man is mistakenly recalled by the Navy.

⑤ **ROCKY KING**—Roscoe Karns
"The Old Mansion." A pair of garden shears is the weapon when the owner of an old house is murdered.

⑦ **TINY FAIRBANKS**—Music
⑨ **MOVIES**—Double Feature

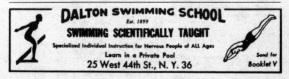
A-29

Morning

Adams	Horwich	Wallace
8:00 a.m. (2)	10:00 a.m. (4) ⑥	11:00 a.m. (2) ⑥

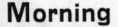

FRI. - MON. - TUES. - WED. - THURS

MORNING

6:55 ④ **Daily Sermonette**

7:00 ④ ⑥ **Today—Jack Lescoulie**
Jack substitutes while Dave Garroway is on vacation in Europe. **Fri.:** Niagara Falls at honeymoon time, from Niagara Falls, N. Y. **Mon.:** "M.D.," medical film on ulcers. Film of Mr. Muggs at the Virginia Beach Sand Festival. **Tues.:** Test run of the British Jet Hydroplane (speedboat), from Long Island. **Wed.:** Gertrude Berg, TV's "Molly Goldberg," is guest. **Thurs.:** Guest: James Michener, author of *Tales of the South Pacific*, discusses his new book *The Bridges at Toko-ri*. *Because Today is a program of current events all listings are subject to change.*

7:20 ② **Prevues and News**

7:25 ④ **Weather—Julia Meade**

7:30 ② **Telecomics—Cartoons**

7:45 ② **Time For Beany—Kids**
Puppet adventures on film.

7:55 ④ **Fix-it—Johnny Stearns**

8:00 ② **Kovacs Unlimited—Variety**
The moustached madman entertains, along with Edith (of *Wonderful Town*) Adams, Eddie Hatrak, Trig Lund, and Andy McKay.

8:25 ④ **Take a Word—Edwards**

8:55 ② **Today in N.Y.—Kathi Norris**

9:00 ② **Margaret Arlen—Interviews**
Fri.: Dr. Robert Lawrence, chiropodist, explains the importance of taking care of the feet.

④ **Morey Amsterdam Show**
Comedy and music with Morey, Milton De Lugg, and vocalist Francey Lane.

9:30 ② **MOVIE—15 Minutes**
Fri.: "Gaiety" (Part II)
Romantic triangle set in Mexico in fiesta time. Alan Mowbray, Bobby Watson.
Mon.-Tues.: "Hayfoot"
Comedy about GI trainees. William Tracy.

Addresses of TV Stations

● **WCBS-TV,** 485 Madison Ave., NYC 22.
● **WNBT,** 30 Rockefeller Plaza, NYC 20.
● **WABD,** 515 Madison Avenue, NYC 22.
⑥ **WNHC-TV,** 1110 Chapel St., New Haven, Conn.
● **WABC-TV,** 7 West 66th St., NYC 23.
● **WOR-TV,** 1440 Broadway, NYC 18.
● **WPIX,** 220 East 42nd St., NYC 17.
● **WATV,** TV Center, Newark 1, N. J.
④⑥ **WICC-TV,** 114 State St., Bridgeport, Conn.

Wed.-Thurs.: "Niagara Falls"
Two pairs of newlyweds, one middle-aged, one young, meet on their honeymoons. Marjorie Woodworth, ZaSu Pitts.

9:45 ② ⑥ **Winston Burdett—News**

10:00 ② **Wheel of Fortune**—(Fri. only.)
Contestants win the opportunity to repay their neighbors' good deeds.

② **Godfrey Time**—(M. thru Thurs.)
Robert Q. Lewis emcees. Arthur may be back Monday, July 6.

④ ⑥ **Ding Dong School** (Chi.)
Nursery school of the air, prepared especially for pre-school youngsters of three to five years of age. Dr. Frances Horwich.

10:30 ④ **Josephine McCarthy**
⑥ **Window Shopper—Malgren**

11:00 ② ⑥ **I'll Buy That—Panel**
Mike Wallace emcees a switch on the old-fashioned auction. Panelists must identify articles offered "for sale" on the show. Panelists are actresses Vanessa Brown and Audrey Meadows, actor Hans Conried, and bridge expert Albert Morehead.

④ **MOVIE—1 Hour**
Fri.: "Murder Is My Business"
Detective Michael Shayne investigates the threatening letters a woman is receiving. Hugh Beaumont, Cheryl Walker.
Mon.: "Rip Roaring Riley"
Federal agent investigates poison-gas manufacture. Lloyd Hughes, Grant Withers.
Tues.: "Loud Speaker"
A radio star loses his popularity. Ray Walker, Jacqueline Wells.
Wed.: "Paradise Isle"
A blind artist, shipwrecked on a South Sea island, is befriended by a beautiful native girl. Warren Hull, Movita.
Thurs.: "Private Snuffy Smith"
The comic-strip character joins the Army to escape the Revenooers. Edgar Kennedy.
⑥ **The Big Picture**—(Thurs.)

11:15 ⑪ **Film Shorts**—(Fri., Mon., Wed.)

11:30 ② ⑥ **Strike It Rich—Hull**
Fri.: John Howard, now appearing in the musical *Hazel Flagg*.

⑦ **MOVIE—1 Hr., 30 Minutes**
Fri.: "Larceny in Her Heart"
Det. Michael Shayne is asked to locate the daughter of a civic crusader. Hugh Beaumont, Cheryl Walker.
Mon. thru Thurs.: To Be Announced

⑪ **Living Blackboard—Educ.**
Fri.: "Art in Your Life." Orestes S. Lapolla.
Mon.: "Face the Facts," social studies.
Wed.: "What's the Big Idea?"

A-16

SATURDAY, JULY 4, 1953

2 WCBS-TV • **4** WNBT • **5** WABD • **6** WNHC-TV CONN. • **7** WABC-TV • **9** WOR-TV
11 WPIX • **13** WATV • **43** WICC-TV CONN.

Saturday

JULY 4

Reese
1:30 p.m. (9)

MORNING

8:55 **4** **Daily Sermonette**

9:00 **4** **Children's Theater—Films**
Roy Forrest narrates children's films.

10:00 **6** **Johns Hopkins Sci. Review**
"Man Will Conquer Space," the first of
three request shows.

10:20 **2** **Prevues and News**

10:30 **2** **F.Y.I.—Film**

4 **Cisco Kid—Western Film**
Duncan Renaldo, Leo Carillo

6 **To Be Announced**

10:45 **2** **Little League Baseball
School** Tommy Henrich plays host to
baseball stars and Little League players.
Guest: Cal Abrams, Pittsburgh Pirates out-
fielder. A salute to the teams of the Lower
East Side, Manhattan

7 **Junior Crossroads—Film**

11:00 **2** **Laughtime—Comedy Films**
Buster Keaton and Charley Chase.

4 **Baseball Hall of Fame**
Lou Fonseca interviews guests.

7 **Space Patrol—Kemmer**
"The Theft of Terra V." When Corry's
space ship is discovered missing, all evidence points to Cadet Happy. (Hollywood)

11:15 **4** **MOVIE—"The Right Man"**
A young man falls in love with his fiancee's
cousin, while an opera singer courts his
fiancee. Alan Ladd, Edith Fellows, Julie
Bishop, Wilbur Evans. Shown three times.

11:30 **2** **Rod Brown of the Rocket
Rangers** "The Phantom Birds of Beloro."
Ranger Brown and Captain Frank Boyd
rescue three gold prospectors on the
planet Beloro from the clutches of an invisible bird.

7 **Sky King—Kirby Grant**
"Formula for Fear." Foreign agents try to
get a poison-gas formula.

AFTERNOON

12:00 **2** **Big Top—Circus**
Snookums, a trained bear; Annell & Brask,
cycling act; Three Royal Rockets, roller-
skating act; Rolando, one-finger balancing
act. Warren Hull substitutes for ringmaster
Jack Sterling.

7 **Italian Cookery—Bontempis**

9 **MOVIE—"Wyoming Wildcat"** Western with Red Barry.

12:20 **5** **Film Shorts**

12:30 **5** **House Detective—Newman**

9 **Merry Mailman—Kids**

13 **MOVIE—1 Hour**

1:00 **2** **Lone Ranger—Western**
John Hart. Film

5 **MOVIE—1 Hr., 30 Min.**

6 **Better Living TV Theater**
"Shining Heart." Story of steel

7 **MOVIES—to 5:30 p.m.**
(1) "Lure of the Islands," Margie Hart
(2) 2:15 P.M., "Mystery of the 13th Guest,"
Helen Parrish. 3—3:15 P.M., "Undercover
Agent," Russell Gleason. 4—4:15 P.M.,
"Police Court," Henry B. Walthall.

9 **Felton's Knothole Gang**

11 **Film Short**

1:10 **11** **Joe E. Brown—Pre Game**

1:15 **5** **MOVIE—"Adventurous
Knights"** "Our Gang" comedy-drama.

4 **Baseball—Dizzy Dean**
Detroit Tigers vs. Cleveland Indian

1:25 **11** **Baseball—Doubleheader**
N. Y. Yankees vs. Philadelphia Athletics

1:30 **2** **Pioneer Playhouse—Film**
"Billy the Kid Outlawed" Bob Steele

9 **Baseball—Doubleheader**
Dodgers vs. Pirates Red Barber, Vince
Scully, Connie Desmond. Sponsored by
the F. & M. Schaefer Brewing Co.

13 **MOVIE—1 Hr., 30 Minutes**

2:00 **2** **Camera Three—Education**
Topic: "Democratic Man." Guest: Dr. Lewis
Leary, Professor of English, Columbia U.

2:30 **5** **MOVIE—"I Cover China-
town"** A girl on her first trip through
Chinatown gets lost and runs into murder.
Norman Foster, Polly Ann Young

2:45 **2** **What's Your Trouble?**
"How to Break the Worry Habit." Dr. and
Mrs. Norman Vincent Peale. (Film)

3:00 **2** **MOVIE—"Amazon Quest"**
(1949) Traveling up the Amazon to collect
evidence proving he has been cheated of
his rightful inheritance, a young man
meets a bandit chief fighting political corruption. Tom Neal, Carole Mathews

4 **Industry on Parade—Film**
(1) War on polio. (2) Pottery. (3) Industrial
lunches. (4) Factory on a college campus.

A-24

121

2 WCBS-TV · 4 WNBT · 5 WABD · 6 WNHC-TV CONN. · 7 WABC-TV · 9 WOR-TV
11 WPIX · 13 WATV · 43 WICC-TV CONN.

Wednesday

| SEPTEMBER 23 |

Backus
8:00 P.M. (4)

6:35 ⑬ **Your Weatherman**
6:40 ⑥ **Weather Report**
6:45 ④ **News—John Wingate**
⑥ **World News Today**
⑤ **News Bulletins**
⑬ **News Bulletins**
6:55 ⑤ **Weather—Janet Tyler**
7:00 ④ **Film Shorts**
⑤ **Captain Video—Kid Serial**
Al Hodge, Don Hastings.
⑥ **Death Valley Days**
⑤ **News—Kevin Kennedy**
⑬ **MOVIE—"Song of the Buck-
aroo"** Western with Tex Ritter.
⑬ **News of the Hour**
7:05 ⑬ **The Cracker Barrel**
7:10 ⑪ **Weather Report**
7:15 ⑨ **Marge & Jeff—Comedy**
⑤ **News—Taylor Grant**
⑪ **Sports—Jimmy Powers**
7:25 ② **Weather—Carol Reed**

THEY'RE BACK!

"MY LITTLE MARGIE"

with Gale Storm and Charles Farrell

NOW 8:30 PM WED.

WNBT

Channel 4

⑦ **Sports—Bill Stern**
⑪ **Telepix News**
7:30 ② **Douglas Edwards—News**
④ **Eddie Fisher—Songs**
⑤ **Weather—Janet Tyler**
⑥ **Connecticut Spotlight**
⑦ **A Date with Judy—Comedy**
⑪ **MOVIE—"The Devil and
Daniel Webster"** For a pittance of
gold and seven years of good luck, a
young farmer makes an unholy pact with
the devil. NY TV debut. Jabez Stone,
James Craig; Mary Stone, Anne Shirley;
Daniel Webster, Edward Arnold; Mr.
Scratch, Walter Huston; Ma Stone, Jane
Darwell; Belle, Simone Simon; Squire Slos-
sum, Gene Lockhart; Miser Stevens, John
Qualen; Justice Hawthorne, H. B. Warner.
⑬ **Studio F—Film**
7:35 ⑤ **News Bulletins**
7:40 ⑤ **Business Report**
7:45 ② **Perry Como—Music**
Perry and the chorus sing "Papaya Mama"
and "Alone Together"; the Fontanes offer
"Falling," and Perry's solo is "Crying in
the Chapel."
④ ⑥ **News—J. C. Swayze**
⑤ **Sports Desk**
⑬ **Poor Richard's Almanac**
8:00 ② ⑥ **GODFREY AND FRIENDS**
Arthur will be seen at his home in Leesburg,
Virginia. Marion Marlowe, Frank Parker,
Janette Davis, Julius La Rosa, Lu Ann Simms,
the Mariners, the McGuire Sisters, Haleloke.
④ **I MARRIED JOAN—Comedy**
Joan's constant lateness almost wrecks her
husband's career. Joan Davis, Jim Backus.
⑤ ⑬ **JOHNS HOPKINS—Science**
Topic: "Symbols of Science." Guests will be
students from foreign countries.
⑦ **TALENT PATROL—Francis**
⑬ **JUNIOR TOWN MEETING**
"Can We Defend the Whole World
Against Communism?"
8:30 ④ **MY LITTLE MARGIE—Comedy**
To prevent Margie from buying an ex-

McNellis and McElhone	Gleason	Pinza	Shriner
7:30 P.M. (7) 43	8:00 P.M. (2)	8:00 P.M. (4) 6	9:00 P.M. (2)

OCT. 3

SATURDAY

❹ ETHEL & ALBERT—Comedy
Albert smells smoke in the Arbuckle house, and takes action, resulting in complete pandemonium. Peg Lynch, Alan Bunce.

❺ TERRY AND THE PIRATES
"Maitland Affair." The Dragon Lady does an about-face and teams up with Terry and Hotshot to protect a shipload of American Relief rice from being pirated.

❻ ARTHUR MURRAY PARTY

❼ LEAVE IT TO THE GIRLS
RETURN A panel of four women engage in a "battle of the sexes" with a male celebrity guest. Panel regulars are Eloise (the "Quiet One") McElhone, actress Vanessa Brown, radio personality Florence Pritchett, and actress Lisa Ferraday. Maggi McNellis presides over the panel. Tonight, George Jessel matches wits with the lovely ladies.

⑪ MOVIE—"Obsessed"
(English; 1951) When an invalid woman is poisoned, her schoolteacher husband and her beautiful secretary-companion, who are in love, suspect each other of the murder. N.Y. TV debut. Geraldine Fitzgerald, David Farrar, Jean Cadell, Roland Culver, Mary Merrall.

8:00 ❷ JACKIE GLEASON—Comedy
Featuring his gallery of characters (Joe-the-Bartender, Reggie, the Honeymooners), the June Taylor dancers, Art Carney, Audrey Meadows. Ray Bloch's orchestra.

❹ ❻ BONINO—Ezio Pinza
Bonino is confronted with a major crisis when his young son Andy balks at going to school. Mary Wickes plays Martha.

❺ 43 PRO FOOTBALL—Wismer
New York Giants vs. Pittsburgh Steelers at Pittsburgh.

❼ ORIENT EXPRESS—Film
"Man of Many Skins," with Eric Von Stroheim. A drama of intrigue and adventure.

⑬ MOVIE—"Marked Men"
A man-hunt gets under way in Arizona when six prisoners make a break for freedom. Warren Hull, Isabel Jewell.

8:30 ❻ AMATEUR HOUR—Mack
The performers vie for the votes of the viewing audience

❼ MADISON SQUARE GARDEN
(1) Pro basketball: Boston Celtics vs. Indianapolis Olympians. (2) College basketball: St John's University vs. Fordham University. (3) Boxing: Ben Travis vs. Danny Taylor. (4) Wrestling: Lu Kim vs. Hardy Kruskamp.

8:55 ⑪ NEWS BULLETINS

9:00 ❷ TWO FOR THE MONEY
Herb Shriner as quiz host, with Dr. Mason Gross, Provost and Professor of Philosophy at Rutgers University, as authority on correctness of contestants' answers.

❹ ❻ ALL-STAR REVUE
RETURN Martha Raye is tonight's star, with guests Margaret Truman, Cesar Romero, Jake La Motta, and Rocky Graziano.

❼ PHILLIES FIGHTS
Phil Kim vs. Arthur Persley, lightweights 10 rounds. See Jimmy Powers' ratings page A-10.

A-25

2 WCBS-TV · **4** WNBT · **5** WABD · **6** WNHC-TV CONN. · **7** WABC-TV · **9** WOR-TV
11 WPIX · **13** WATV · **43** WICC-TV CONN.

Thursday

OCTOBER 22

Bolger
8:30 P.M. (7)

❾ Broadway TV Theater
"The Front Page," by Hecht & MacArthur.

⓫ MOVIE—"Young Widow"
(1946) A young newspaperwoman, widowed by the war, encounters a fresh pilot. NY TV debut. Jane Russell, Louis Hayward, Marie Wilson, Faith Domergue.

⒀ Studio F—Film

7:35 ❺ News Bulletins

7:40 ❺ Business Report

7:45 ❷ Jane Froman—Songs
The setting is Cuba.

❹ ⑥ News Caravan—Swayze

❺ Talk—Rudolph Halley

⒀ Poor Richard's Almanac

8:00 ❷ MEET MR. McNUTLEY
Professor McNutley, an English teacher in a girl's college, puts himself in debt to buy a speedboat, which brings him more trouble than he bargained for. Ray Milland.

❹ ⑥ GROUCHO MARX—Quiz
George Fenneman assists. (Film)

❺ GIANTS—Football News
Professional football information.

❼ QUICK AS A FLASH
Panel: Faye Emerson and Jimmy Nelson.

⒀ PAUL L. TROAST—Talk
Mr. Troast is the Republican candidate for Governor of New Jersey.

⒀ BIG PICTURE—Army Film

8:15 ❷ MOVIE—"Reg'lur Fellers"
A wealthy woman hates children because her son married against her wishes. Roscoe Ates, Sarah Padden, Carl Switzer, Billy Lee.

8:30 ❷ FOUR STAR PLAYHOUSE
Dick Powell in "The Witness." A celebrated

criminal lawyer puts his entire career at stake when he agrees to defend an accused murderer. (Film)

❹ T-MEN IN ACTION—Crime
"Case of the Prosperous Pauper." Agents trail a "poor" newsdealer through 38 banks.

❺ B'WAY TO HOLLYWOOD
Bill Slater is back as host. News features, variety acts. Singer Betty Cox.

⑥ DRAGNET—Jack Webb

❼ WHERE'S RAYMOND?
Ray Bolger stars in a "book musical." Jan Clayton joins him in a singing and dancing role. Allyn Joslyn, Richard Erdman.

⒀ JR. SHOW TIME—Talent

8:55 ❾ NEWS BULLETINS

⓫ NEWS BULLETINS

9:00 ⓫ ⑥ VIDEO THEATER—Drama

❹ DRAGNET—Jack Webb
Joe Friday tries to defend a rookie cop. Alexander plays Officer Frank Smith.

❺ MELODY STREET—Music
Musical variety show drawing on entertainers from its permanent company to sing and dance. Elliot Lawrence, emcee; cast includes Del Hanley, Lynn Gibbs, Art Ostrin, Roberta McDonald, Bob Fitch, Dee Clifford, Jack Krueger, Kay Coster, John Buwen, and the dancing team, Marlene Dell and Don Farnsworth.

❼ DOTTY MACK—Variety
Music and pantomime from Cincinnati.

❾ MOVIE—"Gentleman After Dark" A suave, elusive jewel thief concocts a plan to protect his daughter from the consequences of his crimes. Brian Donlevy, Miriam Hopkins, Preston Foster.

⓫ HOCKEY—[RETURN]
N. Y. Rangers vs. Boston Bruins from Madison Square Garden. Bud Palmer reports.

⒀ ROBERT B. MEYNER—Talk
The Democratic Candidate for Governor of New Jersey.

9:05 ⒀ BOXING—Laurel Garden
Charley Williams vs. Andy De Paul, 8 rounds, welterweights.

9:30 ❷ BIG TOWN—Drama
An escaped convict, using Steve Wilson as his hostage, resorts to time bombs, bullets, and disguises in a bid for freedom.

A-46

WEDNESDAY　　　　JANUARY 6

Ames, John Stephens, Leora Thatcher.

❼ ⒁ JEAN CARROLL—Comedy
Jean hires a maid and dreams of having a business career.

❾ BADGE 714—Film
Jack Webb in a Dragnet re-run.

❶ HOCKEY—Mad. Sq. Gard.
NY Rangers vs. Chicago Block Hawks.

❶ MOVIE—1 Hr., 15 Minutes

9:30 ❹ I'VE GOT A SECRET—Quiz
Panel: Jayne Meadows, Henry Morgan, Bill Cullen, Joan Bennett. Emcee: Garry Moore.

❺ ON YOUR WAY—Collyer
Audience-participation show.

❼ TO BE ANNOUNCED

⑧ MY LITTLE MARGIE—Comedy

❾ DANGEROUS ASSIGNMENT

⒁ TO BE ANNOUNCED

10:00 ❷ ⑧ BOXING—Miami Beach
Bobby Dykes vs. Joey Giambra, middleweights, 10 rounds. See Jimmy Powers Rates the Boxers on page A-2.

❹ THIS IS YOUR LIFE—Edwards
Ralph presents the life story of an interesting person.

❺ STARS ON PARADE
Bobby Sherwood emcees.

❼ BOSTON BLACKIE—Film
Lois Collier and Kent Taylor.

❾ CAPTURED—Film

⒁ MOVIE—"Gangs of Chicago" Seeing his petty-crook father killed by cops, a young man decides to become a criminal attorney and outwit the law. Lloyd Nolan, Barton MacLane, Lola Lane, Ray Middleton, Astrid Allwyn.

10:15 ❹ NEWS BULLETINS

10:30 ❷ DOUGLAS FAIRBANKS, JR.
Drama filmed in England.

❺ THE MUSIC SHOW
Musical half-hour with Robert Trendler's "pops" orchestra and vocalists Jackie Van, Mike Douglas, Eleanor Warner.

❼ FILM DRAMA
"Manhattan Robin Hood," Preston Foster.

❾ HARLEM DETECTIVE

⒀ MOVIE—"Ladies in Distress" Comedy in which a movie actor from Egypt courts a waitress whom he takes to be a disguised duchess casting a

movie she will produce. Alison Skipworth, Polly Moran, Robert Livingston.

10:45 ❷ ⑧ SPORTS—Mel Allen

11:00 ❹ Chronoscope—News Panel

❺ ❼ ❾ ⒁ News Bulletins

❻ News—Van Horn

⑥ The Comedy Hour

⑪ Telepix News

11:10 ❹ Weather—Tex Antoine

❺ Sports—Stern

❼ Weather—Joe Bolton

11:15 ❷ News—Allan Jackson

❹ Sports News

❺ Half Hour Theater—Film
"International Incident," with Wilton Graff, John Baer, and George Wallace. A story of DP's in post-war Germany.

❼ Henry Morgan—Comedy

⑪ Sports—Sam Aro

11:20 ❷ Steve Allen—Variety

❹ Film Short

11:25 ❷ Sports—Jim McKay

11:30 ❷ MOVIE—"Challenge of the Frontier" Late Show. NY TV debut. Western in which a political bully attempts to gain control of another man's ranch. Randolph Scott, Barton MacLane.

❼ Be My Guest—Variety

⑪ Film Shorts

⒀ MOVIE—"Don't Gamble with Strangers" (1946) Gambling partners pose as brother and sister to fleece a small-town banker and take over a local gambling spot. Bernardene Hayes, Kane Richmond, Peter Cookson.

11:45 ❺ News Bulletins

12:00 ❹ News—Kenneth Banghart

❼ MOVIE—30 Minutes

⑧ News Bulletins

12:05 ❷ MOVIE—"Winner Take All"
Rumors say Joe Palooka is washed up. Joe Kirkwood Jr., Elyse Knox.

12:45 ❷ MOVIE—"Women in War"
Late Late Show. Starting time is approximate. A socialite playgirl enters the war nursing-service to escape a prison sentence. Elsie Janis, Wendy Barrie, Patric Knowles, Mae Clarke. Late News follows around 2:00 A.M.

A-42

125

FRIDAY

④ DAVE GARROWAY—Variety
Recording stars Les Paul and Mary Ford (Mr. and Mrs. Paul) are guests.

⑤ FRONT PAGE DETECTIVE
Edmund Lowe in filmed series.

⑦ OZZIE & HARRIET—Comedy
"The Incentive." Ozzie worries that Ricky doesn't appreciate the value of money, so he offers to double any money Ricky has left over from his allowance at the end of the week.

⑩ REQUESTFULLY YOURS
⑬ THE BIG PICTURE—Film

8:30 ② TOPPER—Comedy
Henrietta buys a decidedly modern living-room suite—far too modern for Cosmo's taste. Anne Jeffreys, Robert Sterling, Leo G. Carroll.

④ ⑧ LIFE OF RILEY—Bendix
Chester and Gillis decide to swap houses. It takes some time for them to realize that they ought to consult their wives about the matter.

⑤ MELODY STREET—Variety
Tony Mottola hosts an informal walk down Tin Pan Alley.

⑦ PLAYHOUSE—Arlene Dahl
"Too Gloomy for Pvt. Pushkin," with Robert Strauss, Steve Brodie, and Maurice Kelly. Doing reconnaissance work in a shell-blasted castle on the Italian front in 1943, Private Pushkin is startled to learn that Sergeant Spoda senses the approach of a German patrol.

⑬ JR. SHOWTIME—Talent
8:55 ⑨ NEWS AND WEATHER
① NEWS BULLETINS
9:00 ② PLAYHOUSE OF STARS
William Lundigan in "Give the Guy a Break," with Frances Rafferty. A policeman faces a conflict between loyalty to the force, and loyalty to an old friend who is wanted for murder.

Cast

Jack Fuller . . . . William Lundigan
Maureen . . . . Frances Rafferty
Captain Donaldson . . . . Douglas Kennedy
Desk Sergeant . . . . John Halloran

④ ⑧ BIG STORY—Drama

The story of editor Ed Engledow, of the Lamesa (Texas) *Daily Reporter*, who helped solve the amnesia case of a young woman who couldn't remember who she was or why she was in town. Engledow, who is now on the staff of the El Paso *Times*, will be portrayed by Harry Townes.

⑤ LIFE BEGINS AT 80—Panel
Panelists include Georgiana Carhart and Fred Stein. Guest: Sir Cedric Hardwicke.

⑦ ⑬ PAUL HARTMAN SHOW
Albie tries to prove to his family that he really does enjoy "the finer things in life."

⑨ CAPTURED—Film
"Case of Leo Curt," with Chester Morris.

① MOVIE—"The Courtney Affair" (English; 1951) A domestic love story of three generations of a London family, set against a background of the first 45 years of this century. Michael Wilding, Anna Neagle, Coral Browne.

① WRESTLING—Laurel Garden
9:30 ④ OUR MISS BROOKS—Comedy
Connie and her cohorts kindly agree to sell tickets to a policemen's ball, unaware that the tickets are phony.

③ ⑧ SOUNDSTAGE—Drama
"A Time for Hope," with Carmen Mathews and Malcolm Lee Beggs. A hypochondriac is persuaded by his doctor to take a more reasonable view of life, unaware that he has only a few months in which to make the most of his happiness.

Cast

Marion Merrill . . . . Carmen Mathews
Merrill . . . . Malcolm Lee Beggs
Doctor . . . . Jack Arthur
Specialist . . . . David Orrick

⑤ FILM DRAMA
"The Match," with Gloria Saunders. A spinster is fired after many years at her job, but finds a surprise at the employment agency.

⑦ ⑬ THE COMEBACK STORY
Tonight's subject-guest is Alexander P. de Seversky, aeronautical engineer, designer, and World War I pioneer.

⑨ INNER SANCTUM—[DEBUT]
The celebrated radio series makes its TV

WEDNESDAY, FEBRUARY 10, 1954

WEDNESDAY   FEBRUARY 10

④ KRAFT THEATRE—Drama
"The Barn," by George Lowther. A series of catastrophes wrenches a small-town man from his faith in God, and leads him to believe he can achieve his ambitions without help.

Cast

Mother, Daughter,	
Ghost	Felicia Montealegre
Ben	Edward Binns
Elizabeth	Audra Lindley
Nate	Vaughn Taylor

⑤ FILM DRAMA
⑦ BADGE 714—Film
Jack Webb in a re-run of the *Dragnet* series.
⑪ HOCKEY—Bud Palmer
NY Rangers vs. Detroit Red Wings.
9:30 ② I'VE GOT A SECRET—Quiz
Panel: Jayne Meadows, Henry Morgan, Bill Cullen. Emcee: Garry Moore.
⑤ STARS ON PARADE
⑦ FILM DRAMA
⑧ MY LITTLE MARGIE—Comedy
⑨ DANGEROUS ASSIGNMENT
Dynamite stolen from construction camp. Brian Donlevy as Steve Mitchell.
㊸ CONN. TOWN MEETING
10:00 ② ⑧ BOXING—Hodges
Chico Vejar vs. Jed Black, welterweights, 10 rounds. From Chicago Stadium, Chicago. See *Jimmy Powers'* ratings, pg. A-37.
④ THIS IS YOUR LIFE—Edwards
Ralph presents the life story of an interesting person. (Hollywood)
⑤ THE MUSIC SHOW
Robert Trendler's pop orch., vocalists Jackie Van, Mike Douglas, Eleanor Warner.
⑦ BOSTON BLACKIE—Film
Lois Collier and Kent Taylor.
⑨ CAPTURED—Film
⑬ CLUB CARAVAN—Variety
Bill Cook; The Larks, quartet.
㊸ MOVIE—1 Hour
10:30 ④ DOUGLAS FAIRBANKS, JR.
An old woman helps attract customers to a business. (Film)
⑤ LIFE WITH ELIZABETH
Starring Betty White and Del Moore.
⑦ MY HERO—Cummings
Robert Cummings portrays a happy-go-

lucky real-estate salesman with a penchant for getting into trouble. Julie Bishop.
⑨ HIGH TENSION—Drama
⑬ MOVIE—"Girl from Mandalay" When a man-eating tiger menaces the region, two friendly rivals join forces to get him. Conrad Nagel, Kay Linaker.
10:40 ⑪ NEWS
10:45 ② ⑧ SPORTS—Mel Allen
10:50 ⑪ WEATHER REPORT
10:55 ⑪ SPORTS RESULTS
11:00 ② Chronoscope—News Panel
④ ⑤ ㊸ News Bulletins
⑦ News—Van Horn
⑧ The Comedy Hour
⑨ Man from Times Square
⑪ Treasure Chest—Film
11:10 ④ Weather—Tex Antoine
⑦ Sports—Bill Stern
11:15 ④ News—Allan Jackson
④ Sports News
⑤ Half Hour Theater—Film
⑦ Henry Morgan—Comedy
11:20 ④ Steve Allen—Variety
11:25 ② Sports—Jim McKay
11:30 ② MOVIE—"Dead Men are Dangerous" *Late Show.* (English) An author changes identity with a dead man and becomes the object of hot pursuit by police and criminals. Robert Newton.
⑦ MOVIE—30 Minutes
⑪ Film Short
⑬ MOVIE—"A Desperate Adventure" A Parisian painter attempts to retrieve the painting which brought him fame. Ramon Novarro.
11:45 ⑤ News Bulletins
12:00 ④ News—Kenneth Banghart
⑧ News Bulletins
12:05 ④ MOVIE—"Freddie Steps Out" (1946) *Midnight Movie.* Freddie looks exactly like a crooner who has disappeared, so his teen-age friends announce Freddie is the crooner. Freddie Stewart.
12:45 ② MOVIE—"The Invisible Wall" (1947) *Late, Late Show.* Starting time is approximate. Unable to give up gambling, a criminal gets deeper and deeper into crime. Don Castle.

A-42

127

SUNDAY   FEBRUARY 14

Mulligan'' Comedy about two drug salesmen who join the Army to escape a tax collector, only to find said collector their top sergeant. Nat Pendleton.

5:00 ② Omnibus—Alistair Cooke
(1) The Kabuki Dancers, a Japanese troupe. (2) "Paso Doble," by Budd Schulberg, a drama about a couple honeymooning in Mexico. The girl is attracted to a young man whose father is steering him into a bullfighting career. (Schulberg is author of *What Makes Sammy Run*, *The Disenchanted*, and wrote the screen play of the forth-coming movie *Waterfront*.) (3) "The Whale Who Wanted to Be a Submarine," a series of comic drawings by Leo Salkin. (4) Claude Rains in a scene from T.S. Eliot's *The Confidential Clerk*.

④ ⑧ Hall of Fame—Drama
"Henry Bergh, Crusader Against Cruelty." Henry Bergh creates the American Society For the Prevention of Cruelty to Animals. Sarah Churchill. (Hollywood)

⑦ ⑬ Super Circus—Variety
The Three Hitch Hikers, knockabout act. Rosa Patine, web act. Eddie Fay's Boxing Cats. The Six Mar Vels, tetterboard act.

⑪ Junior Carnival—Cartoons
5:15 ⑪ Cooking Program
5:30 ⑬ Time For Pets
5:45 ⑪ Little Tom-Tom—Puppets

EVENING

6:00 ④ Meet the Press—Panel
The press panel interviews Sen. John F. Kennedy (Dem., Mass.). Panel: Lawrence Spivak, May Craig, Ned Brooks, Marquis Childs. Deena Clark moderates.

⑤ Drew Pearson-Washington
Guest: Igor Gouzenko who was part of a Communist spy ring in Canada.

⑦ Captain Midnight—Film
⑧ Ozzie & Harriet—Comedy
⑨ The Christopher Program
⑪ Ramar of the Jungle
⑬ Harmony Ranch—Variety
Starring Carol Mills, with Johnny Dee Trio.

⑬ Industry on Parade
6:15 ⑤ ⑬ Igor Cassini—Guests
Guests: Basil Rathbone and Edgar Smith, President of the Baker Street Irregulars, a society of Sherlock Holmes fans.

⑨ The Pastor—Religion
6:30 ② ⑧ You Are There—History
"The Hanging of Captain Kidd." He was sentenced to hang on May 23, 1701.

④ Roy Rogers—Western
"Peril from the Past." A bank cashier with a secret past is blackmailed by bandits who plan to rob the Mineral City bank.

⑤ Meet Your Congress
Former Senator Blair Moody moderates.

⑦ ⑬ George Jessel—Variety
A salute to top sports figures. Guests include: Joe Louis; Billy Wells of Michigan State; Al Schacht, "Clown Prince of Baseball"; sports columnist Bill Corum; Dolores Parker, Joe Louis's fiancée; Louis Prima.

⑨ MOVIE—''Valley of Vengeance'' Western with Buster Crabbe.
⑪ Street Corner—Roberts
⑬ Variety Hall
6:45 ⑪ News Bulletins
⑬ News Bulletins
6:55 ⑪ Weather Report
7:00 ② LIFE WITH FATHER
Comedy with Leon Ames and Lurene Tuttle.

④ ⑧ PAUL WINCHELL—Variety
Jerry and Knucklehead use glue instead of varnish when they are hired to polish the chairs in the local town hall.

⑤ AUTHOR MEETS THE CRITICS
Book: *Our Secret Allies*, meaning the common people of the USSR, by Eugene Lyons. Pro Critic: H.V. Kaltenborn, "dean" of

YOU CAN'T BUY
FINER ICE CREAM
THAN
Breyers

SATURDAY, FEBRUARY 27, 1954

MORNING

6:25 ④ Sermonette
6:30 ④ Modern Farmer
7:00 ④ Saturday—Variety
News and features with Herb Sheldon.
7:45 ② Prevues
7:50 ② Give Us This Day
7:55 ② News Bulletins
8:00 ② Laugh Time—Films
"Gold Ghost." with Buster Keaton.
8:30 ② Junior Sports Session
9:00 ② On the Carousel—Kids
(1) Pantomimist Harry Bartron, (2) Ceramic
demonstration, (3) "The Loan's Necklace,"
film, (4) Activities of Junior Achievement.
    ④ Children's Theater—Films
Ray Forrest is the host.
9:30 ⑤ Mr. Wizard—Science
10:00 ② MOVIE—"Bad Men of
Arizona" Western with Buster Crabbe.
    ⑤ Johnny Jupiter—Fantasy
10:15 ② Animal Time—Kids
10:30 ② Here Is the Past
A series on the subject of archeology.
Today: "The Transparent Earth." Professor
Casper Kraemer, NYU, demonstrates the
spectacular results of the archeologist's
use of modern air photography.
    ④ Rocky Jones, Space Ranger
DEBUT Another tale of the future.
Rocky Jones protects his planet.

Cast

Rocky Jones . . . . Richard Crane
Winky . . . . Scotty Beckett
Vena Ray . . . . Sally Mansfield
Bobby . . . . Robert Lyden
    ⑤ Documentary—Film
"We Saw It Happen." History of Aviation.
    ⑦ ⑤ Smilin' Ed's Gang
11:00 ② Winky Dink and You
(1) The young viewers draw a mysterious
object that changes Jack into a clown.
(2) Mike McBean goes on an exploring
trip to the moon and finds some strange
creatures there.
(3) The viewers draw an animal that comes
to life.
    ④ Creative Cookery

Today's recipes: veal scallopini and choco-
late log cookies. . . . Mr. Pope offers his
recipe for peppered steak on page 12.
    ⑦ ⑧ Space Patrol—Kids
"The Blazing Sun of Mercury." Buzz Corry
unmasks the diabolic Mr. Proteus and in so
doing is almost lured to his death.
11:30 ② Rod Brown of the Rocket
Rangers "The Strong Man of Mayron."
A man from Mayron, one of the giant
planets of Alpha Centauri, rescues a rocket
ranger from a cracked-up spaceship.
    ⑤ Tom Corbett, Space Cadet
"Rescue in Space." An exchange student
from Mars brings near disaster to Tom's
crew when he crashes their rocket ship
into a meteor. Frankie Thomas.
    ⑦ Adventures of Blinkey-Kids
    ⑤ Space Cadet
11:45 ⑦ Jr. Crossroads—Kids

AFTERNOON

12:00 ② ⑧ Big Top—Circus
(1) Payo & Mae, Danes, juggling on the
high unicycle.
(2) The Carpis Trio, Greeks, risley act.
(3) The Sensational Kays, Germans, high-
wire duo.
(4) The Arwoods, novelty dog act and
acrobatics.
    ④ Here's Looking at You
Richard Willis gives beauty advice.
    ⑤ MOVIE—"International
Crime" "The Shadow" matches wits with
an international spy ring. Rod La Rocque.
    ⑦ Italian Cookery—Bontempis
Fedora and Pino with recipes and songs.
12:30 ④ Industry on Parade
12:45 ④ MOVIE—"Sailors Three"
Shown several times till 5 P.M. (English)
Comedy in which three drunken sailors try
to capture a German battleship. Michael
Wilding, Claude Hulbert.
1:00 ② Lone Ranger—Western
"Trouble for Tonto." The Lone Ranger and
his Indian friend fall into a trap.
    ⑤ MOVIE—1 Hour
    ⑦ Home Gardner—Aiampi

A-11

TUESDAY, MARCH 16, 1954

Mrs. Novak .... Betty Garde
⑨ SPOTLIGHT—Drama
Please see Monday for details.
⑬ KNOW YOUR STATE
Boy Scouts discuss outdoor manners.
8:30 ② RED SKELTON—Comedy
Character sketches. (Hollywood)
⑤ 43 PANTOMIME QUIZ
Captains: John Barrymore, Jr., Jerry Lester.
Team regulars: Peter Donald, Elaine Stritch,
Dorothy Hart, Jackie Coogan, Mike Stokey
emcees.
⑬ REPORT FROM RUTGERS
Professor Fred Fender discusses Math.
8:55 ⑨ ⑪ NEWS BULLETINS
9:00 ② MEET MILLIE— RETURN
The Manhattan secretary and her comical
mother and friends, return to the video
screens.
Millie .... Elena Verdugo
Mama .... Florence Halop
Mr. Boone .... Roland Winters
Alfred Prinzmetal .... Marvin Kaplan
④ FIRESIDE THEATER—Film
"Ringo's Last Assignment." A veteran reporter learns that he is being pensioned off
and replaced by a young girl. Tom Powers.
⑤ MOVIE—1 Hour
⑦ DANNY THOMAS—Film
Daughter Terry gets a chance at a career
as a child actress. Danny Thomas stars.
Jean Hagen, Rusty Hamer, and Sherry
Jackson are featured.
⑧ PLAYHOUSE OF STARS
⑨ BIG PICTURE—Army Film
⑪ SPORTS EVENT
Professional basketball playoffs. (tentative)
⑬ WOMEN'S WRESTLING
43 MR. & MRS. NORTH—Crime
Richard Denning, Barbara Britton. (Film)
9:30 ② ⑧ SUSPENSE—Drama
"The Fourth Degree." The police attempt
to extract a confession from a criminal
without resort to the third degree.
④ CIRCLE THEATER—Drama
"The Fugitive," by Irene Foley. A teen-age
fugitive hiding in a Roman Catholic or-
phanage must decide whether or not to
sacrifice his freedom to save a child's life.
Postponed from March 2nd.

Cast
Sister Madeleine .... Dolly Haas
Phillippe Grandbois .... Anthony Perkins
Father Desrochers ...... Marcel Hillaire
⑦ 43 U. S. STEEL HOUR
Helen Hayes and Charlie Ruggles star in
"Welcome Home." A loyal and much loved
housekeeper faces dismissal when her em-
ployers decide to close down the house
now that the children have grown up.
Cast
Jenny Libbott .... Helen Hayes
Charles Austin .... Charles Ruggles
Mrs. Charles Austin ... Carmen Matthews
Walter Mason .... Paul McGrath
Mrs. Watson .... Jean Dixon
Sponsored by U. S. Steel Corp.
⑨ THIS IS THE LIFE—Film
**⑪ MOVIE—"Moon Over Mon-
tana"** Western with Jimmy Wakely.
10:00 ② DANGER—Mystery Drama
"Actor." A rugged Hollywood actor re-
turns to his Lower East Side home and finds

TUESDAY · 9:30 PM
THE EXCITING NEW

**U. S. STEEL
HOUR**

**HELEN
HAYES
·
CHARLES
RUGGLES**
in
"Welcome Home"

**CHANNEL
7**

Produced by the Theatre Guild

cealed bottle of nitro-glycerine. He has received a tipoff about a planned prison break. Robert Middleton portrays Deputy Warden Larenz C. Schmuhl.

⑤ THE PLAINCLOTHESMAN

"The Fatal Flaw." A handsome ne'er-do-well is murdered after he has won the affection of a wealthy heiress. Ken Lynch, Jack Orrison.

⑦ ㊸ DR. IQ—Quiz

⑬ BIG PICTURE—Army Film

10:00 ② THE WEB—Mystery

"The Scapegoat." An alibiing businessman really needs an alibi when he is accused of a crime he did not commit.

④ ⑧ LORETTA YOUNG SHOW

"Man's Estate." A young boy attempts to escape from his perfectionist father.

Cast

Roger Stevens..........Douglas Kennedy
Tom......................Bobby Ellis
Stepmother..............Loretta Young

⑤ TWENTY QUESTIONS—Panel

Fred Van Deventer, Florence Renard, Herb Polesie, Bobbie McGuire, Jay Jackson.

⑦ ㊸ BREAK THE BANK

Bert Parks is the host. Bank: $4,000.

⑪ FILM DRAMA

"Feet of Clay." A young prison sociologist learns about the strange events surrounding the life of the warden. Bill Phipps.

⑬ CHAMPIONSHIP BOWLING

Eastern All-Star Classic from Newark.

10:30 ② ⑧ WHAT'S MY LINE?—Quiz

Dorothy Kilgallen, Arlene Francis, Steve Allen, and Bennett Cerf. John Daly moderates as panel tries to guess occupations of guests. A visiting celebrity is featured.

④ I LED THREE LIVES—Film

Philbrick gets involved in a Red scheme to discredit a local citizen. Richard Carlson stars.

⑤ TO BE ANNOUNCED

⑦ MOVIE—"International
Crime"—"The Shadow" matches wits with international spies. Rod La Rocque.

⑪ NEWS BULLETINS

10:40 ⑪ THE WEATHERMAN

10:45 ⑪ SPORT SPOT—McCarthy

㊸ WHAT'S YOUR TROUBLE?

11:00 ② ④ ⑧ News Bulletins

⑤ Film Drama

"One Night Stand." An actor impersonates Wild Bill Hickok. Jeff York, William Lester.

⑦ Maggi McNellis

⑪ Fashions—Ethel Thorsen

⑬ MOVIE—"Angels with
Broken Wings" Three daughters of a charming widow try to help their mother get the man she loves away from his unscrupulous ex-wife. Binnie Barnes.

11:10 ④ Weather—Jon Gnagy

11:15 ② MOVIE—"Tuxedo Junction"

Late Show. A group of migrant boys enter a competition to build a float for a parade. Leon, June, and Frank Weaver.

④ Sports Final—Joe O'Brien

⑦ Dr. Ervin Seale—Talk

⑧ Fireside Theater—Film

11:20 ④ The Cassini Show

11:30 ⑤ News Bulletins

⑪ Film Shorts

11:45 ⑧ Facts Forum

12:00 ④ MOVIE—"The Mysterious
Mr. Davis" Midnight Movie. Comedy about a down-and-out fellow who wangles his way into high financial circles by creating a mysterious and wealthy "partner." Alastair Sim, Kathleen Kelly.

12:15 ⑧ News Bulletins

12:30 ② MOVIE—"Drake of England"

Late Late Show. Starting time is approximate. (English) The story of England's great naval hero of Elizabethan times, Sir Francis Drake. Matheson Lang.

ARB's Top 10 in New York

1.	I Love Lucy.........Mon., 9 P.M.,	②
2.	Talent Scouts.....Mon., 8:30 P.M.,	②
3.	Jackie Gleason.......Sat., 8 P.M.,	②
4.	Dragnet.........Thurs., 9 P.M.,	④
5.	Groucho Marx......Thurs., 8 P.M.,	④
6.	Toast of the Town...Sun., 8 P.M.,	②
7.	This Is Your Life....Wed., 10 P.M.,	④
8.	Godfrey & Friends...Wed., 8 P.M.,	②
9.	Our Miss Brooks....Fri., 9:30 P.M.,	②
10.	Milton Berle.........Tues., 8 P.M.,	④

A-23

SATURDAY, MAY 8, 1954

⑨ Captain Midnight—Film
⑪ The Range Busters
John King, "Crash" Corrigan.
㊽ The Music Box
6:45 ② Art Linkletter and the Kids
Average children discuss worldly topics.
7:00 ② MR. DISTRICT ATTORNEY
The D.A. builds a single clue into a chain of evidence which smashes a gang of hijackers. David Brian. . . . *Sponsored by G. Krueger Brewing Co.*
④ MAN AGAINST CRIME
"Sunset Farm." Mike Barnett comes to the rescue of an old lady who finds herself the only "guest" at a lonely farm retreat. Ralph Bellamy.
⑤ JOE PALOOKA—Film
Joe faces the problem of telling a promising young boxer that he can never fight again. Joe Kirkwood, Jr.
⑦ SUCCESS STORY, U.S.A.
⑧ THIS IS YOUR LIFE
Ralph Edwards and surprise guest.
⑨ MOVIE—"Shaggy"
(1948) The comradeship shared by a father, son, and dog, is threatened when dad remarries. Robert Shayne.
⑪ NEWS—Kevin Kennedy
⑬ NEWS BULLETINS
㊽ TO BE ANNOUNCED
7:10 ② WEATHER—Joe Bolton
7:15 ⑪ COOKING PROGRAM
⑬ MOVIE—"Heartaches"
(1947) A crooner achieves success by using someone else's voice. Sheila Ryan.
7:30 ② BEAT THE CLOCK—Games
Zany stunts and "The Folks Are Fun" flash-photo contest. Bud Collyer. . . . *Did you know that the stunts are pre-tested? See page 8.* . . . *Sponsored by Sylvania.*
④ ETHEL & ALBERT—Comedy
Fun with Mr. and Mrs. Arbuckle. The domestic series is written by Peg Lynch who plays Ethel. Alan Bunce stars as Albert.
⑤ ANNIE OAKLEY—Western
Gail Davis in the title role. . . . *Program is reviewed on page 13.*
⑦ DOTTY MACK—Songs
Pantomime to records. Dotty: Rosemary Clooney's "Lovely Weather for Ducks,"

and Jo Stafford's "Make Love to Me." Bob: Eddie Fisher's "My Mom," for Mother's Day.
⑧ MAN BEHIND THE BADGE
Two junior sheriffs face a showdown with desperate hoodlums.
⑪ MOVIE—"A Millionaire for Christy" *First Show.* NY TV debut. (1951) A legal secretary goes to San Francisco to inform a radio philosopher that he is an heir. Fred MacMurray, Eleanor Parker, Richard Carlson.
8:00 ② JACKIE GLEASON—Comedy
Art Carney, Audrey Meadows, Joyce Randolph, Ray Bloch assist the rotund comic. Ralph (Honeymooner) Cramden is taken in by two swindlers who sell him the "New York interest" in a hair-growing formula.
④ ⑧ SPIKE JONES—Variety
Sketch about the trials and tribulations of the City Slickers when they go on the road for one-night stands. . . . *Final show of the series. Next week: Your Lucky Star, a quiz show with cash prizes.*
⑤ MOVIE—1 Hour, 30 Minutes
⑦ INTO THE NIGHT—Mystery
"Circle of Fire."
⑨ MOVIE—"Swamp Fire"
Time approximate. (1946) A discharged Navy pilot suffering from shock, is helped by his buddies to regain his confidence. Johnny Weissmuller, Buster Crabbe.
㊽ AMERICAN-HUNGARIAN THEATER International variety session.
8:30 ④ ⑧ AMATEUR HOUR—Mack
Tonight Ted introduces talented young-

BEAT THE CLOCK

Join the fun with Sylvania's
Bud Collyer & Roxanne
SAT. **7:30** PM WCBS-TV CHANNEL **2**

A-17

JUNE 14

⑤ **WEATHER—Janet Tyler**

⑦ ⁴³ **FLIGHT NO. 7—** DEBUT
First in a series of travelogues. Tonight:
"Wings to France," a visit to Notre Dame,
the Eiffel Tower, native dances of Brittany,
and the French Riviera. (Film)

⑨ **MOVIE—Drama**
"Oliver Twist." Dickens' classic about the
oppressed orphan boy. Dickie Moore, Irving Pichel, William Boyd.

⑪ **MOVIE—Comedy**
First Show: "The Taming of Dorothy." NY
TV debut. (English; 1950) The misadventures of an Italian bank clerk and an
American gangster who are dead ringers
for each other. Jean Kent, Robert Beatty.

⑬ **MOVIE—Mystery**
"Fog Island." A man seeks revenge on
the people who framed him and killed his
wife. Ian Keith, Veda Ann Borg.

7:35 ⑤ **NEWS—Don Russell**

7:45 ② **PERRY COMO SHOW**

④ ⑧ **NEWS CARAVAN—Swayze**

⑤ **SPORTS DESK—Bob Smith**

8:00 ② **BURNS & ALLEN—Comedy**
Gracie becomes convinced that George is
the "one husband in five" who has a
secret vice. Harry Von Zell, Bea Benaderet,
Larry Keating. (Film)

④ ⑧ **NAME THAT TUNE—Quiz**

⑤ **A DOLLAR A SECOND**
Contestants race against time. Games and
penalty stunts pay them a dollar for each
second they can remain onstage. . . .
Final show of the series.

⑦ ⁴³ **SKY KING—Kirby Grant**
"Operation Urgent." Secret agents unfold
a clever scheme which threatens the security of the country. Kirby Grant as Sky
King, Gloria Winters.

8:30 ② **TALENT SCOUTS—Godfrey**
Three contestants vie for top score.

④ ⑧ **COMMENT—** DEBUT
First in a series of news documentaries
featuring prominent news commentators
interpreting international events.

⑤ **LIFE WITH ELIZABETH**
Domestic misadventures with Betty White.

⑦ ⁴³ **CONCERT—Barlow**

8:55 ⑨ ⑪ ⑬ **NEWS BULLETINS**

9:00 ② ⑧ **I LOVE LUCY—Comedy**
"The Kleptomaniac." Lucy tries madly to
find small objects she can use in the forthcoming church bazaar. Ricky is sure his
wife is a kleptomaniac. Lucille Ball, Desi
Arnaz, Vivian Vance, William Frawley.

④ **DENNIS DAY—Comedy**
Dennis shuttles back and forth between
his apartment and another, trying to have
dinner with his current girl friend and his
former college sweetheart at one and the
same time. Lois Collier, Barbara Ruick, Cliff
Arquette. *Sponsored by RCA Victor.*

⑤ **BOXING—Chris Schenkel**
Preliminaries from St. Nick's.

⑦ ⁴³ **TALENT PATROL—Variety**
Arlene Francis emcees Army talent show.

⑨ **MOVIE—Adventure**
"The Prairie." (1948) Story of a girl captured by the Indians, and her sweetheart
who is unwittingly involved in a murder.
Alan Baxter, Lenore Aubert, Russ Vincent.

⑪ **TENNIS—Don Budge**
National Round Robin Tournament.

A-29

TUESDAY, OCTOBER 19, 1954

④ ⑧ MILTON BERLE—Comedy
Milton, Maxie, Nancy Walker and Arnold Stang in some zany doings.

⑤ ㊸ THE GOLDBERGS
Molly is in a huff. The neighbors are planning a party and she hasn't been invited. What Molly doesn't know is that the ladies plan a surprise nomination of "Berg for president" at the Ladies Auxiliary. Gertrude Berg, Eli Mintz, Robert H. Harris, Arlene McQuade, Tommy Taylor. Final show of the series.

⑦ FILM DRAMA
"Good-bye to the Clown." A little girl creates an imaginary clown to escape the memory of her father's death. Gigi Perreau is starred.

8:30 ② THE HALLS OF IVY
[DEBUT] Ronald Colman and his wife, Benita Hume, star as President and Mrs. Hall of Ivy College. They are at home at Number One, Faculty Row, to all who come with problems of campus life. In the opening episode, Dr. William Todhunter Hall awaits nervously the verdict of the board of governors on a five-year extension of his term as president. As they wait, the Halls recall events from the past. (Film)

Cast
Dr. William Todhunter Hall . Ronald Colman
Vicky . . . . Benita Hume
Alice, the housekeeper . . . . Mary Wickes
Clarence Wellman . . . . . Herb Butterfield
Mr. Merriwether . . . . Ray Collins

⑤ ONE MINUTE PLEASE
The popular panel show features Hermoine Gingold, Marc Connelly, Cleveland Amory and other gifted conversationalists in one-minute discussions of almost anything. John K. M. McCaffery emcees.

⑦ ㊸ 20 QUESTIONS—Panel
Panelists: Fred Van Deventer, Florence Rinard and Bobby McGuire, Jay Jackson moderates. Guest: Harnett T. Kane, noted Southern author and critic. Sponsored by Florida Citrus Commission.

8:55 ⑪ NEWS BULLETINS

9:00 ② ⑧ MEET MILLIE—Comedy
Another turbulent episode. Elena Verdugo.

④ FIRESIDE THEATER
"The Man Who Sold Himself." A young man raffles off his services for one year in order to get a stake and leave a pioneer western town for the big city. The winner is a lovely lady. Dan Barton, Betty Lynn, Sheila Connolly, Kay La Velle, Hugh Sanders, June Evans. (Film)

⑤ STUDIO 57—Film Drama
"The Plot Against Miss Pomeroy." A timid teacher at an exclusive girls' finishing school is asked to do a special favor for a rich pupil. The pupil shows her gratitude by planning a practical joke—the butt, the teacher. Jean Bryon, Natalie Wood, Eleanor Audley, Gay Lynn, Morris Ankrum, Pauline Moore. (Film)

⑦ ㊸ DANNY THOMAS
"Danny Has a Baby." The days of the diaper detail are recalled to Danny when his wife returns home with a neighbor's child. She wants Danny to show off his fatherly talents to Rusty and Terry. Jean Hagen, Rusty Hamer, Sherry Jackson. (Film)

Is it animal... vegetable... or mineral?

Watch **20** QUESTIONS
8³⁰ - 9 P.M. Channel 7

A-41

134

Sydney Tafler, Bill Owen, Charles Farrell, Robert Adair.

⑬ MOVIE—Crime
"Black Market Babies." A gangster takes over the mob's leadership, makes a doctor his partner, and begins operating the racket of black-marketing babies. Ralph Morgan, Kane Richmond.

7:45 ② JANE FROMAN—Songs
Jane salutes the electric light and Thomas Edison with: "Diamonds Are a Girl's Best Friend," "The Old Lamplighter," "I Believe in Miracles" and "I Believe."

④ ⑧ NEWS—John C. Swayze

8:00 ② RAY MILLAND—Comedy
Peggy McNulty's sub-teen-aged cousin enrolls in Comstock College as a freshman. Prof. McNulty finds that matching wits with a boy genius can be anything but pleasant. Ray Milland, Phyllis Avery, David Stollery. (Film)

④ ⑧ GROUCHO MARX—Quiz
Groucho interviews contestants from the studio audience and quizzes them.

⑤ THEY STAND ACCUSED
Another courtroom drama from Chicago.

⑦ ㊷ THE MAIL STORY
The series explores the various departments of the U. S. Post Office. Investigations of attempts made to misuse the U. S. mails will be featured.

8:30 ② CLIMAX—Drama
"Casino Royale" stars Barry Nelson, Linda Christian and Peter Lorre in Anthony Ellis's adaptation of the recent Ian Fleming novel. James Bond is assigned by the Secret Service to destroy the power of "Le Chiffre" ("The Chief"), a powerful Communist agent in France. Bond's plan is to force "Le Chiffre" to lose a large sum of money at the gambling tables of Monte Carlo's famed Casino Royale.

Cast

James Bond	Barry Nelson
Valerie	Linda Christian
Le Chiffre	Peter Lorre

Sponsored by Chrysler Corp.

④ JUSTICE—Drama
"The Safecracker." A convicted safecrack-

THURSDAY NIGHT, OCT. 21

"CLIMAX!"

FULL HOUR OF TENSE DRAMA DIRECT FROM TELEVISION CITY, HOLLYWOOD

PETER LORRE
LINDA CHRISTIAN
BARRY NELSON
IN
"Casino Royale"

PRESENTED BY

CHRYSLER CORPORATION

Plymouth • Dodge • DeSoto
Chrysler • Imperial

8:30 to 9:30 EST
CHANNEL 2

A-55

WEDNESDAY OCTOBER 27

made rifle. Liphalet Remington builds his own rifle on his father's forge and enters a shooting match to prove that the American made goods can compete with the more expensive European made items. George Nader, Kathleen Crowley. (Film)

⑨ COWBOY G-MEN—Kids
"Safecrackers," Russell Hayden.

⑪ NEWS—Kevin Kennedy
⑬ NEWS, SPORTS, WEATHER
7:10 ⑪ **WEATHER—Joe Bolton**
7:15 ⑤ **NEWS—Morgan Beatty**
⑦ ㊸ NEWS ANALYSIS
John Daly at the microphone.

⑪ NEWS—John Tillman
⑬ PHOTO QUIZ
7:20 ② **N.Y. DEMOCRATIC COMM.**
7:25 ② **WEATHER—Carol Reed**
⑪ SPORTS—Jimmy Powers
7:30 ② **NEWS—Commentary**
Douglas Edwards is the reporter.

④ EDDIE FISHER—Songs
Axel Stordahl and his orchestra accompany Eddie, and Fred Robbins emcees.

⑤ LIFE WITH ELIZABETH
⑦ ㊸ DISNEYLAND
[DEBUT] The veteran animator of cartoons and producer of realistic nature films comes to TV with a weekly series. Tonight he introduces both the show, Disneyland, and the amusement park of the same name now under construction in California. Show begins with a magic-carpet journey to the Walt Disney Burbank studios, a trip around the world and then a trip to outer space. Then Mickey's creator will salute the great mouse by presenting the memorable cartoon, "The Sorcerer's Apprentice," about a magic broom that gets out of control. Music is the famous score by Paul Dukas, and Mickey is the frantic apprentice. For a description of the series, turn to our story starting on page 5.

⑧ CONNECTICUT SPOTLIGHT
⑨ MOVIE—Drama
Million Dollar Movie: "The Private Affairs of Bel Ami." NY TV debut. (1947) De Maupassant's story of a man who uses women as tools in his attempt to climb to the top of Parisian society. George Sanders, Angela Lansbury, Ann Dvorak, Warren William, Frances Dee, Marie Wilson, John Carradine, Hugo Haas, Albert Basserman, Katherine Emery.

⑪ MOVIE—Western
First Show: "Stampede." NY TV debut. (1949) When a group of settlers in the West discover they have no water supply, a conflict develops with two neighboring cattlemen whose land is fertile.

Cast
Mike .... Rod Cameron
Connie .... Gale Storm
Tim .... Don Castle
Sheriff Ball .... Johnny Mack Brown
Stanton .... Don Curtis

⑬ HARRISON WILLIAMS—Talk
7:45 ② **PERRY COMO—Music**
④ ⑧ NEWS—John C. Swayze
⑬ MOVIE—Crime
"The Old Homestead." A woman mayor poses as a crook in order to fight crime. Anne Jeffreys, Dick Purcell, Leon Weaver, Frank Weaver, June Weaver.

8:00 ② ⑧ **GODFREY & FRIENDS**
Frank Parker, Marion Marlowe, Janette Davis, Haleloke, Lu Ann Simms.

④ I MARRIED JOAN—Comedy
Brad's colleague, Judge Cushing, has a wife who helps him with his work, so Joan decides to favor Brad with her able assistance. In order to do this, our earnest but dizzy heroine enrolls in secretarial school. Joan Davis, Jim Backus, Emlen Davies, Paul Keast. (Film)

⑤ COUNTERPOINT
"The Money." A harried husband, worried over the imminent arrival of his second child and a stack of unpaid bills, suddenly finds himself in possession of $20,000. John Doucette, Elizabeth Fraser. (Film)

8:30 ④ **MY LITTLE MARGIE**
"The Shipboard Story." Vern is on his way to London to settle the estate of a new earl. Margie wants to go along, so she pretends, with the help of a handsome stranger, that she is going to elope. Gale Storm, Charles Farrell, John Lupton. (Film) How old is Mrs. Odetts. See page 15.

A-46

For Week Beginning December 11th, 1954

SATURDAY
MORNING

6:25 ④ **Sermonette**
6:30 ④ **Modern Farmer**
6:45 ② **Previews**
6:55 ② **Give Us This Day**
7:00 ② **Cartoon Carnival**
"Five Little Pups," "Mouse Trapper," "Canine Commandos."
  ④ **Saturday—Herb Sheldon**
7:30 ② **MOVIE—Western**
"Bad Men of Nevada." A bandit decides to leave his gang and go straight. Jean Parker, Russ Hayden.
8:30 ② **Junior Sports Session**
Frankie Frisch presents track team of St. John's Prep School.
9:00 ② **On the Carousel—Kids**
Paul Tripp presents his version of the story of Rip Van Winkle, with thespians Ruth Anders, Joe Silver and Ted Tiller. "The Little Ballerina," a story with dance, is performed by three little ballerinas coached by Alma Kaaber. Ted Kazimiroff, boy naturalist, presents his collection of bugs, snakes, toads, etc. We see old-fashioned toys from the Museum of the City of New York and exhibits from the current Science Fair. Take a Tripp on the carousel in next week's TV GUIDE.
  ④ **Children's Theater-Forrest**
  ⑦ **MOVIE—Western**
"Caravan Trail," Eddie Dean.
9:15 ⑧ **Mr. Wizard—Science**
Highlights of the most interesting experiments done by Mr. Wizard (Don Herbert) and Buzz (Bruce Podewell) in the past will be repeated. (Film)

9:45 ⑧ **Barker Bill's Cartoons**
10:00 ② **Time for Beany—Puppets**
  ④ **Happy Felton—Kids**
Happy shows the children films of Krone's Seals; The Two Luvas, high wire act; The Three Lesters, trampoliners.
  ⑦ **Rin Tin Tin—Drama**
"The Outcast of Fort Apache." Rusty's good friend, Lt. Masters, is accused of cowardice and drummed out of the Army. Lee Aaker, James Brown. (Film)
  ⑧ **Captain Midnight—Kids**
"The Deserters." The young members of the Secret Squadron tell Captain Midnight that they are being ejected from their headquarters. Richard Webb. (Film)
10:30 ② **Winky Dink and You**
The kids at home help Jack and Winky Dink by drawing a grandfather's clock. Adventures of Mysto the Magician.
  ④ **Winchell & Mahoney**
A thunderstorm strikes while the gang is holding a meeting at the clubhouse. The members soon realize that the roof leaks like a sieve. The Milton De Lugg Trio.
  ⑤ **MOVIE—Western**
"Romance of the West," Eddie Dean.
  ⑦ ⑧ **Smilin' Ed's Gang**
Smilin' Ed relates the story of Gunga and his pet dog, Nabu, who go after a wounded leopard terrorizing the village.
11:00 ② **Captain Midnight—Kids**
"The Electrified Man." Professor Berglund of Captain Midnight's laboratory is working on an electric charge which will immunize humans against radioactive dust. An accident during the experiment turns him into a danger to the United States. Richard Webb, Ian Keith, Olan Soule.

⑥ Film Short
6:45 ⑪ Telepix News—McCarthy
⑬ News Report
6:55 ⑪ Weather Report
7:00 ② LASSIE—Drama
"The Fawn." Jeff and Gramps find a two-weeks-old fawn in the woods. Jeff wants to keep the helpless animal, but Gramps and Doctor Wilson insist that dogs and deer don't mix. Tommy Rettig, George Cleveland, Jan Clayton. (Film)

④ ⑧ PEOPLE ARE FUNNY
Art releases some more helium balloons with the $2000 riddle in them. A married couple is invited to a fashionable party under unusual stipulations. Art determines whether a man is more loyal to his wife or to his best friend. (Film)

⑤ HEART OF THE CITY
⑦ ⑥ YOU ASKED FOR IT
Art Baker presents sequences from the new movie, "The Bridges at Toko-Ri." Footage will consist of fighter planes taking off and landing aboard a carrier, and the activities of a landing-signal officer, portrayed by actor Robert Strauss. Film clips of tennis star Jack Kramer.

⑪ HY GARDNER CALLING
⑬ GOV. MEYNER'S REPORT
The governor discusses legislative proposals made in his annual message.

7:30 ② JACK BENNY—Comedy
Jack welcomes back his old violin-playing partner, singer Gisele MacKenzie. The Hit Parade thrush has come West for tonight's show . . . only to find that she's implicated in the theft of Jack's precious violin. Jack and Rochester join in the search. Gisele sings "Mr. Sandman," accompanying herself at the piano. Eddie "Rochester" Anderson, Don Wilson, Mel Blanc. (Hollywood)

④ MR. PEEPERS—Wally Cox
Robinson's ritzy English shipmate urges him to join him in a trip up to the first-class deck. In progress, there, are formal festivities celebrating the last night aboard ship. Cyril Ritchard.

⑤ OPERA CAMEOS
Giovanni Martinelli introduces Donizetti's

"Don Pasquale," in which Don Pasquale prevents his nephew's marriage to Norina. In revenge, Norina charms the wealthy Don into marriage and spends all his money. Salvatore Baccaloni, Florence George Crosby, Charles Anthony.

⑦ ⑥ PLAYHOUSE—Bergen
There's terror in "The House on Judas Street," where a schoolteacher has rented a room with a most unusual family. Jean Byron, Peter Votrian. (Film)

⑨ MOVIE—Melodrama
Million Dollar Movie: "LuLu Belle." See Sat. 7:30 P.M. (9) for details.

⑪ DATELINE EUROPE—Drama
"The Troop Train." Robert Cannon changes his travel plans when at a chance meeting at a Frankfort airport he encounters an ambitious West German agitator. Jerome Thor. (Film)

⑬ N. J. LEGISLATIVE REPORT
8:00 ② ⑧ TOAST OF THE TOWN
Ed's guest list is headed by actor Robert Taylor, who introduces a film clip from "Many Rivers to Cross," in which he stars with Eleanor Parker and Victor McLaglen. Metropolitan Opera star Roberta Peters and William Warfield offer a singing salute to Jerome Kern. Also on the agenda: The dancing of Tony and Sally DeMarco; Eugenie Leontovich and Viveca Lindfors playing their big scene from the Guy Bolton drama about the missing Russian Czarist princess, "Anastasia"; the antics of comedienne Sybil Bowen and the music of the 30-youngster Alfredi's all-accordion symphony band, which plays the Kern melodies, "Who" and "I've Told Ev'ry Little Star." Ed discovered a new secret for success: "Get yourself insulted regularly." Page 5.

④ COMEDY HOUR—MacRae
Host Gordon MacRae introduces Gloria Vanderbilt and Franchot Tone, who star in a TV adaptation of the Irwin Shaw story, "In the French Style." An American arranges a cafe rendezvous with the girl he forsook to travel in Egypt. He realizes now, that he has always loved her. Those appearing in tonight's variety acts are:

TUESDAY, FEBRUARY 8, 1955

TUESDAY
FEBRUARY 8

4 Kit Carson—Adventure
"Devils Renueda," Bill Williams.

5 Magic Cottage-Pat Meikle
"A Valentine for Two Gun Tim" continues. Timothy's gun is of no value to him when he comes up against a problem of the heart.

7 Gloria Swanson Show
Gloria Swanson stars in "Short Story." A famous woman novelist becomes intrigued with a struggling young commercial artist. When they marry, the newspapers have a field day and label her young husband "The Cinderella Man."

8 Stage 8—Drama

9 Merry Mailman—Kids

11 Ramar of the Jungle

13 MOVIE—Western
"Desert Bandit," Don Barry.

43 News Report

6:05 2 Feature—Bill Leonard
43 Sports—Manning Slater

6:10 2 Sports—Jim McKay

6:15 2 MOVIE—Mystery
Early Show: "Terror by Night." (1946) Sherlock Holmes is called upon to protect a fabulous diamond against jewel snatchers. Basil Rathbone, Nigel Bruce, Alan Mowbray.
43 Wallie Dunlap—Variety

6:25 7 Film Short

6:30 4 Sky's the Limit—Quiz
5 The Old-Timer—Kids
7 Files of Jeffrey Jones
"The Evil Ones." An exotic woman hires Jeff Jones to secretly investigate the strange behavior of her sister. Don Haggerty, Frank Sully. (Film)
8 Sports—Syd Jaffe
11 Liberace—Music
"Bumble Boogie," "Tales of the Vienna Woods," "I Miss You So," "Blue Tango."

6:40 8 Weather Forecast

6:45 4 News—John Wingate
5 News—Larry McNamara
9 News—Lyle Van
43 Family Rosary—Religion

6:55 4 Weather—Tex Antoine
5 Weather—Janet Tyler
9 Sports—Jack O'Reilly

7:00 4 ELLA RAINES—Drama
"The Hutchins Case." Janet Dean discovers her charge, a young woman who attempted to commit suicide, has given her baby to a wealthy woman for adoption under unusual circumstances.
5 43 CAPTAIN VIDEO-Serial
The Video Rangers pursue Clipper Evans.
7 KUKLA, FRAN & OLLIE
8 WILD BILL HICKOK
"Jingles' Disguise." Jingles disguises himself as a peddler's wife to find a gang of outlaws. Guy Madison. (Film)
9 HOLLYWOOD PREVIEW
11 NEWS—Kevin Kennedy
13 MOVIE—War Drama
"The Purple V." An American flyer is lost behind German lines. John Archer.

7:10 11 WEATHER—Joe Bolton

7:15 5 NEWS—Morgan Beatty
7 43 NEWS—John Daly
11 NEWS—John Tillman

7:25 2 WEATHER—Carol Reed
11 SPORTS—Jimmy Powers

7:30 2 NEWS—Douglas Edwards
4 DINAH SHORE—Music
Back in Hollywood again, Dinah and the Skylarks sing "I've Got My Love to Keep Me Warm," "We Just Couldn't Say Goodbye," "Hooray for Love" and "Old Man River."
5 WATERFRONT—Drama
"Cap'n Christopher." Cap'n John tangles with the officers of a foreign ship over a Polish stowaway seeking freedom in America. Preston Foster. (Film) *Series reflects careful handling in all phases, says our reviewer on page 21.*
7 43 CAV'LC'DE OF AMERICA
"New Salem Story." The legendary tale of Abe Lincoln and his great love for Ann Rutledge. Young Lincoln, just starting in politics, falls tragically in love with a woman who is fated to die before the two can wed. (Film)

Cast
Abe Lincoln .... James Griffith
Ann Rutledge .... Jeff Donnell

9:00 ④ ACADEMY AWARD NOMINATIONS

SPECIAL Jack Webb takes a short holiday from homicide to emcee an hour-and-a-half program to select the nominees for the Academy Awards ceremony, March 30. TV camera crews will cover the excitement in the motion-picture studios, noted supper-clubs and hotels for the announcement of the chosen on this fateful evening. From the slate of glittering names selected tonight will come the eventual Oscar winners for the

Webb

year. This is the first time that Hollywood's nominations have been televised nationally.

Viewers will be taken through the studios and other locations in the film capital. Plans also include a behind-the-scenes look at set designers at work, special effects, a tour of the Disney studios.

Outstanding personalities of the motion-picture industry will participate in the festivities from NBC's studios in Burbank, Cal.

The following are some of the "most likely to succeed" in the Acting, Directing and Production categories:

BEST PERFORMANCE BY AN ACTRESS

Garland   Hepburn

Kelly   Saint   Taylor

Judy Garland.................."A Star is Born"
Audrey Hepburn........................"Sabrina"
Grace Kelly................."The Country Girl"
Eva Marie Saint........."On the Waterfront"
Elizabeth Taylor..."The Last Time I Saw Paris"

BEST PERFORMANCE BY AN ACTOR

Bogart   Brando

Crosby   Ferrer   Holden

Humphrey Bogart............"The Caine Mutiny"
Marlon Brando.............."On the Waterfront"
Bing Crosby................."The Country Girl"
José Ferrer................."The Caine Mutiny"
William Holden............."The Country Girl"

BEST PERFORMANCE BY A SUPPORTING ACTRESS

Foch   Ritter   Sterling

Nina Foch...................."Executive Suite"
Thelma Ritter..................."Rear Window"
Jan Sterling.........."The High and the Mighty"

BEST PERFORMANCE BY A SUPPORTING ACTOR

Cobb   Malden   O'Brien

Lee J. Cobb................."On the Waterfront"
Karl Malden................."On the Waterfront"
Edmond O'Brien........."The Barefoot Contessa"

BEST DIRECTOR

George Cukor.................."A Star is Born"
Alfred Hitchcock..............."Rear Window"
Elia Kazan..............."On the Waterfront"
George Seaton..............."The Country Girl"
William Wellman.."The High and the Mighty"

BEST PRODUCTION

"A Star is Born"..................Warner Bros.
"On the Waterfront"..................Columbia
"Robinson Crusoe"............United Artists
"The Egyptian"............20th Century-Fox
"The High and the Mighty"......Warner Bros.

MONDAY, MARCH 7, 1955

7:30 ❹ ❽ PRODUCERS' SHOWCASE

MARY MARTIN *in* "PETER PAN"

SPECIAL COLOR Two-hour musical version of Sir James M. Barrie's fantasy about the Lost Boys who live in Neverland, where they never grow up, and the Darling children's adventures there.

ACT I
Scene 1. NURSERY
"Tender Shepherd"..........Mrs. Darling, Wendy, John, Michael
"I've Got to Crow".....................................Peter
"Neverland" ....Peter
"I'm Flying".........Peter, Wendy, John, Michael

Scene 2. FLIGHT TO NEVERLAND

ACT II
Scene 1. NEVERLAND
"Pirate Song"..............................Hook, Pirates
"A Princely Scheme" ....Hook, Pirates
"Indians!".........................Tiger Lily, Indians
"Wendy".................................Peter, Boys
"Another Princely Scheme" ....Hook, Pirates
"Neverland Waltz"...............................Liza

Scene 2. FOREST PATH
"I Won't Grow Up".........................Peter, Boys
"Mysterious Lady"..................Peter, Hook

Scene 3. NEVERLAND
"Ugg-a-Wugg" } ----------- { Peter, Tiger Lily,
"Powwow Polka"   Children, Indians
"Distant Melody".................................Peter

ACT III
Scene 1. PIRATE SHIP
"To the Ship"...................Peter, Ensemble
"Hook's Waltz"....................Hook, Pirates
"The Battle"..............Peter, Hook, Ensemble

Scene 2. FOREST PATH
Reprise: "I've Got to Crow"
Peter, Liza, Ensemble

Scene 3. NURSERY
Reprise: "Tender Shepherd"
Wendy, John, Michael
Reprise: "I Won't Grow Up"
Darling Family, Lost Boys

Scene 4. NURSERY, YEARS LATER
Reprise: "Neverland".............................Peter

CAST

Peter Pan...Mary Martin
Mr. Darling } ....Cyril Ritchard
Capt. Hook }
Mrs. Darling...........................Margalo Gillmore
Wendy....................................Kathy Nolan
John...............................Robert Harrington
Liza.................................Heller Halliday
Michael..............................Joseph Stafford
Nana (the Dog) } ....Norman Shelly
Crocodile }
Smee.....................................Joe E. Marks
Tiger Lily.................................Sondra Lee

Mary Martin

ALSO . . . Lion: Richard Wyatt. Kangaroo: Carle Erbele. Ostrich: Joan Tewkesbury. Slightly: David Bean. Tootles: Ian Tucker. Curly: Stanley Stenner. Nibs: Paris Theodore. First Twin: Jackie Scholle. Second Twin: Darryl Duran. Starkey: Robert Vanselow. Cecco: Richard Winter. Noodler: Frank Lindsay. Mullins: James White. Jukes: Frank Bouley. Wendy (grown-up): Ann Connolly. Jane: Kathy Nolan.

CREDITS . . . Music: Mark Charlap, Jule Styne. Lyrics: Carolyn Leigh, Betty Comden, Adolph Green. Guest Producer: Richard Halliday.

141

MONDAY, MARCH 28, 1955

MONDAY   MARCH 28

⑪ **New York Calendar**
⑬ **Shop, Look, Cook—Bean**
12:45 ② **Guiding Light—Serial**
⑧ **Electric Show—Baker**
Guests: Joanne Burnett, from the John Robert Powers School, will give advice on selecting Easter hats, and Gertrude Norcross speaks about Easter seals.

1:00 ② **The Inner Flame—Serial**
Staley triumphs in the arrest of Walter.
④ **News—Kenneth Banghart**
⑤ **Glamor Secrets—Mann**
⑨ **Public Service Film**
⑪ **Travel Film**
⑬ **Musical Jackpot—Quiz**

1:05 ④ **Norman Brokenshire Show**
Harry Snow replaces Bill Hayes.

1:15 ② **Road of Life—Serial**
1:30 ② ⑧ **Welcome Travelers**
④ **Beauty Advice—Willis**
⑤ **Food for Thought**
⑧ **MOVIE—Drama**
"Lost Youth." (Italian; 1949) A young detective poses as a college student in order to infiltrate a gang of youthful hoodlums. Massimo Girotti.
⑪ **Opinion Please-Interview**

2:00 ② ⑧ **Robert Q. Lewis-Variety**
[COLOR] "We Love Color TV" .... All "Hard-Hearted Hannah" .... Robert Q. "Friendship" Jaye P. Morgan, Don Liberto
④ **Big Matinee—Drama**
"Magic Interlude." A man falls in love with an attractive girl at a sales convention. Damian O'Flynn. (Film)
⑤ **Maggi McNellis—Women**
Host of the week: James Strauss shows several types of tea. Instructor John Vassoe of the Silvermine Guild Art School and two students show paintings.
⑪ **Dione Lucas—Cooking**
⑬ **MOVIE—To Be Announced**

2:30 ② **Linkletter's House Party**
Guest: Dr. Paul Popenoe.
④ **Jinx Falkenburg's Diary**
Guests: "Tony" Award winners.
⑤ **Letter to Lee Graham**
⑧ **Guiding Light—Serial**
⑨ **Liberace—Music**
"Stormy Weather," "Over the Rainbow."

2:45 ⑧ **News—Joe Burns**
3:00 ② ⑧ **The Big Payoff—Quiz**
Bess Myerson parlayed a winning figure into six figures a year. Story on page 15.
④ **The Greatest Gift—Serial**
Phil hears tragic news about Eve.
⑤ **Paul Dixon Show—Music**
⑦ **Romantic Interlude**
"The Trail." The description of an outlaw-murderer tallies with that of the son of an upright citizen. Douglas Kennedy.
⑨ **Ted Steele—Variety**
Ceil does reducing exercises.
⑪ **Bob Kennedy—Variety**
3:15 ④ **Golden Windows—Serial**
Julie fears that she has shot Carl.
3:30 ② **Bob Crosby—Music**
④ ⑧ **One Man's Family**
Father Barbour hears news of Cliff.
⑦ **Memory Lane—Franklin**
⑬ **Fun Time—Kids**
3:45 ④ ⑧ **Miss Marlowe—Serial**
Belle Mere refuses Barbara's request.
3:55 ⑪ **News—Kevin Kennedy**
4:00 ② **Brighter Day—Serial**
Skip visits Rev. Dennis.
④ ⑧ **Hawkins Falls—Serial**
Mitch Fredericks enlists the help of Millie.
⑤ **MOVIE—Mystery**
"The Case of Charles Peace." (English; 1949) Late-Victorian drama about a series of burglaries and the murder of a police constable. Michael Harve.
⑦ **Hopalong Cassidy**
"Cassidy of Bar 20." Hoppy gets himself convicted on trumped-up charges in order to smoke out a crook. William Boyd. (Film)
⑪ **Bob Kennedy—Variety**
⑬ **MOVIE—Western**
4:15 ② ⑧ **Secret Storm—Serial**
④ **First Love—Serial**
Andrews warns Zach. Val Dufour.
4:30 ② **On Your Account—Quiz**
④ **The World of Mr. Sweeney**
"The Prize." Charles Ruggles is Sweeney. Ed Herlihy, announcer on the show, is his own best customer. Details on page 10.
⑧ **Meet the Stars—Variety**
4:45 ④ **Modern Romances**
"The Doubt," by Jesse Sandler, is intro-

THURSDAY, SEPTEMBER 1, 1955

⑪ TROTTING—Yonkers
Tonight's feature race is the Yonkers Futurity Trot, trotting's second most lucrative race, a mile-and-a-sixteenth contest for a purse of approximately $75,000.

9:30 ② FOUR STAR PLAYHOUSE
Dick Powell stars in "The Returning." A professor at a school in Tokyo falls in love with a Eurasian, just prior to World War 2. She is not accepted socially and he is forced to choose between his love for her and his career. (Film)

Cast

Capt. Avery	Dick Powell
Laura	Joan Elan
Maj. Evans	Christopher Dark
Co-Pilot	Bob Stevenson
Navigator	Bill Boyett
Cory Howard	Walter Kingsford

④ ⑧ FORD THEATRE—Drama
Charles Coburn stars in "Pretend You're You." An honest boy meets and falls in love with an engaged girl. His not-so-honest family decides to "help" the romance along. (Film)

Cast

Uncle Henry	Charles Coburn
Freddy Remington	Keith Andes
Sue Smith	Lucy Marlow
Aunt Clara	Nana Bryant
Amy Remington	Joan Hovis

⑦ GREATEST SPORTS THRILLS
⑬ SONGS OF YESTERYEAR
Jimmy Shearer sings: "Alice, Where Art Thou Going?" "You're a Million Miles from Nowhere," "Oh, You Million Dollar Doll," "I'm Waiting for Ships That Never Come In," "Red Sails in the Sunset."

④⑬ TOP SECRET—Drama
"Feedback." A company, engaged in making equipment for rocket research, reports that its top machinery has gone haywire. Powell of the Bureau of Science Information checks. Paul Stewart. (Film)

9:45 ⑬ INDUSTRY ON PARADE
1. A visit to the modern oil city of Maracaibo, Venezuela.
2. Building airplanes that fly very slowly.
3. Making heat-resistant brick.
4. Making the masks for masquerades.

10:00 ② JOHNNY CARSON—Variety
Johnny's sketches include a satire on mental wizards. Johnny can't remember the name of the product but he can remember how the commercial goes. Jill Corey is the featured singer. (Hollywood)
Johnny holds forth with some of his comedy ideas in next week's TV GUIDE.

④ ⑧ VIDEO THEATER
"Kenny." A returned Marine and former high school sports star, is looked on as a hero by his father and mother. But Kenny's wife and mother are exasperated when he won't work, cheats and steals. Kenny finds it hard to explain when he becomes involved in a case of murder. (Hollywood)

⑦ OUTSIDE U.S.A.—Howe
[DEBUT] Analyst Quincy Howe in a new half-hour series devoted to the international scene.

SATURDAY   SEPTEMBER 10

⑬ GERMAN VARIETY

8:30 ④ THE DUNNINGER SHOW
Dunninger's guest is U.E. Braughman, chief of the U.S. Secret Service. The master mentalist will attempt to take Braughman out of the "skeptics' corner."

⑬ BIG PICTURE—Army Film

9:00 ② TWO FOR THE MONEY
Hoosier humorist Herb Shriner is back from a summer of easy livin' to take over emcee chores. Dr. Mason Gross judges the accuracy of the answers given by tonight's contestants as they battle the ticking clock.

④ ⑧ MUSICAL CHAIRS—Quiz
Singer-comedienne Rose Marie sits in with panelists Johnny Mercer, Mel Blanc and Bobby Troup. Bill Leyden emcees.

⑤ COUNTERPOINT—Drama
"The Old Man." An old man, losing contact with his friends and neighbors, draws attention to himself by making nocturnal trips deep into the woods, armed with all manner of mining equipment. Houseley Stevenson.

⑦ LAWRENCE WELK—Music
"Jolly Caballero"	Myron Floren
"Many-Splendored Thing"	Jim Roberts
"School Days"	Alice Lon
"Tina Marie"	Bob Lido
"Angel Bells"	Alice
What's the score on this musical show? Review on page 14.

⑨ MOVIE—Drama
"Derby Day." (English; 1955) It's the big day at Epsom Downs, and to it come: an attractive widow who scandalizes her household by going to the Derby; a newspaper cartoonist who doesn't care about horses; and an elegant film star who's been "won" in a raffle as escort to Mrs. Harbottle-Smith. Anna Neagle, Michael Wilding, Googie Withers.

⑪ MOVIE—Western
"Six-Gun Serenade," Jimmy Wakely.

⑬ CARNAVAL HISPANO

9:30 ② IT'S ALWAYS JAN
DEBUT Janis Paige stars as a widowed night-club entertainer who shares an apartment with her 10-year-old daughter

and two friends, a secretary and a model. In the first episode Jan is upset about her daughter's behavior. The girl seems too mature for her age. The only solution seems to be to move out of the apartment. Jan sings "Nothing Could Be Finer." (Film)

Cast
Jan Stewart	Janis Paige
Josie Stewart	Jeri Lou James
Pat Murphy	Patricia Bright
Val Marlowe	Merry Anders
Stanley Schreiber	Arte Johnson

④ ⑧ DONALD O'CONNOR
Donald feels its time his career got a boost, and hires a publicity agent. In a flashback sequence we meet some of Donald's ancestors. As a riverboat gambler Don sings "He's My Don" and "Li'l Ole You." Sidney Miller and Donald are seen in a satire on British commercials. Don sings "Aren't People Nice?" Also in the cast are Lisa Davis, Douglas Fowley, Joyce Smight and Nestor Paiva. (Film) *This is the last Donald O'Connor Show.*

⑤ AMERICAN BARN DANCE
Uncle Tom invites us in to another session of country songs and dances. Emcee Bill Bailey introduces tonight's cast: singers Mary Jane, Kay Brewer, Tex Williams, Homer and Jethro, Doc Hopkins, the Candy Mountain Girls; and comic "Cousin Alvin." Tunes include: "Pretty Little Pink," "I Can't Help It," "Erie Canal," "So Long No. 2." (Film)

10:00 ② GUNSMOKE—Western
DEBUT A new Western adventure series starring James Arness as Dodge City Marshal Matt Dillon. A gunman, hunted by Texas authorities, seeks refuge in Dodge City in "Matt Gets It." Dillon goes into action against this outlaw who shoots before he talks. (Film)

Cast
Matt Dillon	James Arness
Dan Grat	Paul Richards
Kitty	Amanda Blake
Cheater	Dennis Weaver
Doc	Milburn Stone

MONDAY, SEPTEMBER 19, 1955

11 LIBERACE—Music
"Hi Neighbor," " Malagueña," "Canadian Capers," "Nearness of You." (Film)

5 FILM SHORTS

7:45 **2 JULIUS LA ROSA—Songs**

4 8 NEWS—John C. Swayze
Mr. Swayze makes a strong case for TV's
benefits to children. Story on p. 20.

8:00 **2 BURNS & ALLEN—Comedy**
George is somewhat taken aback when
Gracie begins interviewing applicants for
the next Mrs. George Burns! Gracie mistakenly believes she is not long for this
world, and wants to see that George will
be taken care of. Daughter Sandra
Burns is featured. (Film)

8:00 4 8 PRODUCERS' SHOWCASE
OUR TOWN by Thornton Wilder
starring

| Frank SINATRA | Eva Marie SAINT | Paul NEWMAN | Paul HARTMAN | Ernest TRUEX | Sylvia FIELD |

music by James VAN HEUSEN • lyrics by Sammy CAHN

[COLOR] "Our Town," 17 years old now,
has become a musical. David Shaw has adapted
the script. The songs have been written by
James Van Heusen and Sammy Cahn.

Sinclair Lewis familiarized a generation of
Americans with a satiric view of the American
small town. Wilder's play reverses the trend,
and takes a loving look at the inhabitants of
the typical town of Grover's Corners. The plot
is pegged on the lives of a boy and a girl in
the town, and follows them through romance,
marriage and death.

Act 1: It's May 1901. The act opens at dawn
on a street in Grover's Corners. The cameras
rove through the Gibbs' and Webbs' homes
and the office of the Grover's Corners Sentinel.
Songs: "Our Town"..."Grover's Corners"...
"The Impatient Years."

Act 2: Now it's July 1904. Again we enter the
Gibbs' and Webbs' homes, then travel to an
area near the school, the drugstore soda foun-
tain and the church.
Songs: "Love and Marriage"..."A Perfect
Married Life"..."The Impatient Years" (reprise)
..."Wasn't It A Wonderful Wedding?"

Act 3: Nine years have passed. It's summer
1913. The scene is the cemetery. Flashbacks
take us back to 1899. We return to the town
in 1913, where the play closes.
Song: "Look to Your Heart."

Paul Newman

Frank Sinatra

Eva Marie Saint

Cast

Stage Mgr. (Mr. Morgan)	Frank Sinatra
Emily Webb	Eva Marie Saint
George Gibbs	Paul Newman
Dr. Gibbs	Ernest Truex
Mrs. Gibbs	Sylvia Field
Editor Webb	Paul Hartman
Mrs. Webb	Peg Hillias
Mrs. Soames	Carol Veazie

Credits: PRODUCER: Fred Coe. DIRECTOR:
Delbert Mann. CHOREOGRAPHY: Valerie Bettis.
SETS: Otis Riggs. COSTUMES: Robert Campbell.

take care of his young son. Lee Marvin and Joanne Davis star. (Film)

11 ELLERY QUEEN—Mystery

[RETURN] A film series featuring the adventures of the famed society detective and his father, Inspector Queen, of the New York Police force. Tonight: "Stranger in the Dark." A blind girl returns to her apartment too late to prevent her roommate's murder. But she's in time to brush against the fleeing murderer. Hugh Marlowe is Ellery Queen. Florenz Ames is Inspector Queen.

13 EVANGEL HOUR—Religion
43 FILM SHORTS

9:30 2 ALFRED HITCHCOCK

[DEBUT] Director Alfred Hitchcock, master of cinematic mystery, brings to television the techniques of inducing suspense that won him renown in the movies. He personally has directed some of the 30-minute film dramas to be shown in the series. Ralph Meeker and Vera Miles star in the first Hitchcock-directed thriller, "Revenge." An aircraft expert quits his job to care for his wife, the victim of a recent nervous breakdown. He arrives home to find her unconscious, the victim of a beating. When she revives they scour the city in a suspenseful search for the guilty party. (Film)

Cast

Carl Spann	Ralph Meeker
Elsa Spann	Vera Miles
Mrs. Fergerson	Frances Bavier
Man in Grey Suit	Ray Montgomery

**THE WHISTLER
10:30 PM
WPIX 11**

Doctor .... John Gallaudet
Lieutenant .... Ray Teal
Sergeant .... Norman Willis
Cop .... John Day

5 STORY THEATER

"Why Thomas Was Discharged." Two young fortune hunters check into a fashionable resort hotel in search of rich wives. When they find their prey, they decide to toss dice to see how they should pair off. Mark Daniels, Marcia Jones, Gil Stratton, Jr. (Film)

7 8 LIFE BEGINS AT 80

The panel from *Juvenile Jury* matches wits with the oldsters. Here are: Mark Swartz, Douglas Stuart, Nancy Mitman and Carol McDonald. The panel of over-80's includes: Georgiana Carhart and Mr. and Mrs. Thomas Clark. Jack Barry.

9 CONFLICTS—Drama

"Center Ring." Tale of circus jealousy in which a big circus talent-scout shows up at a small circus to catch the acrobatic act of a young couple. Jane Darwell, James Mitchell. (Film)

11 CITY DETECTIVE—Drama

"The Beautiful Miss X." A girl suffering from amnesia, an attempted murder, mixed identities and an engagement ring add up to a baffling plot for Bart Grant to unravel. Rod Cameron and Lynn Bari are featured. (Film)

13 SPANISH SHOW—Variety

Drama: a group of actors rehearsing find their own lives tangled in a real-life counterpart of "Othello." Othello: Onix Baez. Iago: Edmundo Larra. Desdemona: Josefina Claudio. Hosts Don Passante and Miguel Machuca introduce the entertainment: "Jiboro" group Los Tipicos, doing typical Puerto Rican country music; Cuban Alma Tropical Orchestra; Muan and Elena, mambo and cha cha interpreters; comedian Abigail Narvaez; and Mexican singer Gilberto Mendoza. "The Sergeant and the Recruit" get into more trouble.

43 FILM SHORTS

10:00 2 APPOINTMENT WITH ADVENTURE—Drama

"The Allenson Incident," by Richard

TUESDAY

OCTOBER 4

ick, including the story of Patrick Sarsfield. Music is provided by tenor Michael Joyce and the Sarsfield Pipe Band.

8:30 ❷ YOU'LL NEVER GET RICH
"WAC." Bilko's bilked in his attempt to "con" an outside sergeant named Hogan into refusing the job of armory guard. Just when Bilko thought he had the position tied up, Hogan appeared on the scene. (Film)

Cast

Sgt. Bilko	Phil Silvers
Rocco	Harvey Lembeck
Henshaw	Allan Melvin
Hogan	Elisabeth Fraser
Mildred	Jane Dulo
Sgt. Sowici	Harry Clark
Sgt. Pendleton	Ned Glass
Colonel	Paul Ford

❼ ⑧ WYATT EARP—Western
"Wyatt Earp Comes to Wichita." A local dancehall queen falls into disfavor with some dangerous outlaws because of a favor she granted Wyatt Earp. But no one wants to argue with the fast-drawing marshal when he lets it be known the woman is his girl friend. Hugh O'Brian, Collette Lyons, Don Haggerty, House Peters. (Film) *The Marshal has his six-shooter checked by our reviewer in next week's TV GUIDE.*

⑪ FILM DRAMA
"Wedding Morning." Pre-wedding excitement almost results in the cancellation of the ceremony. But love is the factor that keeps everything on schedule. James Lydon, Allene Roberts.

⑬ MOVIE—Spy Drama
"Storm over Lisbon." A Lisbon café becomes the meeting place of a dancer and an American correspondent carrying secret information. Erich von Stroheim.

㊸ FILM SHORTS

9:00 ❷ MEET MILLIE—Comedy

❹ FIRESIDE THEATER—Drama
Hostess Jane Wyman presents Keenan Wynn and Jayne Meadows as stars of "The Sport." A conservative young busi-

nessman buys a dazzling foreign sports car. He soon discovers that the orderliness of his family and office routine has disappeared. Herbert Heyes, Reginald Denny, Harlan Warde. (Film)

❺ PLAY OF THE WEEK
"House of Death." A gentleman on intimate terms with monsters and goblins finds a new complication in his life when his beautiful niece comes to live with him. Boris Karloff. (Film)

❼ ⑧ DANNY THOMAS
A new next-door neighbor, who also works nights, causes a bit of trouble for night-club entertainer Danny Williams. Daughter Terry steps in to lend a helping hand. Frank Faylen and Virginia Gregg play the neighbors; Sherry Jackson plays Terry. (Film)

❾ DATELINE EUROPE—Drama
"The Sleeper Village." Robert Cannon is hunting for a story in a small mining town and discovers some international thievery. Jerome Thor stars. (Film)

⑪ MO▨▨ Melodrama
"The Diar▨▨▨▨▨▨ of th▨▨" A domineering Fre▨▨▨▨▨ 19th Century tries to ▨▨▨▨▨ hold on her son by involving him with the chambermaid. Paulette Goddard, Burgess Meredith, Hurd Hatfield, Francis Lederer, Judith Anderson, Reginald Owen, Irene Ryan.

㊸ FILM SHORTS

9:30 ❷ RED SKELTON—Comedy
[COLOR] Red's special guest is Jackie Gleason. The two comedians do a historical sketch tracing the origin of Skelton's character, Freddy the Freeloader. We see Freddy and his pal (played by Gleason) trying to crash their way into the World Series. David Rose conducts the orchestra. Jackie Gleason plays his Poor Soul character in the sketch with Red. David Burns plays a man who sells ball-park tickets.

❹ PLAYWRIGHTS '55
[DEBUT] First in a series of alternate-week dramatic shows produced by Fred

WEDNESDAY

⑤ MOVIE—Drama
"The Guilty." (1947) Twin sisters' love for the same man leads to violence. Bonita Granville, Don Castle.

⑪ CASES OF SCOTLAND YARD
"Silent Witness." A detective-inspector has reason to doubt a man's story that his wife's death was caused by a fall. Ivan Craig. (Film)

⑬ HOUSE DETECTIVE

8:30 ④ FATHER KNOWS BEST
"Father Is a Dope." Protesting about a TV series in which the father is portrayed as a gullible fool, Jim Anderson says that real life is quite the opposite. To prove his point he refuses to be "tricked" into canceling his fishing trip. Robert Young, Jane Wyatt, Elinor Donahue, Billy Gray and Lauren Chapin play the Andersons. Robert Foulk plays Ed Davis, Harry Antrim plays Dr. Conrad. (Film)

⑦ ⑧ M-G-M PARADE
Host George Murphy presents film clips from "The Philadelphia Story," 1940

movie starring Katharine Hepburn, Cary Grant and James Stewart; "The Tender Trap," 1955 movie with Frank Sinatra singing the title song; and Victor Moore and Edward Arnold in a sketch called "Pay the Two Dollars." (Hollywood) Our reviewer takes a look at Leo the Lion's TV efforts in next week's TV GUIDE.

⑪ BADGE 714—Jack Webb
"The Big Threat." Friday and Smith have a real problem on their hands when a druggist who has been beaten and robbed refuses to identify his assailants. Ben Alexander. (Film)

9:00 ② THE MILLIONAIRE
James Daly stars in "The Story of Tom Bryan." An ex-convict tries to regain the confidence of his son. The boy learns that his dad's time "overseas" was actually spent in jail. (Film)

Cast

Michael Anthony	Marvin Miller
Tom Bryan	James Daly
Tommy Bryan, Jr.	Billy Chapin
Myrtle Haskins	Gladys Hurlbut
Betsy Bryant	June Vincent

Spending a million dollars a week, even scriptwise, is not easy. Story on P. 18.

④ THEATER—Drama
"Number Four with Flowers," a comedy by Louis Pelletier. The bullet-riddled body of a gangster is brought to the morgue. Since no one appears interested

MONDAY

NOVEMBER 14

murder, falls in with a renegade thief. George "Gabby" Hayes.

⑪ Ramar of the Jungle
"The Mask of Kreenoh." Greed and selfishness invade the jungle when a white trader and his scheming wife rob a sacred jewel from a native chief. Jon Hall, James Fairfax.

6:35 ㊸ NEWS
6:40 ⑧ Weather—St. George
6:45 ④ Weather—Tex Antoine
   ⑧ NEWS—Joe Burns
   ㊸ Family Rosary
6:50 ④ NEWS—Kenneth Banghart
6:55 ⑦ Weather—Janet Tyler
7:00 ④ HIGHWAY PATROL
Dan Mathews learns that five migratory workers have died under similar circumstances. The insurance policies of these men name different beneficiaries—but all list the same address. (Film)

   ⑤ NEWS—Mike Wallace
   ⑦ KUKLA, FRAN AND OLLIE
   ⑧ MR. DISTRICT ATTORNEY
Mr. D.A. investigates a gas station robbery marked by unwarranted violence and destructiveness. David Brian. (Film)

   ⑪ NEWS—Kevin Kennedy
   ⑬ MOVIE—Detective
"The Old Homestead." A woman mayor poses as a thief in order to fight crime. Anne Jeffreys, Dick Purcell.

   ㊸ TELE-COMICS—Kids
7:10 ② WEATHER—Carol Reed
   ⑪ WEATHER—Joe Bolton
7:15 ② NEWS—Douglas Edwards

⑤ TEX McCRARY'S M.I.P.
Tex's guest tonight is Abba Eban, the Israeli delegate to the UN.

⑦ NEWS—John Daly
⑪ NEWS—John Tillman
㊸ FILM SHORTS
7:25 ⑤ LES PAUL & MARY FORD
7:30 ② ROBIN HOOD—Adventure
"Queen Eleanor." Robin Hood comes to the aid of Queen Eleanor of Aquitaine, mother of Richard the Lionhearted, who is touring England to raise ransom for the release of her captured son. (Film)

Cast
Robin Hood .... Richard Greene
Queen Eleanor .... Jill Esmond
Mair Marian .... Bernadette O'Farrell
Friar Tuck .... Alexander Gauge
The Sheriff .... Alan Wheatley
Review of the beau-and-arrow saga, p. 21.

④ TONY MARTIN—Songs
Tony presents his version of "Guys and Dolls." He sings "Luck Be a Lady" and a medley from the film. The Interludes offer "Guys and Dolls" and "Bushel and a Peck." Tony also sings "Same Old Saturday Night" and "Someone You Love." (Hollywood)

⑤ LONE WOLF—Adventure
"Memo: The Smuggling Story." Mike Lanyard is a busy sleuth when after being kidnaped by thugs police call him in for questioning about a $250,000 jewel robbery. Louis Hayward. (Film)

⑦ TOPPER—Comedy
"Legacy." An unscrupulous lawyer tries to cheat Marion and George out of the huge legacy left them by grandpa Augustus Kerby. The lawyer has discovered the will's stipulation that if George and Marion should die, the money goes to charity. (Film)

Cast
Topper .... Leo G. Carroll
George .... Robert Sterling
Marion .... Anne Jeffreys

⑧ THIS IS YOUR WORLD
RETURN Prof. Albert Burke, of Yale University, returns to discuss the basic principles of why people throughout the

MONDAY, JANUARY 16, 1956

fender Bart Matthews tries to help an ex-convict who is suspected of robbing his boss. Reed Hadley, Chris Drake. (Film)

⑬ REPORT FROM RUTGERS
First of a new series entitled, "A Short Course in Ornamentals and Gardening." Today's guest is Prof. Raymond Korbobo.

8:30 ② TALENT SCOUTS—Godfrey
⑦ ⑧ VOICE OF FIRESTONE
Guest tonight is tenor Eugene Conley.

Program
Mr. Conley
"All Day on the Prairie" .. David Guion
"Another Mile" .... Dana Suesse
"Solenne in Quest' Ora"
  ("La Forza") .... Verdi
"My Heart Stood Still" ..Richard Rodgers
Orchestra
"O Susannah" .... Stephen Foster
"Arkansas Traveler" .. Old Dance Tune
Fourth Mvt., Fifth Symphony
  .... Tchaikovsky
Chorus
"Across the Wide Missouri"..Jimmy Shirl.

⑪ SAN FRANCISCO BEAT
"The Rewrite Pilot." Two hired gunmen, shoot a lawyer down for a $5000 fee. Warner Anderson, Tom Tully. (Film)

⑬ WESTERN JAMBOREE

9:00 ② ⑧ I LOVE LUCY—Comedy
When the Ricardos and the Mertzes board the ship for their trip to Europe, Lucy insists on bidding farewell to Ricky, Jr., once more. Lucy really has a problem when the liner pulls away from the dock without her. Lucille Ball, Desi Arnaz, Vivian Vance, William Frawley. (Film)

④ MEDIC—Drama
"The Laughter and the Weeping" is the story of a young boy who gives up college in order to support his bedridden father. He turns to professional wrestling, and later must undergo plastic surgery. Richard Boone is seen as Dr. Styner. (Film)

Cast
Harold Lord Percival .. Michael Ansara
Stan Thompson ...... Charles Delaney
Luther Jackson .... Otis Greene
Dr. Chalmers .... Hal Gerard

Chuck Strong ... . Will White
Eddie .... Hank Weaver

⑤ FIGHT PRELIMS—St. Nick's
⑦ DOTTY MACK—Pantomime
Dotty, Bob Braun and Colin Male pantomime to records. (Cincinnati)
Dotty: "Sailor Boy"..Rosemary Clooney
Bob: "Natives Are Restless"..Ray McKinley
Dotty, Bob: "Aba Daba Honeymoon"
  ..Debbie Reynolds, Carlton Carpenter
Colin: "Lot More Layin'
  Down" .... Phil Harris
Dotty: "Taboo" .... June Valli

⑨ BIFF BAKER—Adventure
"Detour to Cairo." Biff and Louise go to Cairo to look up an old friend, and find unexpected adventure. (Film)

⑪ THE TRAP—Drama
"Blind Man's Bluff." A beautiful woman finds herself torn between her love for her husband and for his doctor. (Film)

⑬ COLLEGE BASKETBALL—Adelphi vs. Rider
Adelphi College is host to the Rough Riders of Rider College. Marty Glickman.

Rider
23 Faas	54 Pratt	58 Karp
50 Krol	55 Burke	59 Solana
51 Adams	56 Sheil	60 Simon
52 Chester	57 Piotrowski	63 Anderson
(Coach: Tom Leyden)

Adelphi
4 Funk	9 Bedell	15 Halfond
5 Pachter	11 Ehrlich	16 Soderstrom
7 Gordon	12 Sternstreim	17 Scott
8 Wellens	14 McDonough	
(Coach: George Faherty)

㊸ BIG PICTURE—Army Film
"Defense of Japan." (Film)

9:30 ② DECEMBER BRIDE
Lily calls on son-in-law Matt to appear in her club's annual play. When Matt receives rave reviews, he decides he is destined for a movie career. (Film)

Cast
Lily .... Spring Byington
Matt .... Dean Miller
Ruth .... Frances Rafferty
Pete .... Harry Morgan
Hilda .... Verna Felton
Cast continued on page A-42

FEBRUARY 14

Pauline, at first successful in employing beauty and charm to pass through the Southern lines, is eventually exposed, arrested, tried and sentenced to death. But even then, her feminine ingenuity helps her plan a means of escape. (Film)

Cast

Pauline Cushman . . . . Gertrude Michael
Lt. Young . . . . Michael Hall
Lt. Sinclair . . . . Fred Beir
Union Lieutenant . . . . Richard Bauman

9 STRANGE STORIES—Drama
"Bright Boy." A young man tries to double-cross his two partners in crime by making off with the loot. Tom Drake. (Film)

10:00 2 8 $64,000 QUESTION
As of the 35th show, contestants have won $448,608 and six luxury automobiles. This week Hal March has more than one excuse for excitement. On Friday he will marry Candy Toxton in Las Vegas.

5 THE HUNTER—Drama
"The Iron Capitol." Bart Adams is involved with an artist's model, a coded message and a Czech patriot as he tries to promote freedom behind the Iron Curtain. Barry Nelson. (Film)

7 43 TOMORROW'S CAREERS
"The Rocket Engineer" is an actual career a youngster can consider. Present in the Baltimore studios to describe it are engineers from the Martin Company, the firm that is developing the satellite to be launched by the U. S. They will discuss the training and opportunities in their field and chances of flying to the Moon one day. Lynn Poole acts as host.

9 MOVIE—Musical-Biog.
Million Dollar Movie: "The Great Gilbert and Sullivan." See Mon., 7:30 P.M. 9.

10:30 2 DO YOU TRUST YOUR WIFE?
Edgar Bergen is host to contestants. Returning for the second time tonight will be Bernard and Kathy Brawner of Beverly Hills, Cal. Other contestants are: 1. Charles and Betty Hawley, restaurant operators of Sherman Oaks, Cal. 2.

Richard and Janice Pearson of North Hollywood, Cal. Richard is a director of Forest Lawn Cemetery. 3. Jim and Beth Dannaldson of Tarzana, Cal. The Dannaldsons train snakes, alligators and the like for motion pictures and TV. (Film)

4 BIG TOWN—Mark Stevens
"Hung Jury." Despite overwhelming evidence, the trial of a racketeer ends in a hung jury. Steve Wilson is sure that the one holdout juror has been "fixed" and sets out to prove it. Mark Stevens, Barry Kelley, Doe Avedon. (Film)

Cast

George Constantine . . . Frank de Kova
Edward Stacy . . . . Harry Antrim
Klinger . . . . James Todd

5 STAR PLAYHOUSE—Drama
"Manhattan Robin Hood." A writer masquerades as a high school basketball coach to get a story on a gang of hoodlums. Preston Foster. (Film)

7 BORIS KARLOFF—Mystery
"The Devil Sells His Soul." Although

tonight...
PONTIAC PRESENTS PLAYWRIGHTS '56
A Fred Coe Production

TV's Greatest Dramatic Series

9:30 to 10:30
Channel 4

Brought to you by
YOUR PONTIAC DEALER

THURSDAY, MARCH 15, 1956

43 TELE-COMICS—Kids
7:10 **2** WEATHER—Carol Reed
11 WEATHER—Joe Bolton
7:15 **2** NEWS—Douglas Edwards
5 TEX McCRARY'S M.I.P.
7 NEWS—John Daly
8 SPORTSMEN'S CLUB
11 NEWS—John Tillman
43 FILM SHORTS
7:25 **5** LES PAUL & MARY FORD
7:30 **2** SGT. PRESTON—Adventure

A trapper takes a dangerous shortcut through the mountains to find a faster way of staking a gold claim. But two tramp miners waylay him, steal his supplies, and leave him to freeze. (Film)

Cast

Sgt. Preston .... Richard Simmons
Beaver Louie .... Dick Wilson
Slocum .... Gene Roth
Hod .... Ralph Neff
Brady .... Ed Dearing
Zack Wilson .... Harry Tyler

4 DINAH SHORE—Songs

Dinah and the Skylarks pay a musical tribute to St. Patrick's Day.

5 THE GOLDBERGS
7 8 THE LONE RANGER

"Jornada del Muerto." The Lone Ranger and Tonto invade an Apache Indian stronghold to seek the evil white man responsible for stirring up many a bloody Indian uprising. Clayton Moore, Jay Silverheels. (Film)

9 MOVIE—Drama

Million Dollar Movie: "Fighting Father Dunne." See Mon., 7:30 P.M. **9**.

11 I LED THREE LIVES

Counterspy Herb Philbrick is ordered by the Communist party to impersonate an FBI counterspy. Richard Carlson. (Film)

7:45 **4** NEWS—John C. Swayze
8:00 **2** BOB CUMMINGS—Comedy

Bob and Margaret differ on what college Chuck should attend. Both ply him with favors in an effort to win him to the "right" choice. Rosemary DeCamp, Ann B. Davis, Dwayne Hickman. (Film)

4 GROUCHO MARX—Quiz
5 LIBERACE—Music

RETURN Tonight's theme is candelabra through the ages, with Liberace's own collection on display. He plays Liszt's "Hungarian Rhapsody No. 2," "Ave Maria," "I'm in the Mood for Love" and an Italian medley of "O Sole Mio," "Ciribiribin," "Marie," "Sorrento." He will dance "The Emperor Waltz" in a Vienna court setting.

7 8 BISHOP FULTON SHEEN
11 CAPTURED—Drama

"Joseph McCann." The story of one of the country's most daring criminals, who operated with a flagrant disregard for his own safety. William Haade. (Film)

13 WESTERN JAMBOREE
8:30 **2** SHOWER OF STARS

COLOR Jack Benny, Fredric March, Peggy Lee and Elsa Lanchester are the stars of tonight's show. One of the features of the show is George Kelly's one-act play, "The Flattering Word." The wife of a college professor receives an unexpected visit from a former suitor, now an actor. The professor is somewhat carried away by the actor's insistence that the teacher should be on the stage.

Cast

Loring Rigley .... Jack Benny
Eugene Tesh .... Fredric March
Mary Rigley .... Peggy Lee
Mrs. Zooker .... Elsa Lanchester

To round out the show, Jack is featured

SATURDAY, MARCH 17, 1956

SATURDAY   MARCH 17

9:00 ④ TELEVISION "EMMY" AWARDS

SPECIAL   The eighth annual awards will be presented from the Pan-Pacific Auditorium in Hollywood, and the Waldorf-Astoria Hotel in New York City. "Emmy" statuettes will be presented to winners in each of 41 categories. Check your choices below and see how they compare with the judges' selections.

Best Actor
(Series Performance)
1. Cummings 2. Gleason
3. Silvers 4. Thomas
5. Young

Best Actress
(Series Performance)
1. Allen 2. Arden
3. Ball 4. Hagen
5. Sothern

Best Actor
(Single Performance)
1. Bellamy 2. Ferrer
3. Nolan 4. Sullivan
5. Sloane

Best Actress
(Single Performance)
1. Harris 2. Martin
3. Saint 4. Tandy
5. Young

Art Linkletter will be master of ceremonies for the show from Hollywood, with John Daly as New York emcee. Jimmy Durante, Bob Cummings and George Gobel will make the Hollywood presentations. Sam Levenson, Hal March and Phil Silvers will do the same in New York.

Nominated for best single program of the year are Davy Crockett and the River Pirates, Caine Mutiny Court Martial, Peter Pan Meets Rusty Williams, Peter Pan, The Sleeping Beauty, No Time for Sergeants and The American West.

The players nominated for best single performance were for these shows: Ralph Bellamy, Fearful Decision; José Ferrer, Cyrano de Bergerac; Lloyd Nolan and Barry Sullivan, Caine Mutiny Court Martial; Everett Sloane, Patterns; Julie Harris, Wind from the South; Mary Martin, Peter Pan; Eva Marie Saint, Our Town; Jessica Tandy, The Fourposter; and Loretta Young, Christmas Stopover.

Among other nominations in top categories are: best comedian, Jack Benny, Sid Caesar, Art Carney, George Gobel and Phil Silvers; best comedienne, Gracie Allen, Eve Arden, Lucille Ball, Nanette Fabray and Ann Sothern. The show is scheduled to last 90 minutes.

[5] Hollywood Preview
Conrad Nagel interviews his guests, actress Nancy Kelly and dancer-producer Don Loper.

[13] Junior Frolics-Fred Sayles

4:55 [9] NEWS

5:00 [2] MOVIE—Comedy
"Sleepy Lagoon." Racketeers are having a field day until an all-female reform ticket takes over the city government. Judy Canova, Dennis Day.

[4] Pinky Lee—Kids

[5] Food for Thought-Graham
Henri Bendell conducts a fashion show.

[7] [8] Mickey Mouse Club
(1) Newsreel: water wizard; hee-haw; a sky train; the tale of no shirt; girl saves dolls. (2) Mouseketeers: songs are "Humphrey Hop" and "Green Grass Grows." (3) Spin and Marty: "A Surprise Decision." (4) Cartoon: "Delivery Boy."

[9] Teen Bandstand—Steele
Catholic charities: Kennedy Home, Bronx.

5:25 [11] NEWS

5:30 [4] Howdy Doody—Kids
[COLOR] Bob Smith entertains.

[5] Commercial Film

[11] Cartoon Comics

[13] Super Serial—Adventure
"Radio Patrol." (Part 11) Grant Withers.

5:45 [5] Tim McCoy—Western

5:55 [2] Les Paul and Mary Ford

EVENING

6:00 [2] WORLD NEWS-Robt. Trout

[4] Superman—Adventure
"Around the World with Superman." Superman is faced with problems that tax his ingenuity as a result of a promotion stunt. George Reeves, Noel Neill, Jack Larson.

[5] Capt. Video Cartoons

[7] Gene Autry—Western
"Blazeaway." Gene, as a Cavalry officer, helps to prove the innocence of an Indian tribe that has been wrongly blamed for a series of tomahawk murders. Gene Autry, Bob Bice.

[8] Stage 8—Drama
"Lady Investigator." A pretty schoolteacher inherits a private detective agency in New York. Gene Raymond, Bonita Granville.

[9] Cartoon Parade
Once again, Ray Heatherton is seen as the Merry Mailman.

[11] Clubhouse Gang Comedies

[13] MOVIE—Western
"Lawman Is Born." Johnny Mack Brown.

[43] Film Shorts

6:05 [2] LOCAL NEWS—Cochran

6:10 [2] Feature—Bill Leonard

6:15 [2] MOVIE—Mystery
Early Show: "The Unholy Four." (1954) An amnesia victim returns after four years to find his wife still surrounded by his old rivals. When one of the rivals is murdered, the husband is suspected of the crime. Paulette Goddard, William Sylvester.

6:30 [4] Inside NBC
Bill Cullen is visited in *Cullen's Corner* by guests with unusual merchandise to sell home viewers.

[5] Looney Tunes—Cartoons

[7] Cisco Kid—Western
A rancher, who is working the old water toll racket on unsuspecting cattle drivers, meets unexpected resistance from a woman rancher. Duncan Renaldo.

[8] SPORTSCOPE—Syd Jaffe

[9] MOVIE—Western
"Romance on the Range." Bandits rob trappers of furs brought into the trading post. Roy Rogers.

[11] Ramar of the Jungle
"Valley of No Return." The Afro-Continental Railroad is plagued by mishaps, accidents and desertions which almost doom its existence. Jon Hall.

[43] NEWS—Harry Downie

6:35 [43] Film Shorts

6:40 [8] Weather—Ed Caputo

6:45 [4] [COLOR] Weather-Antoine

[8] NEWS—Joe Burns

[43] Family Rosary

6:50 [4] NEWS—Ken Banghart

6:55 [7] Weather—Janet Tyler

WEDNESDAY   MAY 16

while Arthur Godfrey is vacationing. Songstress Jaye P. Morgan and singer-dancer
Bobby Van are featured. Jaye sings "Play
for Keeps" and "Swanee." Bobby does a
song-and-dance routine to "Too Darn Hot."

❹ DIRECTORS PLAYHOUSE
George Sanders, Sal Mineo and Patricia
Morison star in "The Dream," directed by
Hugo Haas. In France in 1887 an im-
pressionable youth gives his mother cause

9:00 ❹ KRAFT THEATRE — Drama

A Profile
in Courage:
Edmund G. Ross

May 12, 1868. Ross (seated,
center) in Senate chamber.

by Sen. John F. Kennedy

The U.S. House of Representatives has the
power to impeach the President. On Feb. 24,
1868, the House, by a huge vote, adopted a
resolution of impeachment of
President Andrew Johnson.
Only the Senate can try a
President who has been im-
peached. On March 5, 1868,
the Senate began trial of
President Johnson under
11 Articles of Impeachment.

Kennedy

The first vote of the trial came on May 16.
It marked the climax of a public tug-of-war
between President Johnson and the Radical
Republican members of Congress, a battle on
which hung tremendous issues involving the
relation of the Executive and the Legislative
branches of the Government. The Radicals had
prepared their ground carefully, and had only

brought the issue to trial when they were confident they had the two-thirds majority to assure
conviction.

It looked, at first, as if they had just the needed
number. Then the alarm went out: One Senator
they thought they could count on, Edmund G.
Ross, of Kansas, was "shaky." Between March
5, the start of the trial, and May 16, date of the
first vote, Ross became the object of nearly overwhelming pressure. His own political future,
the fate of the Radicals and the shape of the
U.S. Government hinged on his vote.

Tonight's one-hour play was adapted by
Wendell Mayes from a chapter of the current
best-seller "Profiles in Courage," by Sen. John
F. Kennedy of Massachusetts. Direction is by
Fielder Cook, who directed "Patterns" both for
Kraft Theatre and the movies.

blance, then learn they are twin sisters. Written, produced and directed by Emeric Pressburger. Yolande and Charmion Larthe, Hugh Williams, Elizabeth Allan.

9:00 ② ⑧ CHARLIE FARRELL

A visiting French gourmet and his beautiful daughter nearly cause a disaster at Charlie's Racquet Club. Charlie's chef is so enamored of the daughter he is ready to forsake all and return to France with his lady love. (Film)

Cast

Charlie	Charlie Farrell
Anatole Boulanger	Marcel Dalio
Mignin Boulanger	Maria Palmer
Pierre	Leon Askin
Dad Farrell	Charles Winninger

Sherman Hull	Richard Deacon
Mrs. Papernow	Kathryn Card

④ MEDIC—Drama

"When I Was Young" tells the story of a middle-aged woman going through the change of life. Her irritability and nervousness prompt quarrels when her family fails to realize the cause of her moodiness. (Film)

Cast

Dr. Konrad Styner	Richard Boone
Gwen Kellogg	Mae Clarke
Jim Kellogg	Ray Bailey
Kathy Kellogg	Martha Randall
Dave Kellogg	Don Gardner
Madge	Marlo Dwyer

⑤ FIGHT PRELIMS—St. Nick's

9:30   ② ④ ⑦ ⑧   DEMOCRATIC CONVENTION

SPECIAL Highlighting the convention's opening day is tonight's keynote speech, to be delivered by Gov. Frank Clement of Tennessee. A second address will be given by Mrs. Franklin Delano Roosevelt, widow of the president. Also on the evening program is a documentary film depicting the history of the Democratic Party. The film's narrator and commentator will be Sen. John F. Kennedy of Massachusetts. Sen. Kennedy's best-selling book "Profiles in Courage" sketched a number of political heroes in U.S. history. The first talk of the evening will be given by Paul M. Butler, Chairman of the Democratic National Committee.

Kennedy

Highlights of today's convention program:

DAY SESSION

Opening Remarks	Party Chairman Butler
Welcome	Mayor Richard Daley, Chicago
Call for the Convention	Dorothy Vredenburgh
Address	Sen. Paul Douglas (Ill.)

EVENING SESSION

Address	Paul M. Butler
Film	Democratic Party
Film Narrator	Sen. John F. Kennedy
Keynote Address	Gov. Frank Clement
Address	Eleanor Roosevelt
Songs	Frank Sinatra

Clement

MONDAY   AUGUST 20

⑧ Breakfast Playhouse
Host Tom Romano introduces the films: "Papa Goes to the Ball"; "Present for Sarah"; "Security"; a cartoon, "Bosco the Shepherd"; "Dancing Tree."

9:30 ② Amos 'n' Andy—Comedy
"Cousin Effie's Will." Kingfish looks around for someone to adopt, in order to get an inheritance for new members of the family. Tim Moore, Spencer Williams co-star.

10:00 ② Of All Things—Variety
④ Ding Dong School—Kids
Artist Waylande Gregory guests.
⑤ MOVIE—Drama
Tune in Anytime Theater: "The Girl from

Manhattan." (1948) A model tries to outwit a realtor threatening to forceclose on her uncle's house. Dorothy Lamour, Charles Laughton, George Montgomery.

⑦ MOVIE—Drama
Afternoon Show: "Lost Moment." (1947) A publisher visiting an old woman who was once loved by a famous poet is attracted to the woman's niece. Robert Cummings, Susan Hayward.

10:30 ② Peter Lind Hayes—Variety
RETURN Peter Lind Hayes again steps in to take over for the vacationing Arthur Godfrey. With the youthful comedian is his "child bride," Mary Healy, and the Toppers, a male quartet.

2:00 ② ④ ⑦ ⑧ ④ REPUBLICAN CONVENTION

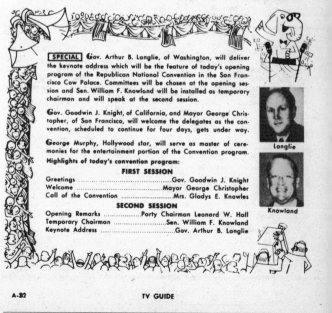

SPECIAL Gov. Arthur B. Langlie, of Washington, will deliver the keynote address which will be the feature of today's opening program of the Republican National Convention in the San Francisco Cow Palace. Committees will be chosen at the opening session and Sen. William F. Knowland will be installed as temporary chairman and will speak at the second session.

Gov. Goodwin J. Knight, of California, and Mayor George Christopher, of San Francisco, will welcome the delegates as the convention, scheduled to continue for four days, gets under way.

George Murphy, Hollywood star, will serve as master of ceremonies for the entertainment portion of the Convention program.

Highlights of today's convention program:

FIRST SESSION

Greetings	Gov. Goodwin J. Knight
Welcome	Mayor George Christopher
Call of the Convention	Mrs. Gladys E. Knowles

SECOND SESSION

Opening Remarks	Party Chairman Leonard W. Hall
Temporary Chairman	Sen. William F. Knowland
Keynote Address	Gov. Arthur B. Langlie

Langlie

Knowland

SUNDAY NOVEMBER 18

[11] **FOREIGN LEGIONNAIRE**
"As Long as There Will Be Arabs." Accused of murdering a village wise man, a European seeks aid from the Legion.

[13] **GOV. MEYNER'S REPORT**
"Air Pollution Control." The Governor's guests are William R. Bradley, chairman of the N. J. Air Pollution Control Committee; Dr. Miriam R. Sachs, chief of the Bureau of Adult and Occupational Health; and Fitzhugh W. Boggs, president of the N.J. Assoc. for Retarded Children, Inc.

[43] **THIS IS THE LIFE—Religion**
"Too Good for Heaven." A wealthy old woman is annoyed to find herself in conflict with her gardener. Steven Geray.

7:00 [2] **LASSIE—Drama**
"Fish Conservation." Gramps has the Miller lake stocked with fish. Then Jeff discovers poachers are fishing in the lake and using methods that could contaminate the water. Tommy Rettig. (Film)

[4] **BENGAL LANCERS—Adven.**
"The Traitor." Lts. Rhodes and Storm are sent to lead the honor guard of a British caravan traveling through Mohmand territory. They arrive at the caravan to find the commissioner dead and the money carried by the caravan gone. Phil Carey, Warren Stevens. (Film)

Cast

Lance-Daffadar Rahman	Michael Ansara
Patricia Carnovan	Millie Doff
Yakoob Ali	Douglass Dumbrille
Lal Dad	Richard Avonde

[5] **RACKET SQUAD—Police**
"Front Man." This is the story of an oil swindle. Reed Hadley. (Film)

[7] [8] **YOU ASKED FOR IT**
1. The Amazing Randi's record-making submersion in a steel tomb. 2. London Dixieland band. 3. George and Igor Rudenko, Russian jugglers. 4. Grand Coulee Dam. Art Baker, host. (Film)

[11] **CRUNCH AND DES—Adven.**
"Hookey, Line and Sinker." A young man whose job as a bank clerk provides his only income jeopardizes his career by going fishing. Forrest Tucker. (Film)

[13] **MOVIE—Drama**
All-star Movie: "The Foxes of Harrow." See Sat., 7 P.M., Ch. 13 for details.

[43] **FILM SHORTS**

7:30 [2] [8] **JACK BENNY—Comedy**
A major crisis develops for Jack when his ancient and beloved Maxwell automobile is stolen. Rochester, Don Wilson and the Sportsmen. (Hollywood)

[4] **CIRCUS BOY—Adventure**
"The Proud Pagliacci." Old Fritz, once a circus headliner and now a stable-hand, tries to impress the circus gang by giving the impression that he owns the farm. But when Joey and Corky go to visit the farm, the spoiled daughter of the real owner destroys the illusion. (Film)

Cast

Corky	Mickey Braddock
Joey	Noah Beery
Fritz	Otto Waldis
Marcy	Judy Short
Big Tim	Robert Lowery

Our reviewer visits the big top next week.

[5] **CAVALCADE OF PROGRESS**
Today's quests are executives of a company that manufactures analog computing equipment. (Film)

[7] **AMATEUR HOUR—Talent**
Ted Mack salutes Grand Rapids, Mich. Talent includes: vocalist Connie Webber, Columbus, O.; Three Guys and a Rose, instrumental group, Bronx, N.Y.; James Raglan, vocalist, White Plains, N.Y.; the Backers Sisters, vocal trio, Weathersfield, Connecticut.

[11] **VICTORY AT SEA—Docum.**
"Return of the Allies." The Allies come back to the Philippines, and a bitter struggle ends. (Film)

FRIDAY

DECEMBER 14

8:00 2 WEST POINT—Drama
"Heat of Anger." A cadet is confined to quarters the week-end of a big dance, so he asks his roommate to escort his girl. The other boy falls in love with her, which creates a major conflict between the two boys. Host: Cadet Thompson.

Cast

Joe Simpson	Henry Silva
Will Adler	Brian Hutton
Meg	Pat Crowley
Capt. Holmes	Larry Thor
Cadet Carter	James Hickman

The cadets are marching, coast to coast. See our story on page 4.

4 LIFE OF RILEY—Comedy
"Honeybee's Mother." Gillis hasn't seen his mother for several years, and Riley decides to remedy the situation. William Bendix, Marjorie Reynolds, Tom D'Andrea, Gloria Blondell. (Film)

5 COUNT OF MONTE CRISTO
"The Luxembourg Affair." A duchess borrows money from the count. (Film)

7 8 JIM BOWIE—Western
"The Swordsman." Jim Bowie goes to New Orleans to buy an estate, but loses sight of his objective after he meets the owner. (Film)

Cast

Jim Bowie	Scott Forbes
Liane Trudeau	Lilyan Chauvin
Armand De Nivernais	Michael Landon
Count De Nivernais	Richard Avonde
Maurice Toulouse	Ivan Triesault

11 FEDERAL MEN—Police
"The Case of the Man Outside." Convicts make counterfeit money while serving time. Walter Greaza. (Film)

8:30 2 ZANE GREY—Western
Dick Powell stars in "Courage Is a Gun." A gunslinger is wounded by the marshal in a shooting while waiting to settle a feud with an enemy. He then plots to kill the marshal as well. (Film)

Cast

Marshal Jess Brackett	Dick Powell
Ellen	Beverly Garland
Johnny	Robert Vaughn
Bert	James Westerfield
Collins	Claude Akins
Walter Blake	Leonard Penn

4 WALTER WINCHELL—Variety
COLOR Singer Nat "King" Cole joins Winchell tonight. (Hollywood)

Our reviewer covers this show on page 20, and Winchell talks about it in next week's TV GUIDE.

5 ETHEL BARRYMORE TH'TER
"Winter and Spring." An elderly man who has never held a job becomes a baby sitter. Charles Coburn stars. (Film)

7 8 CROSSROADS—Drama
"Tenement Saint," with Cecil Kellaway. An Episcopal minister tries to help a crippled woman. She has been forced out of her home by her daughter and son-in-law. (Film)

Cast

Father Hunsicker	Cecil Kellaway
Michael	Gerald Mohr
Mary Donovan	Elizabeth Patterson
Ann	Andrea King
Jerry	Billy Chapin

11 UNCOVERED—Drama
"Snapshot." A chance photograph leads to a case of mistaken identity. (Film)

9:00 2 CRUSADER—Adventure
"The Boy on the Brink." The son of a banker is despondent when he becomes convinced his father committed suicide over the theft of a million dollars. (Film)

Cast

Matt Anders	Brian Keith
Dick	Mike Landon
Jim Reynolds	Carl Betz
Ruth	Jorja Curtright

4 ON TRIAL—Drama
"The Jameson Case," starring Everett Sloane. Comm. Andrew Jameson is called before a court-martial in 1852. While at sea, he was responsible for the hanging of three seamen suspected of plan-

SATURDAY DECEMBER 29

⓫ BUFFALO BILL, JR.—Western
"Kansas City Lady." A ten-year-old boy is surprised and resentful on the arrival of his new stepmother. Dick Jones, Nancy Gilbert, Harry Cheshire.

⓭ POLKA SHOW—Gronet
Regina Kujawa sings "Johnny Oberek" and "Danny Boy." She and Ginger Smith duet "Oh! Mama Oberek."

6:05 ❷ LOCAL NEWS—Ron Cochran

6:10 ❷ WEATHER—Carol Reed

6:15 ❷ VINCENT LOPEZ—Music
Vincent builds tonight's show around the music of 1956. He solos "Poor People of Paris" and "Autumn Leaves." The band will play "I Could Have Danced All Night" and "Picnic." Judy Lynn sings "Married I Can Always Get."

6:30 ❺ LOONEY TUNES—Cartoons
❼ MOVIE—Adventure
"Trail of the Yukon." (1949) A Mountie sets out on the trail of bank robbers. Kirby Grant, Chinook.

❽ ANNIE OAKLEY—Western
"Justice Guns." A vicious young gunman provokes a proud old marshal into action. Annie, realizing that the brave old man is no match for the outlaw, tries to prevent a tragedy. Gail Davis.

❾ WAR IN THE AIR—Docum.
"The Cold Dawn." The final bombing of German communications and materials, winter and spring, 1944-45.

⓫ SHEENA—Adventure
"Rival Queen." A dangerous man, who manages to escape from his guards while working in the jungle, seeks revenge on Sheena. Irish McCalla and Buddy Baer are starred.

⓭ IRELAND'S HERITAGE
"County Fermanagh" and the names McGuire and McManus will be discussed by Rita Murphy and Maurice O'Sullivan. Irish tenor Pat Reynolds also appears.

7:00 ❷ BEAT THE CLOCK—Stunts
The producers "upped" the jackpot and "Beat the Clock" kept right on ticking. See next week's TV GUIDE.

❺ LONG JOHN SILVER—Adven.
"Sword of Vengeance." Taunted by their kidding, a young man whose looks are deceptive, is ready to take on all the sailors in port when Long John intervenes. Robert Newton. (Film)

❽ ROBIN HOOD—Adventure
"Outlaw Money." Sir William's taxes threaten to reduce the entire town of Lotham to poverty. Robin comes to the aid of the villagers by providing them with silver from his own supply. (Film)

Cast

Robin Hood	Richard Greene
Henry	Sidney James
Sheriff	Alan Wheatly
Derwent	Victor Wolf

❾ IT'S FUN TO TRAVEL

⓫ ABBOTT AND COSTELLO
"Uncle from New Jersey." When their landlord threatens to evict them for nonpayment of rent, the boys invent a mythical millionaire uncle to appease him. Bud Abbott, Lou Costello. (Film)

⓭ MOVIE—Western
All-star Movie: "Belle Starr." (1941) The Southwest's most notorious female bandit poses as a Missouri slave owner. Randolph Scott, Gene Tierney, Dana Andrews, Chill Wills, Louise Beavers.

7:30 ❷ BUCCANEERS—Adventure
"Gentleman Jack and the Lady." Dan Tempest gets into a battle royal with female pirate Anne Bonny over a captured Spanish galleon. Anne makes things even more difficult when she hides her identity in the guise of a foppish male. (Film)

Cast

Dan Tempest	Robert Shaw
Anne Bonny	Hazel Court
Beamish	Peter Hammond
Gaff	Brian Rawlinson
Taffy	Paul Hansard
Dickon	Wilfrid Downing

❹ PEOPLE ARE FUNNY—Stunts
A man who once was hoaxed into believing he would appear on the show, tonight actually does appear. Art Linkletter also meets Bill Harper and Dorilla Dufresne, a couple brought together by Univac, who play anagrams for a possible $10,000. Sandy Lowenberg returns from Las Vegas to report on how he fared using his "system." (Film)

❺ CRUSADE IN EUROPE

❼ FAMOUS FILM FESTIVAL
"School for Secrets." (English; 1947) TV Debut. Two civilian research experts and an RAF officer risk their lives to develop

FRIDAY

MORNING

6:45 **(8) SACRED HEART**—Religion

6:50 **(2) PREVIEWS**

6:55 **(2) GIVE US THIS DAY**—Religion
Rabbi Karl Weiner.

(4) DAILY SERMONETTE
Rabbi Moshe Weiss.

7:00 **(2) GOOD MORNING!**—Rogers

(4) TODAY—Dave Garroway
Tom Ewell and Nancy Olson chat about "The Tunnel of Love," the new stage comedy in which they will star.

(8) CARTOON CARNIVAL—Kids

7:25 **(8) NEWS AND WEATHER**

7:30 **(8) SHORT SHOW**—Drama
"Blazing the Trail." (1912) The trek west to California and a tragic Indian massacre are given realistic treatment.

7:55 **(8) NEWS AND WEATHER**

7:56 **(2) MORNING PRAYER**—Religion
Rabbi Israel Gerstein.

8:00 **(2) CAPT. KANGAROO**—Kids
The Captain talks about water, and Grandfather Clock has an appropriate poem. Mr. Moose shows how to play his new game, antler toss.

(7) TINKER'S WORKSHOP—Kids
Tinker Tom and the puppets do pantomimes to children's records. Cartoons: "Beau Best," "Tit for Tat."

(8) HAPPY THE CLOWN—Kids
Happy reveals a surprise.

8:30 **(5) FEATURAMA**

(8) BREAKFAST PLAYHOUSE
"So Help Me." A night club entertainer tells the millionaire who wants to marry her about the other men in her life.

8:45 **(5) SANDY BECKER**—Kids

8:55 **(4) MOVIE**—Comedy
First Feature: "Sons of the Desert." (1934) This time Stan and Ollie are lodge members who have a high old time trying to get away from their wives for the annual convention. Stan Laurel, Oliver Hardy, Dorothy Christy, Mae Busch.

(8) NEWS

9:00 **(7) STU ERWIN**—Comedy
"Baby Knows Best." Stu offers to baby-sit for his daughter, Joyce.

(7) DRAMA OF LIFE
"Segment." A woman's nagging drives her husband berserk when she threatens to tell their daughter she is an adopted child. William Bendix, Ward Bond, Joanne Woodward, Rosemary De Camp.

(8) THIS, OUR FAITH—Religion
This Is Judaism: Moderator Samuel Markle continues the current series, "Great Moments in Jewish History" with today's episode, "Jeremiah."

9:30 **(2) AMOS 'N' ANDY**—Comedy
"The Adoption." Sapphire and Kingfish want to adopt a child, but end up only with a long babysitting job instead. Tim Moore, Spencer Williams, Ernestine Wade, Alvin Childress.

(4) MOVIE—Drama
Morning Feature: "Force of Evil." See Mon., 9:30 A.M. for details.

(8) UNIVERSITY OF THE AIR
The World Behind the News: Today's topic: "City Planning."

(9) OUR CHILDREN—Docum.

10:00 **(2) GARRY MOORE**—Variety
Varvel and Bailly with Les Chanteurs de Paris, French singing group, visit the show. The whole cast does its version of grand opera. Denise Lor sings "Pigalle" and Ken Carson does "Baby Doll."

(4) HOME—Arlene Francis
Camp Kilmer, N.J., is today's hometown, U.S.A., as Arlene and Hugh Downs interview some of the Hungarian refugees who are living there.

(5) MOVIE—Comedy
Tune In Anytime Theater: "The Galloping Major." (English; 1951) A group of English village people decides to buy a race horse, but their expert returns with the sorriest nag imaginable. Basil Radford, Jimmy Hanley, Janette Scott.

(8) MY LITTLE MARGIE—Comedy
"Vern's Winter Vacation." Margie's boyfriend wins a trip to Palm Beach on a TV program and Margie tries to trick Vern into taking her there, too. Gale Storm, Charles Farrell.

(9) CARTOON TIME—Kids

10:25 **(4) WINDOW**—Lind
COLOR Toiletries for the home.

10:30 **(4) HOME**—Arlene Francis

(8) MOVIE—Drama
"Two Alone." (1934) An orphan is adopted by a tyrannical farmer who treats her

FRIDAY

scored knockout victories over John Holman and Nino Valdez, and he beat Johnny Summerlin in his last start. Eddie is a strong puncher, but if he allows Maxim to set the pace he could be upset. Scoring by points.
Compiled by Nat Fleischer (The Ring)

7 8 43 RAY ANTHONY—Music
Musical selections include "Lucky, Lucky Me," "Moonlight Gambler," "Lucky in Love," "You Are My Lucky Star," "Gambler's Guitar," "Cool Water," "Wind and Sand," "I'm Looking Over a Four Leaf Clover," and "I've Got Beginner's Luck."

9 MOVIE—Drama
Million Dollar Movie: "Cry the Beloved Country." See 7:30 P.M., Ch. 9.

11 MOVIE—To Be Announced

13 MOVIE—Mystery
All-star Movie: "Quiet Please, Murder." See Monday, 7 P.M., Ch. 13 for details.

10:30 2 PERSON TO PERSON
Ed Murrow visits actor Paul Douglas and his wife, Jan Sterling, in their Manhattan apartment. The second guest is Herbert Lehman, former U.S. Senator from New York State.

8 MOVIE—Mystery
"Lady in the Lake." (1946) When a private detective tries to sell his experiences to a horror magazine he discovers several murders.

Cast
Phillip Marlowe	Robert Montgomery
Adrienne Fromsett	Audrey Totter
Lt. DeGarmot	Lloyd Nolan
Capt. Kane	Tom Tully

10:45 4 COLOR SPORTS—Red Barber

11:00 2 NEWS—Ron Cochran

4 NEWS—John McCaffery

5 NIGHT BEAT—Mike Wallace
Mike's guests are Norman Thomas, leader of the Socialist Party; Sam Levenson, schoolteacher turned humorist.

7 NEWS—Cecil Brown

11:10 2 WEATHER AND SPORTS

4 COLOR WEATHER—Antoine

7 MOVIE—Mystery
Night Show: "Nocturne." See Monday, 11:10 P.M., Ch. 7 for details.

11:15 2 MOVIE—Drama
Late Show: "Saturday's Children." (1940) NY TV Debut. A couple discovers their early days together are not a bed of roses

when they have financial difficulties.

Cast
Rims Rosson	John Garfield
Bobby Halevy	Anne Shirley
Mr. Halevy	Claude Rains
Willie Sands	Roscoe Karns

4 COLOR HY GARDNER

11:30 4 TONIGHT—Steve Allen
Steve Allen, Skitch Henderson, Gene Rayburn and vocalists Pat Kirby, Eydie Gorme, Steve Lawrence and Andy Williams all say goodbye to viewers with a big farewell party. Next week there will be a complete change of format in the "Tonight" show with Jack Lescoulie acting as emcee.

9 MOVIE—Drama
"Ace of Aces." (1933) A war hero has strange reactions to his prowess as a killer of enemy soldiers. Richard Dix, Elizabeth Allan, Ralph Bellamy.

11 NEWS—John Tillman

13 HOLLYWOOD HALF HOUR
"Closeup." See Monday, 11:30 P.M., Ch 13 for details.

12:00 5 MOVIE—Mystery
"Hidden Corpse." A murder is committed under mysterious and unusual circumstances. ZaSu Pitts.

8 NEWS

13 FRONT PAGE DETECTIVE
"Friend of the Corpse." See Monday, 12 Midnight, Ch. 13 for details.

12:30 4 OLD OLD SHOW—Movies
Charlie Chase stars in "Bad Boy."

12:40 7 EVENING PRAYER—Religion
Rev. Eugene Houston.

1:15 2 MOVIE—Comedy
Late Late Show: "Eve Knew Her Apples." (1945) Time approximate. A radio singer takes a vacation to get away from her manager, her press agent and all her fans. Ann Miller, William Wright.

1:00 4 SERMONETTE
Rev. Charles McManus.

1:30 5 PREVIEWS; CALL TO PRAYER

2:30 2 NEWS

2:35 2 GIVE US THIS DAY—Religion
Rabbi Sanford Shapiro.

Stations reserve the right to make last-minute changes.

MONDAY, FEBRUARY 11, 1957

MONDAY   FEBRUARY 11

Chorus
" 'S Wonderful" ....Gershwin
"Song in My Heart" ....Idabelle Firestone

Orchestra
"Belle of the Ball" ....Leroy Anderson
"Our Love" ....Tchaikovsky

⓫ SAN FRANCISCO BEAT–Police
"Dennis Case." The trail of a suspected
murderer leads to San Francisco's famed
Coit Tower. (Film)

⓭ HOLLYWOOD HALF HOUR
"Coals of Fire." A young woman finds a
way to humanize her domineering, pen-
ny-pinching father. Carol Thurston. (Film)

9:00 ❷ ❽ I LOVE LUCY–Comedy
The Ricardos are settled in their new
Connecticut country home, but can't get
used to the peace and solitude. Lonesome
for their friends the Mertzes, Lucy and
Ricky start out for New York just about
the same time Fred and Ethel leave to
visit the Ricardos in Connecticut. Lucille
Ball, Desi Arnaz, Vivian Vance. (Film)

❹ TWENTY-ONE–Quiz
The biggest winner in TV-radio history,

Charles Van Doren, who has accumulated
$122,000, will tell whether he is going
on to try for higher stakes or call it
quits with his present winnings.

❺ RACKET SQUAD–Police
"His Brother's Keeper." The police department investigates a racket consisting
of organized beggars. (Film)

❼ BISHOP SHEEN–Talk
Bishop Sheen's topic tonight is "The Best
City in the World When You Are Broke."

❾ MOVIE–Mystery
Who Dunit Theater: "Ellery Queen, Master Detective." (1940) NY TV Debut.
A young girl, prevented from marrying
the man she loves, mysteriously disappears.

Cast
Ellery Queen ....Ralph Bellamy
Barbara ....Marsha Hunt
James Rogers ....Michael Whalen

⓫ FABIAN OF SCOTLAND YARD
"Murder in Soho." Three young hoodlums
kill a man who tries to block their
escape. Bruce Seton. (Film)

captured by the Indians because they think a Christian can save them from starvation. Jan Merlin, Tamar Cooper star. (Film)

⑪ STAGE 7—Drama
"Happy New Year." A beautiful girl holds an internationally famous jewel thief captive to give her accomplices an opportunity to steal some gems. The jewel thief had his eye on these particular gems. Cesar Romero, Virginia Field are featured. (Film)

㊸ FILM SHORTS

8:25 ⑬ HOLLYWOOD HALF HOUR
"Rewrite for Love." A Hollywood producer gets a surprise when he offers a movie contract. Wanda Hendrix. (Film)

8:30 ❷ TALENT SCOUTS—Godfrey
While Arthur is on vcacation, the third of the "Talent Scout" shows is presented, featuring two singers and a vocal trio. Mark Stuart, from Buffalo, N.Y., sings "You're My Everything." The three dolls —Patti Fuller, Bonni Haycock and Jacki Cronk—sing "Baby, Won't You Please Come Home." Lorraine Forman of Vancouver, B.C., sings "From This Moment On." (Film)

❹ STANLEY—Comedy
Stanley and Celia get involved in a lovers' quarrel, but neither wants to be the first to give in. Buddy Hackett plays Stanley, Carol Burnett plays Celia.

This is the final show of the series. Next week at this time: "Wells Fargo," a western adventure series.

❺ JUDGE ROY BEAN—Western
"Katcina Doll." A group of Americans steal the doll which is used in the religious ceremony by the Mexicans and hold it for ransom. Edgar Buchanan stars. (Film)

❼ ❽ VOICE OF FIRESTONE
Brian Sullivan, tenor, is the soloist, and he is seen in a record-shop setting for tonight's musical salute to St. Valentine's Day and St. Patrick's Day. Howard Barlow conducts the orchestra and chorus.

Mr. Sullivan
"Open Thy Blue Eyes" ....Massenet
"Magic Is the Moonlight" ..Maria Crever
"Come un Bel Di" ("Andrea Chenier")
....Giordano
Irish Medley ....Various

Orchestra
"March" ("Nutcracker Suite")
....Tchaikovsky
Finale, Symphony No. 5 ......Beethoven
Chorus
"Love's Old Sweet Song" ..James Molloy

⑪ SAN FRANCISCO BEAT-Police
"The Radio Case." Two ingenious thieves jam the police radio. Warner Anderson and Tom Tully star. (Film)

8:55 ⑬ FRONT PAGE DETECTIVE
"The Deadly Curio." A girl goes to visit her friend and finds the apartment empty except for a corpse. (Film)

9:00 ❷ ❽ I LOVE LUCY—Comedy
When their brood of chicks is traded in at a loss, in exchange for laying hens, Ricky is even more discouraged by the small amount of egg production. In an attempt to help matters along, Lucy buys five dozen eggs, intending to place them in the nests without Ricky's knowledge. Lucille Ball, Desi Arnaz, William Frawley, Vivian Vance, Richard Keith. Bruce: Ray Ferrell. (Film)

❹ TWENTY-ONE—Quiz
Charles Van Doren, winner of $143,000 to date, faces Mrs. Vivienne Nearing, attorney, for the third time. Tonight, every point is worth $2000. Jack Barry is the quizmaster.

For a word portrait of Mrs. Nearing, turn to page A-1.

The Terry twins are pictured in next week's TV GUIDE.

❺ RACKET SQUAD—Police
"Lady Luck." Capt. Braddock tracks down and exposes the practices of card sharpers. Reed Hadley, Roy Roberts, John Hubbard, Dawn Addams. (Film)

TUESDAY

APRIL 30

8:00 **2** PHIL SILVERS—Comedy
"Sgt. Bilko, the Marriage Broker." A new man, Lt. Wallace, is put in command of the motor platoon and has definite ideas about how things should be done in the most efficient way Bilko decides this eager beaver must be calmed down, and feels the best way is to get him interested in a girl. Phil Silvers, Maurice Gosfield, Harvey Lembeck Allan Melvin, Jack Healy. Lt. Wallace: Biff McGuire. Miss Wigman: Constance Ford. (Film)

4 ARTHUR MURRAY—Variety
COLOR Kathryn Murray welcomes actress Helen Hayes and her actor son, James MacArthur, who is currently being seen in the movie, "The Young Stranger"; comedian Sam Levenson; and actor Don Ameche, now starring in the Broadway show "Holiday for Lovers."

5 UNCOMMON VALOR—Marines
"Korea, Act 1." Gen. Holland M. Smith narrates the story of the Marines and their drive in North Korea. (Film)

8:30 **2** PRIVATE SECRETARY
"How to Handle the Boss." Susie has a problem when a scathing article carrying her by-line is accepted by a magazine publisher. Ann Sothern, Don Porter, Ann Tyrrell. (Film)

4 PANIC!—Drama
James Mason and Pamela Kellino (Mrs. Mason) star in "Marooned." A family with two young children is stranded 22 floors up in an unfinished apartment building. They face a long three-day holiday week-end with no food or water and no way of obtaining any. Also appearing are the two Mason offspring, Portland, 8, and Morgan, 21 months. (Film)

5 ENTERTAINMENT PANEL
Special guest is Broadway and TV actress Kim Stanley. Panelists include Al Morgan.

7 **8** WYATT EARP—Western
"Beautiful Friendship." Wyatt's new friend Doc Holliday gets himself involved in a Dodge City gun battle. Wyatt is forced to throw him into jail, and Doc's wife shows up to get him released. Hugh O'Brian. (Film)

Cast

Doc Holliday .... Douglas Fowley

Frank Loving .... Dennis Moore
Dog Kelley .... Paul Brinegar

43 MOVIE—To Be Announced

9:00 **2** TO TELL THE TRUTH—Panel
Mary Healy, substituting for Polly Bergen, joins regular panelists Kitty Carlisle and Hy Gardner. Bud Collyer emcees.

4 JANE WYMAN—Drama
Jane Wyman and Craig Stevens star in "The Man in the Car." Just after moving into a partly furnished home, a husband leaves his wife alone to make an out-of-town business trip. An unexpected visitor and the half-empty house create strange fears in the wife's mind. (Film)

Cast

Martha Carroll .... Jane Wyman
Danny Carroll .... Craig Stevens
Jonas .... Ainslie Pryor
Terry Carroll .... Charles Herbert
Pharmacist .... Jason Johnson

5 MR. AND MRS. NORTH
"The Third Eye." A dress shop is the scene of a murder in which the Norths are involved. Richard Denning. (Film)

7 **8** BROKEN ARROW—Western
"The Archeologist." Indian agent Tom Jeffords guides two archeologists to the Apache reservation to record primitive Indian customs for their museum. But when the two archeologists start probing into things the Indians regard as sacred, the Apaches take up arms against them. John Lupton, Michael Ansara. (Film)

Cast

Morgan .... Arthur Hanson
Regis .... Antony Eustrel
Katena .... Richard Hale

13 MOVIE—Drama
Command Performance: "Kiss of Death." See Mon., 9 P.M. Ch. 13, for details.

9:30 **2** RED SKELTON—Comedy
COLOR Arnold Stang and Lina Romay join Red in a sketch entitled "Cookie in Cuba." Sailor buddies Cookie and Snorke manage to get into trouble in Havana when they meet a pretty cafe singer named Carmen. (Hollywood)

4 CIRCLE THEATER—Drama
"Night Court," by Art Wallace, dramatizes a six-hour session in a New York night court, where viewers meet vagrants, shoplifters, a phony palmist, a sharp bookie, a spoiled young college boy and a

SUNDAY, MAY 12, 1957

Randolph C. Pate, Marine Corps; and Vice Adm. Alfred C. Richmond, Coast Guard. They will speak from the Pentagon. Other stops on the tour include: **Fort Sill, Okla.:** platoon nailed down by enemy action. **Ft. Carson, Col.:** mountain troops climbing steep bluffs. **Norfolk, Va.:** the Boston, guided-missile cruiser; the FDR, big new carrier; refueling a destroyer; firing of the Regulus guided missile from a submarine. **Miami, Fla.:** frogmen destroy an underwater obstacle. **George Air Force Base, Victorville, Cal., and Luke Air Force Base, Phoenix, Ariz.:** gunnery training and planes attacking enemy group.

7 COLLEGE NEWS CONFER'NCE
Sen. Stuart Symington (D., Mo.) is interviewed by a panel of college students. Ruth Geri Hagy moderates. (Wash., D.C.)

13 JUNIOR CARNIVAL—Kids

4:30 2 NEWS ROUNDUP—Sevareid

7 8 MEDICAL HORIZONS
"Infectious Diseases." Dr. William Frye, dean of Louisiana State Medical School, New Orleans, and his staff report on the treatment of diseases caused by virus infections. They will explain their research program to isolate the various strains by virus, to discover the particular disease each can cause and to develop antibiotics to fight them.

9 STRANGE STORIES—Drama
"Hotel at the Beach." A prison doctor attempts to complete a case that calls for some private investigation.

4:50 11 SPORTS—Frankie Frisch

5:00 2 ODYSSEY—Collingwood
"The Kremlin." See page A-23 for details.

5 GANGBUSTERS—Police
"The Lawrence Case." The police are on the trail of a murderer who killed a local meat market manager.

7 43 DEAN PIKE—Religion
Today Dean Pike tells his four children, Cathy, Jimmy, Connie and Christy, "The Story of Jamestown, Va." This year marks the 350th anniversary of the first permanent English settlement in Jamestown, Va.

8 MAN BEHIND THE BADGE
"Most Dangerous Game." A prison warden discovers that the hunter can become the hunted. Charles Bickford hosts.

9 MOVIE—War Drama
Million Dollar Movie: "Marine Raiders." See Sat., 7:30 P.M. Ch. 9.

11 MOVIE—Western
"Trail of Kit Carson." Allan Lane

13 COMEDY CORNER

5:30 4 OUTLOOK—Chet Huntley
Chet Huntley rounds up late news and presents film features. Today there will be filmed reports on the U.S. guided missile program and on the Israeli Army as recently covered by "Outlook" cameramen.

5 THREE MUSKETEERS
"Petticoat Castle." The King orders the Musketeers to rescue a duchess.

7 43 OPEN HEARING-Secondari

8 CARTOONS—Kids

13 EVANGEL HOUR—Religion
Lane Adams sings "How Great the Art and "He's Got the Whole World in His Hands." Rev. Ronald Fleming directs.

EVENING

6:00 2 MY FRIEND FLICKA
"Growing Pains." New to the McLaughlin Ranch are a little girl and a prize bull. Ken learns that both are extremely unpredictable. Johnny Washbrook, Gene Evans, Anita Louise. Susan: Reba Waters. Gus: Frank Ferguson.

4 MEET THE PRESS—Panel
Today's guest is Gen. Maxwell D. Taylor, Chief of Staff of the U.S. Army. The panel consists of May Craig, Portland (Me.) Press-Herald; Martin Agronsky, NBC; Yates McDaniel, Associated Press; and Richard Wilson, Cowles Publications. Ned Brooks moderates the discussion from Washington, D.C.

5 GREAT GILDERSLEEVE
"The Nightmare." Gildy becomes such an avid devotee of pocket mysteries that he re-enacts all the plots in his dreams. Willard Waterman stars. (Film)

7 CORLISS ARCHER—Comedy
Mr. Archer finds it impossible to make a long-distance phone call from his house. Ann Baker, Bobby Ellis.

8 BADGE 714—Jack Webb
Legal restraints which limit police officers in searching and arresting suspects are examined. Ben Alexander.

SATURDAY, JULY 6, 1957

tonight and tries to talk Jimmy into getting his nose bobbed in order to get more romantic movie roles. In a dream sequence, Jimmy imagines that he goes through with the operation. His songs include "It's the Clothes That Make the Man," "Pike's Peak or Bust" and "Battle of the Bands." Also appearing are Eddie Jackson, Jack Roth, Jules Buffano, Angie Dickinson, puppeteer Allen Henderson and Jackie Barnett. (Film)

④ JULIUS LA ROSA—Variety
[COLOR] Julius welcomes comedian Stubby Kaye, dancer Bobby Van, the singing Fontaine Sisters and the Piero Brothers, a novelty act, as his guests tonight. Julius, Stubby Kaye and Bobby Van sing "You Gotta Have Heart." Julius offers "Just One of Those Things," "Everything Happens to Me," "You've Changed," "You Took Advantage of Me" and "I'll Know."

⑤ MOVIE—Mystery
"Charlie Chan and the Chinese Ring." (1947) When a Chinese princess comes to the U.S. to purchase planes, she is murdered in the den of Charlie Chan. Roland Winters, Mantan Moreland, Warren Douglas, Victor Sen Yung.

⑦ ⑧ BILLY GRAHAM—Religion
⑪ MOVIE—Drama
"Another Man's Poison." (1952) A ruthless, self centered woman who will stop at nothing, believes she is in love with her secretary's fiance. Bette Davis, Gary Merrill, Emlyn Williams, Anthony Steele.

8:30 ② TWO FOR THE MONEY—Quiz
Tonight's teams include the two cabin boys from the Mayflower II, Joe Meany and Graham Nunn; Virginia Payne, radio's "Ma Perkins," and Don MacLoughlin, radio's Dr. Brent of "Road of Life"; and the Lashley quadruplets, 16-year-olds from Leitchfield, Ky.

9:00 ② OH! SUSANNA—Comedy
"The Chimpanzee." Capt. Huxley of the S.S. Ocean Queen has had enough of Susanna's shenanigans. He advises her that she must not get involved in any more problems or she'll be fired! Gale Storm, ZaSu Pitts, Roy Roberts, Jimmy Fairfax. (Film)

④ GEORGE SANDERS—Mystery
"The Call," by Bethel Laurance. A nurse

begins to receive a series of mysterious phone calls in the middle of the night. The unidentified caller keeps warning her she is going to die, until the nurse almost has a nervous breakdown. TV adaptation by Halsted Welles. George Sanders is the host. (Film)

Cast
Ann Higginson ....Toni Gerry
Lewis ....James Gavin
Tom McElroy ....Adam Williams
Police Captain ....Marshall Bradford

⑤ N.O.P.D.—Police
"The Goobers Case." Detectives Beaujac and Conroy try to crack down on a narcotics-smuggling ring. Stacy Harris, Lou Sirgo. (Film)

⑦ ⑧ LAWRENCE WELK
Featured are regulars Alice Lon, the Lennon Sisters, Larry Dean, Dick Kesner, Bob Lido, Tiny Little, Jr., Rocky Rockwell and Larry Hooper. (Film)

Highlights
"Fourth R—Religion" ....Hooper
"Taking a Chance on Love" ....Lon
"Stardust" ....Dean
"Two Different Worlds" ....Lennons
"Moonlight and Roses" ....Kesner
"I Love a Piano" ....Lido, Little
"Mocking Bird Hill" ....Lennons
"I Wanna Be a Cowboy" ....Rockwell

⑨ MOVIE—Drama
"Green Promise." (1949) Story of a farmer who acts like a tyrant to his four motherless children, and of a young agricultural agent who comes into their lives. Marguerite Chapman, Walter Brennan, Robert Paige, Natalie Wood.

⑬ ZERO 1960—Religion
Dominic Reis describes the "miracle of the Sun," which took place at Fatima, Portugal, 40 years ago.

Postponed from last week.

9:30 ② S.R.O. PLAYHOUSE—Drama
Will Rogers, Jr., in "On a Dark Night." A college professor is unexpectedly caught up in a dangerous situation. The professor and his friend Max give a good-looking blonde a lift, only to wind up in a gangster's hideout. The professor guesses that Max, who's always telling tall stories about himself, staged the episode, but he's dead wrong! (Film)

Continued on next page

MONDAY

AUGUST 19

his nurse from a blackmailing charge. Macdonald Carey stars. (Film)

8:30 ❷ TALENT SCOUTS—Godfrey
Tonight's talent includes the Hampton Sisters, vocal trio, Lowell, Mass.; Virginia Hauer, vocalist, Dorchester, Mass.; and Charles Castleman, violinist, Braintree, Mass. Arthur Godfrey is host. (Film)

❹ ACTION TONIGHT—Drama
Ralph Bellamy stars in "The Payoff." Joe McQuade is a respected and honest detective, who is in line for a promotion. Suddenly his daughter needs an operation. Joe needs money, and a racketeer offers to supply it under certain conditions. (Film)

Cast
Joe McQuade	Ralph Bellamy
Otto "Bitsy" Lamb	Nat Pendleton
Mary	Marianne Stewart
Henry	Lewis Charles
Frank	Bob Bice
Augie	Harry Tyler

❺ CONFIDENTIAL FILE—Docum.
"Exceptional Children." There are 1,-500,000 mentally retarded persons in the U.S. and only 10 per cent of them are institutionalized, Paul Coates reveals in this film. The other 90 per cent live at home and have to put up with the day to day problems of the real world. (Film)

❼ ❽ BOLD JOURNEY—Docum.
"Lost." Milton Farney narrates the rescue of a party of hikers in the wilds of Baja California. Highlights include scenes of the ground-and-air rescue team as they try to find the teen-age students and their professor in the wild country. John Stephenson is the host. (Film)

Spine tingling! INNER SANCTUM 9:30 WPIX 11

⓫ SAN FRANCISCO BEAT-Police
"The Wildcat Case." It looks like a senseless crime when a robbery-murder nets only seven dollars. (Film)

8:40 ⓭ THRILLS IN SPORTS

9:00 ❷ THOSE WHITING GIRLS
Barbara concocts an unusual plan to convince her mother and Margaret they should buy her a car. Her first step is to acquire an ancient automobile from their gardener. Margaret and Barbara Whiting, Mabel Albertson. Penny: Kathy Nolan. Gardener: Robert Kino. (Film)

❹ TWENTY-ONE—Quiz

❺ RACKET SQUAD—Police
"C.O.D. Honeymoon." An auction racket wherein a wealthy woman bids a high price for an art object. When she discovers she hasn't sufficient money for payment the auctioneer agrees to deliver the article C.O.D. Reed Hadley. (Film)

❼ PASSPORT TO DANGER
"Havana." Diplomatic courier Steve McQuinn is disturbed to discover that a friend is piloting passengers into the country illegally. Cesar Romero. (Film)

❽ OPPORTUNITY UNLIMITED

❾ MOVIE—Mystery
"The Saint Meets the Tiger." (1943) Murder complicates matters in this tale of gold thieves. Hugh Sinclair, Jean Gillie, Gordon McLeod.

⓫ CITY DETECTIVE—Police
"Thirteen O'Clock." When the college campus clock strikes 13, a student mysteriously disappears. Rod Cameron. (Film)

⓭ MOVIE—Mystery
All-star Movie: "Bermuda Mystery." See 7:30 P.M. Ch. 13, for details.

9:30 ❷ RICHARD DIAMOND-Mystery
"The Chess Player." Diamond is hired as a bodyguard to a wealthy financier without the man's knowledge. The man's wife is worried because he refuses to take threats against his life seriously. David Janssen. (Film)

Cast
Mildred Tyler	Catherine McLeod
Julian Tyler	Vaughn Taylor
Norman Devitt	Ross Elliott
Colin	Jonathan Hole
Warburton Flagge	Larry Dobkin
Arminta Tyler	Madeline Holmes
Thaddeus Tyler	George Baxter

SUNDAY

the Huron tribe is on the warpath against the British Fort, Col. Thorpe sends out his chief scout for reinforcements. John Hart, Lon Chaney, Lili Fontaine.

⑨ SCIENCE FICTION THEATER
"Time Is Just a Place." A couple discover that their new neighbors possess a "sonic broom." Don DeFore, Marie Windsor.

⑬ GOV. MEYNER'S REPORT
The Governor and his guests discuss Civil Defense Week.

㊹ THIS IS THE LIFE—Religion
"The Hateful Heart." A teen-age boy tries to cope with a bitter uncle and an ex-convict father. Robert Arthur.

7:00 ② LASSIE—Drama
"The Suit." Young Timmy accidentally ruins a brand-new suit and is afraid to tell Ellen he is responsible. He decides to place the blame on Lassie. Jan Clayton, Tommy Rettig, George Cleveland, Donald Keeler. Timmy: Jon Provost. (Film)

⑤ DR. BRUNO FURST
Dr. Bruno Furst, director of the School of Memory and Concentration, and his students demonstrate how imagination and association can be used to acquire a reliable memory.

⑦ ⑧ YOU ASKED FOR IT
1 Russia's famous performing bears. 2. World's largest mechanical monster. 3 Mexico's amazing kid riders. 4. Hiroshima bombing witness, Nobuko Sakoda, showing the city as it is today. Art Baker is the host. (Film)

⑨ EDDIE CANTOR—Variety
Eddie Cantor presents Joe E. Brown, who finds out on tonight's show that a practical joke can backfire. (Film)

⑬ CARTOON COMICS—Kids
㊹ FILM SHORTS
7:20 ⑪ SPORTS—Red Barber
7:30 ② BACHELOR FATHER—Comedy
DEBUT John Forsythe stars in this half-hour comedy series, which will alternate with the "Jack Benny Show" in this time spot. Forsythe, plays Hollywood attorney Bentley Gregg, whose life becomes complicated when he takes over the care of his teen-age niece Kelly, played by Noreen Corcoran. Tonight: "Uncle Bentley and the PTA." Kelly is anxious for her uncle to be active in the PTA at her school. But Uncle Bentley has other things on his mind—a series of dates with a glamorous Italian movie actress. (Film)

Cast

Bentley Gregg	John Forsythe
Kelly	Noreen Corcoran
Peter	Sammee Tong
Angela Giovanni	Sheila Rudy
Vickie	Alice Backes

④ SALLY—Comedy
DEBUT Joan Caulfield stars in this half-hour comedy series as Sally Truesdale. companion to globetrotting Mrs. Myrtle Banford, played by Marion Lorne. In tonight's episode Sally, employed as a sales-girl, believes that an elderly matron is a penniless shoplifter and tries to help her Actually, the matron is one Myrtle Banford, a wealthy widow. Viewers will recall Marion Lorne as Mrs. Gurney in the "Mr. Peepers" TV series. (Film)

Cast

Sally Truesdale	Joan Caulfield
Myrtle Banford	Marion Lorne
Quincey	Parley Baer

WEDNESDAY, SEPTEMBER 18, 1957

⑨ MOVIE—Drama
Million Dollar Movie: "The Boy with the Green Hair." See Mon., 7 P.M., Ch. 9.

⑪ NEWS—Kevin Kennedy

⑬ BOATING—Lewis King

7:05 **②** LOCAL NEWS—Ron Cochran

7:10 **②** WEATHER—Carol Reed

⑦ WEATHER—Janet Tyler

⑪ WEATHER—Joe Bolton

7:15 **②** NEWS—Douglas Edwards

⑦ NEWS—John Daly

⑪ NEWS—John Tillman

7:30 **②** I LOVE LUCY—Comedy
"The Matchmaker." Lucy hopes that the Ricardos' display of wedded bliss will induce her friends Dorothy and Sam to marry. Lucille Ball, Desi Arnaz, Vivian Vance, William Frawley. (Film)

④ WAGON TRAIN—Western
DEBUT The pioneers' trek from the midwest to the Pacific is recounted in this weekly one-hour series. Details, below.

Television saddles up for a big season. See the article on page 17.

⑤ MICKEY ROONEY—Comedy
"Scoop Mulligan." Mickey aspires to move up the ladder of success and vies for a position on the local newspaper because of his experience with the editorial section on the network paper. (Film)

7:30 ④ WAGON TRAIN — Western

Robert Horton   Ward Bond   Ernest Borgnine

DEBUT Ward Bond and Robert Horton co-star in this weekly series of one-hour Westerns. It recounts the adventures of a group of pioneers traveling in covered wagons from the Midwest to California.

Ward Bond will be seen each week as the wagonmaster, Seth Adams, and Robert Horton portrays the guide, Flint McCullough. Their job is to lead the wagon train and its passengers through hostile Indian territory.

Tonight Ernest Borgnine and Marjorie Lord are seen in the first episode, "The Willy Moran Story." Seth Adams rescues a man from a street brawl and is surprised to find that he is Willy Moran, under whom Adams served in the Civil War. Moran, once a great fighter, is now an alcoholic, unable to fight or to earn an honest living. Adams offers Moran a job with the wagon train, but warns him to stay away from liquor during the trip. A widow traveling with the train is attracted to Moran and tries to help him. She is confident that Moran has conquered his craving for liquor, even though the other members of the wagon train try to tell her otherwise. (Film)

Cast

Seth Adams	Ward Bond
Flint McCullough	Robert Horton
Willy Moran	Ernest Borgnine
Mary Palmer	Marjorie Lord
Brady	Andrew Duggan
Susan	Beverly Washburn
Robinson	Donald Randolph
Palmer	Richard Hale
Fabor	John Harmon
Ben	Michael Winkleman
Lansing	Kevin Hagen

men leave on furlough while waiting for reassignment. Then the Comanche chief, Okoma, takes over the fort and joins forces with Chief Black Cloud to raid the surrounding territory. Maj. Swanson is ordered to retake Fort Apache. Lee Aaker, James Brown, Joe Sawyer.

⑧ WATERFRONT—Adventure
"Backwash." A former suitor of a young woman sells her husband a diesel motor booster. Preston Foster.

⑪ BRAVE EAGLE—Western
[RETURN] "Valley of Decision." A stubborn settler refuses to let Brave Eagle's tribe pass through his property to harvest their corn. Keith Larsen.

⑬ HOUSE DETECTIVE—Newman

6:30 **⑤ LOONEY TUNES—Cartoons**

⑦ ANNIE OAKLEY—Western
"The Mississippi Kid." A young man feels that he must demonstrate his

7:30 ❷ PERRY MASON — Mystery

Raymond Burr

Barbara Hale

[DEBUT] ❶ne of the best-known fictional sleuths is attorney Perry Mason, the brainchild of author Erle Stanley Gardner, and the hero of 52 different novels. Raymond Burr portrays Mason in this one-hour series dramatizing the famous attorney's adventures. Barbara Hale is seen as Mason's girl Friday, Della Street.

Tonight: "The Case of the Restless Redhead." Mason receives a call for help from a redhead, Evelyn Bagby, who finds a gun that doesn't belong to her in her apartment. After questioning her, Mason learns that Evelyn was recently tried for diamond theft, but was acquitted for lack of evidence.

Later, Evelyn is chased by a hooded man and shoots him. When she is picked up by the police and accused of murder, Mason comes to her defense and begins a dramatic courtroom battle to save her life. (Film)

Cast

Perry Mason	Raymond Burr
Della Street	Barbara Hale
Paul Drake	William Hopper
Evelyn Bagby	Whitney Blake
Lt. Arthur Tragg	Ray Collins
Hamilton Burger	William Talman
Mr. Boles	Vaughn Taylor
Mrs. Boles	Jane Buchanan
Mervyn Aldrich	Ralph Clanton
Sgt. Holcomb	Dick Rich
Helene Chaney	Gloria Henry
Judge Kippen	Grandon Rhodes

TUESDAY

OCTOBER 22
Evening

Uncle Hugo ....Horace Cooper
Martin ....Leonard Elliott
Aunt Jennifer ....Eda Heinemann

Dina Merrill, society girl who has been featured on this show, is written up in next week's TV GUIDE.

④ GEORGE GOBEL—Variety

COLOR Actress Hedy Lamarr is George's guest. The theme of tonight's show is "Glamour Through the Ages." In a comedy sketch, George brings his dog into a poodle parlor presided over by Hedy Lamarr. The Johnny Mann Singers. Regulars include Eddie Fisher, Shirley Harmer, Barbara Bostock, Howard McNear and John Scott Trotter and the orchestra.

⑤ UNCOMMON VALOR—Marines

"Between the Wars." Gen. Holland M. Smith discusses the problems the Marine Corps met due to the rapid demobilization after World War 2. (Film)

⑪ DEEP SEA ADVENTURES

"Hull Down." An elderly businesswoman feels old and useless when she suffers a heart attack. Forrest Tucker. (Film)

8:30 ② EVE ARDEN—Comedy

Liza Hammond decides to return to her home town, about which she has written a best-selling book. But Liza is shocked to find that the townspeople, far from welcoming her back as a local celebrity, are up in arms against her. Allyn Joslyn, Frances Bavier, Karen Greene, Gail Stone. (Film)

Cast

Headwaiter ....Gil Stuart
Waiter ....Jimmy Cross
Cigaret Girl ....Gloria Marshall
Customers ....Ray Kellogg, Mike Ross

⑤ GOLDBERGS—Comedy

"Fledermaus." Molly organizes a production of "Fledermaus" to aid her ladies' auxiliary. Gertrude Berg. (Film)

⑦ ⑧ WYATT EARP—Western

"Warpath." Wyatt steps in when a band of renegades sells several cases of forbidden guns to a tribe of Indians. The Army intelligence officers sent in to prevent an Indian uprising find themselves in trouble. Hugh O'Brian. (Film)

Cast

Lt. Clark ....Donald Murphy
Brave Bull ....Monte Blue
Young Wolf ....Michael Carr

Mr. Cousin ....Rico Alaniz
Mr Brother ....Rodd Redwing

⑪ FOOTBALL HIGHLIGHTS

㊸ MOVIE—To Be Announced

8:45 ⑬ MOVIE—To Be Announced

9:00 ② TO TELL THE TRUTH—Panel

④ MEET McGRAW—Mystery

"Mojave." McGraw, a stranger in a desert town, is accused of killing the woman owner of a cafe in a robbery. The townsfolk refuse to believe McGraw, who claims he is innocent. A lynching party forms. Frank Lovejoy. (Film)

Cast

Andy Plummer ....Robert Armstrong
Jim Bennet ....Claude Akins
Lila Doyle ....Carol Kelly
Broderick ....Kay Kuter
Joe Cappo ....Ralph Moody

⑤ MOVIE—Musical

"Casbah." (1948) To avoid capture by the police, a jewel thief hides in the Casbah section of Algiers. Yvonne DeCarlo, Tony Martin, Peter Lorre, Marta Toren.

⑦ ⑧ BROKEN ARROW-Western

"Ghost Sickness." A young Apache boy sets out to clear the reputation of his dead father, who is regarded as a traitor. He feigns sickness in an attempt to expose the real traitor. John Lupton, Michael Ansara, Armand Alzamora. (Film)

⑨ FAVORITE STORY—Drama

"Crime of Sylvestre Bonnard." A French professor wants to help the young daughter of a childhood sweetheart. Ralph Morgan, Ellen Corby. (Film)

⑪ INNER SANCTUM—Mystery

"Killer's Choice." A man relies on human nature to revenge the cold-blooded murder of his brother. Warren Stevens. (Film)

9:30 ② RED SKELTON—Comedy

COLOR Red's guest is actor Cesar Romero. Romero, as an advertising-agency executive, is given the task of finding a cowboy television star to match that of a rival advertiser's TV show. In a saloon in the wild and woolly West he discovers Deadeye. David Rose conducts the orchestra. (Hollywood)

④ BOB CUMMINGS—Comedy

"Bob Hires a Maid." While Margaret is away Bob decides to hire a maid. Paul Fonda, who thinks it's an excellent idea, is determined to be the first to make a

FRIDAY

N O V E M B E R   2 2
Evening

⑬ ZERO 1960!—Religion

9:00 ② MR. ADAMS AND EVE
"The Artist." Eve is flattered and Howard skeptical when an egotistical artist asks Eve to sit for a portrait so that he can paint the "real woman" within her. Ida Lupino, Howard Duff, Olive Carey, Hayden Rorke, Jay Novello. (Film)

④ M SQUAD—Police
"Killer in Town." A man sought for another murder is responsible for the death of an off-duty Chicago policeman. Lt. Ballinger has only a few leads to the man's favorite haunts. Lee Marvin. (Film)

Cast
Rex Lang ....Lee Farr
Det. Cooper ....Russ Conway
Olga ....Gail Kobe
Capt. Grey ....Paul Newlan
Police Driver ....John Hiestand
Counter Man ....Roy Glenn

⑤ TV READER'S DIGEST
"The Gigantic Banknote Swindle." A Portuguese swindler is convinced that dishonesty is the best policy and sets out to prove it. Victor Jory. (Film)

⑦ ⑧ FRANK SINATRA—Variety
Erin O'Brien is Frank's guest. She sings "I'm Glad There Is You." She and Frank duet "Let's Get Away from It All." Frank's songs are "Wrap Your Troubles in Dreams," "I Get Along Without You Very Well," "My Funny Valentine," "I Wish I Were in Love Again," "Write Myself a Letter" and "One for My Baby." Nelson Riddle Orchestra. (Film)

Frank won't starve if he loses his voice. See the article which begins on page 17.

⑨ I'M THE LAW—Police
"Fight Fix." A girl and her boss try to fix a fighter by blackmailing him with the accidental killing in his past. (Film)

⑪ DAVID NIVEN—Drama
"Man of the World." A bored husband looks for romance with a girl he meets on a train. David Niven. (Film)

⑬ HARLEM SHOWCASE—Bostic
Appearing Tonight: The Willows vocal group, dancer Bunny Briggs, the Orlandos Group, and singer Bobbe Caston.

9:30 ② SCHLITZ PLAYHOUSE—Drama
Myrna Loy stars in "No Second Helping." A woman suspects her husband of having a romance with the young daughter of one of his former flames. In an attempt to show her husband the error of his ways, the wife invites the girl out for a weekend. (Film)

Cast
Nancy Knox ....Myrna Loy
George Knox ....Tom Helmore
Eloise Moore ....Jill St. John

④ THIN MAN—Mystery
"A Ring Around Rosie." Giving a wedding party for friends proves explosive for Nick and Nora when their car is demolished by a bomb. Peter's mother, Lady May Lawford, plays a society reporter. Phyllis Kirk, Peter Lawford. (Film)

Cast
Porky Bunker ....Joey Faye
Lt. King ....Tol. Avery
Miss Lamont ....Lady May Lawford
Rosie Muldoon ....Lisa Gaye

⑤ BIG STORY—Drama
"Shield 21." Ralph Hennings, South Bend (Ind.) Tribune reporter, is eye witness

SUNDAY, DECEMBER 1, 1957

[13] **MOVIE—Mystery**
All-star Movie: "Strange Triangle." (1946) This story involves a beautiful, unscrupulous woman, embezzlement and murder. Signe Hasso, Preston Foster.

[43] **FILM SHORTS**

7:30 [2] **JACK BENNY—Comedy**
"Filming Jack's Life Story." Van Johnson and Buddy Adler, production head of 20th Century-Fox Studios, are Jack's guests. Over Adler's protests, Jack wants to produce, direct, write and star in the film. When Van Johnson is suggested as the star, Jack blows up. Mary Livingstone, Don Wilson and Eddie (Rochester) Anderson. (Film)

[4] **SALLY—Comedy**
In North Africa, Sally does a favor for a young girl. Result—she's kidnaped and taken to an Arab's desert camp. Joan Caulfield, Marion Lorne. (Film)

[5] **MICKEY ROONEY—Comedy**
"Ghost Story." Mickey's folks decide to rent a cottage near a lake for their vacation. Regis Toomey. (Film)

[7] [8] **MAVERICK—Western**
"The Wrecker," based on the novel by Robert Louis Stevenson and Lloyd Osbourne. During an auction, curiosity causes Bret and Bart Maverick to outbid a man for the cargo of a ship, mysteriously beached on Midway Island in the South Pacific. Bart travels to the island, where his life is endangered by an unknown assassin. James Garner, Jack Kelly, Patric Knowles. (Film)

[9] **FILM DRAMA**
"Deception." An overprotective mother learns that her son, reported killed in Korea, is alive and coming home.

[11] **VICTORY AT SEA—Docum.**
"The Fate of Europe." Some unusual sequences of the Russian fleet in action in the Black Sea are shown. (Film)

[43] **MOVIE—To Be Announced**

8:00 [2] **ED SULLIVAN—Variety**
Ed's guests are actor Douglas Fairbanks, Jr.; singer Polly Bergen; the Glenn Miller Band led by Ray McKinley; Collier's all-American football team now taken over by General Mills; recording stars Sam Cooke, the Rays, the Crickets and Bobby Helms; comedienne Jean Carroll; and Tony and Sally DeMarco, ballroom dance team.

Songs

"You Send Me" ....Cooke
"Silhouettes" ....The Rays
"That'll Be the Day" ....The Crickets
"My Special Angel" ....Helms

[4] **STEVE ALLEN—Variety**
[COLOR] Steve's guests are comedienne Martha Raye; actor Errol Flynn; singer Jimmy Dean; the captains and stars of the Army and Navy football teams; comedian Don Adams; and singer Jennie Smith. The Army and Navy players will talk over yesterday's game with Steve. Tom Poston, Don Knotts, Louis Nye, Skitch Henderson, Gene Rayburn.

Highlights

"Come Rain or Come Shine" ....Raye
Satire on Panel TV Shows ....Flynn
"Sandy Sleighfoot" ....Dean
"Sometimes I'm Happy" ....Smith

[5] **NEW HORIZONS—Pearson**
Guest business executives are Maurice Chaffin, decorative plastic coverings; Robert Franklin, abrasive grinding wheels; Alex Manoogian, faucets. (Film)

TONIGHT at 6 PM
Ann Baker stars in
MEET CORLISS ARCHER
channel **9**

Is fact stranger than fiction?
Tune in
SCIENCE FICTION THEATRE
and see.
TONIGHT at 6:30 PM
channel 9

THURSDAY

Phoebe ....Gail Kobe
Max Birnbaum ....Benny Rubin
Guard ....Mark Scott
Inspector ....David McMahon

④ GROUCHO MARX—Quiz
Contestants tonight include columnist Mike Jackson, Laguna Beach, Cal., with Mrs. Marie Spangler, Glendale, Cal.; and Mr. and Mrs. Anil Hutheesing, UCLA exchange students from India. (Film)

⑤ RAY MILLAND—Comedy
"Ray's Other Life." Sheila Castle, a glamorous actress and ex-girl friend of prof. McNulty, arrives in town to register her daughter at Lynnhaven. Ray Milland. Phyllis Avery, Gordon Jones. (Film)

⑦ ⑧ ZORRO—Adventure
"Agent of the Eagle." An outlaw leader known as the Eagle orders members of his gang to extort cash from the frightened citizens to aid the unscrupulous magistrate of Los Angeles. Zorro decides to unmask the outlaw. Guy Williams, Henry Calvin, Gene Sheldon. (Film)

Cast
Eagle ....Charles Korvin
Ortega ....Anthony Caruso
Magistrate ....Vinton Hayworth

8:30 ② VERDICT IS YOURS—Drama
In tonight's re-enacted trial 19-year-old Thomas Crane is charged with murder. A grocery store was robbed and its owner shot by George Winslow, who confessed and was given a life sentence. Crane, who was the driver of the getaway car, is charged with playing the same part in the crime as Winslow. Jim McKay.

④ DRAGNET—Jack Webb
Joe Friday and Frank Smith try to track down a con man who has been swindling women by posing as a European count. Jack Webb, Ben Alexander. (Film)

Cast
Count H. Bukary ....Roger Till
Glenda Cardell ....Ann Morrison
Lucille Montrose ....Dodie Wright
Jed Stapes ....Charles Courtney
Sgt. Anders ....Craig Duncan

⑤ DOUGLAS FAIRBANKS
"Personal Call." A young American couple are in London on business. When the wife starts getting telephone calls at all hours, their marriage seems headed for the rocks. Phil Brown, Diane Hart. (Film)

⑦ ⑧ REAL McCOYS—Comedy
"Luke Gets His Freedom." Grampa McCoy talks Luke into going fishing without asking Kate, to prove he isn't henpecked. Kate retaliates, but the boys schedule a stag poker party to even the score. Walter Brennan, Richard Crenna, Kathy Nolan, Lydia Reed, Michael Winkelman. (Film)

Cast
Pepino ....Tony Martinez
George MacMichael ....Andy Clyde
Mac Maginnis ....Willard Waterman
Lila Maginnis ....Shirley Mitchell

9:00 ③ DR. HUDSON'S JOURNAL
"Caroline Story." Dr. Hudson saves the life of a young girl injured in an automobile accident. John Howard. (Film)

④ PEOPLE'S CHOICE—Comedy
An old boy friend of Mandy's drops in for a visit. He tells them how rich and successful he's become, so Sock, not to be outdone, invents a tale of his own importance. Then J.B. Barker, Sock's boss, arrives. Jackie Cooper, Pat Breslin, Paul Maxey. Roger Crutcher: John Stephenson. J.B. Barker: Addison Richards. (Film)

⑤ WRESTLING—Wash., D.C.

⑦ ⑧ ANDY WILLIAMS
Andy introduces "Promise Me, Love," a new song by Kay Thompson. Dick Van Dyke, Bob Hamilton Trio.

⑨ NIGHTMARE—Mystery
"Change of Heart." A famous criminal lawyer is approached by a beautiful girl at a party and asked to defend her brother who is accused of murder. Stephen McNally, Barbara Hale. (Film)

⑬ JAZZ PARTY—Art Ford

9:30 ② PLAYHOUSE 90—Drama
Tab Hunter and Geraldine Page in "Portrait of a Murderer," based on a real murder case and written by Leslie Stevens. Handsome 29-year-old Donald Bashor leads an exemplary life during the day and is a respected member of the community. But at night a change comes over the young man, and he frequents bars in another neighborhood. After committing a robbery, Bashor is arrested, though the police don't suspect that their prisoner may be connected with two recent murders. Tonight's script is based on remarks made by Bashor after his conviction and

SATURDAY

audience. 2. A woman who works for a travel agency is given a prize for each city she correctly locates on a map of the U.S. having no state boundaries. 3. Corrine Clark tries for $10,000 in the phone game. Art Linkletter hosts. (Film)

⑤ RANCH PARTY—Tex Ritter
Tommy Duncan sings "Dusty Skies." Smiley Burnette: "Way Down Low." (Film)

⑦ DICK CLARK—Music
Dick's guests are singers Ruth Brown, Tommy Edwards and Bobby Denton, and the Quin-Tones, vocal group.

Songs
"This Little Girl's Gone Rockin' " ..Brown
"It's All in the Game" ....Edwards
"Down the Aisle of Love" ....Quin-Tones

⑧ SWORD OF FREEDOM
"The School." Marco searches for a spy who is reporting Republican sympathisers. Edmund Purdom. (Film)

⑨ MOVIE—Western
Million Dollar Movie: "Station West." (1948) A Federal agent investigates robbery and murder in the Old West. Dick Powell, Jane Greer, Agnes Moorehead.

⑪ TOMAHAWK—Adventure
"The Betrayal." After being thrown in jail on their homecoming, Radisson and Medard appeal to the king. (Film)

⑬ MOVIE—Comedy
First Inning Theater: "Mister Universe." (1950) A sidewalk hawker turns an ex-war buddy into a wrestler. Jack Carson, Bert Lahr, Janis Paige, Robert Alda.

8:00 ④ BOB CROSBY—Variety
[COLOR] The guests tonight are singers Billy Eckstine and Eileen Rodgers, song-and-dance man Eddie Foy, Jr., and singing comedian Gary Morton. Gretchen Wyler, Peter Gennaro Dancers, Clay Warnick Singers, Carl Hoff Orchestra.

This is the last show of the season. "The Perry Como Show" returns next week.

⑤ CRUSADE IN THE PACIFIC
"The War in the China, Burma, India Theater." The film shows the importance of this area for checking Japanese aggression in Asia.

⑦ JUBILEE, U.S.A.—Variety
Red Foley's guests are country-music composer Cindy Walker and ragtime pianist Bob Darch. Red and the gang sing some

of Cindy's songs, including "Night Watch." Bobby Lord, Suzi Arden, Tall Timber Trio, Slim Wilson's Jubilee Band.

⑧ SHIRLEY TEMPLE'S STORY-BOOK—Fairy Tale
[SPECIAL] King Watkins may be a king, but he is also very poor. In fact he works for a living, as a mere junior clerk for miserly Mr. Gingery. The pretty Princess Alicia, eldest of his children, does her best to stretch his earnings to feed her little brothers and sisters. Barry Jones, Lisa Daniels, Estelle Winwood. (Film)

⑪ FILM DRAMA

8:30 ② WANTED—DEAD OR ALIVE
[DEBUT] This new weekly half-hour Western series concerns the adventures of Josh Randall, bounty hunter, who tracks down wanted men and brings them in—dead or alive—to collect the rewards offered for them. Tonight: Two outlaws shoot down the sheriff of a small Western town and their crime is witnessed by Josh Randall. The outlaws escape and a reward is offered for their capture. Josh Randall sets out after the fugitives. (Film)

Cast
Josh Randall ....Steve McQueen
Carl Martin ....Michael Landon
Andy Martin ....Nick Adams
Louise Martin ....Jenifer Lea

⑤ MR. AND MRS. NORTH
The Third Eye." Pam and Jerry visit an exclusive dress shop. (Film)

⑪ AMOS 'N' ANDY—Comedy
"The Kingfish at the Ballgame." A rich widow accidentally drops her diamond ring into Kingfish's crackerjax box. (Film)

8:45 ⑬ SPORTS—Bert Lee, Jr.
Guest is Frank Bruggy, ex-catcher with the Philadelphia Phillies.

8:55 ⑬ BASEBALL—Dodgers
St. Louis Cardinals vs. Los Angeles Dodgers, from St. Louis. Jack Buck.

9:00 ② GALE STORM—Comedy
"Diamonds Are a Girl's Best." Susanna, who usually arranges romances for other people, falls in love with Air Force Lt. Arnold Van Dyke. Gale sings "Soon I'll Wed My Love" and "South of the Border." Gale Storm, ZaSu Pitts, Roy Roberts. Van Dyke: John Agar. (Film)

SUNDAY

(8) HARBOR COMMAND—Police
Racketeers force fishermen to pay for "insurance" or else suffer sudden violence. Wendell Corey.

(9) FARMER AL FALFA—Cartoons

(13) GOVERNOR MEYNER—Panel
"N.J.'s School Problems."

(43) THIS IS THE LIFE—Religion
"The Poisoned Banquet." A jealous woman invites trouble when she pries into a newcomer's business. The woman threatens her position in the community.

7:00 (2) LASSIE—Drama
June Lockhart and Hugh Reilly join Lassie and Jon Provost as Timmy's parents Ruth and Paul Martin. In tonight's episode Timmy's friend Boomer brings his dog Mike to meet Timmy and Lassie. The boys decide to assign Lassie to teach Mike the duties of a farm dog. (Film)

Cast
Timmy .... Jon Provost
Ruth Martin .... June Lockhart
Paul Martin .... Hugh Reilly
Uncle Petrie .... George Chandler
Boomer Bates .... Todd Ferrell
Fred Bates .... George Cisar

(3) 26 MEN—Western
"Apache Water." A religious group camped on Apache land nearly causes an Indian uprising. Tris Coffin. (Film)

(4) NOAH'S ARK—Drama
COLOR "The Guide." Noah and Sam try to help an injured dog and his owner. The owner suffers from a lack of confidence in his own abilities. Paul Burke, May Wynn, Vic Rodman. (Film)

Cast
Doug Conner .... Paul Brinegar
Florence Bixel .... Lillian Powell
Fred .... Jim Horan

(5) LILLI PALMER—Drama
"Just Off Piccadilly." In a house reputed to be haunted, a woman unexpectedly meets her lover of many years ago. (Film)

(7) (8) YOU ASKED FOR IT
The entire program today is devoted to a tour of the late William Randolph Hearst's estate, San Simeon, which has been converted into a public institution. Jack Smith hosts. (Film)

(9) CARTOONS—Kids

(10) POLITICAL TALK

(43) FILM SHORTS

7:15 (13) JUNGLE—Documentary
"The Scorpion." Life cycle of the scorpion of the American Southwest. (Film)

7:20 (11) SPORTS—Red Barber

7:30 (2) BROTHERS—Comedy
"Prisoners of Love." Three woman fugitives and a hoodlum friend of theirs hide out at the Boxes' place. Gale Gordon, Bob Sweeney, Nancy Hadley, Hope Emerson, Kathleen Freeman, Ellen Corby. (Film)

Last show of the season.

(3) MOVIE—Drama
"The Baron of Arizona." (1950) An ambitious land-office clerk turns up fraudulent papers to "prove" that the whole of Arizona belongs to his wife. Vincent Price, Ellen Drew, Beulah Bondi.

(4) NO WARNING!—Drama
"Twenty-six Hours to Sunrise." A youngster is bitten by a rabid coyote and must receive quick treatment. A stagecoach driver takes the boy and begins the ride to town, but is held up by a bandit. (Film)

Cast
Boy .... Michael Winkelman
Sam Travis .... Peter Hansen
Arty Hoyt .... Steve Brodie

Final show of the series. Beginning next week at this time: "Northwest Passage," an adventure series based on Kenneth Roberts' novel of the French and Indian War.

(5) MICKEY ROONEY—Comedy
"Lion Hunt." Mickey unwittingly allows

TUESDAY

SEPTEMBER 9
Afternoon-Evening

he believes belonged to Alladin. He rubs it and wishes for a horse—the horse appears. Buster Crabbe.

⑧ NEWS—Dave Kiernan

⑨ JET JACKSON—Space
"Trapped Behind Bars." Jackson and Ikky are planted in a prison to quell an expected riot. Richard Webb, Sid Melton.

⑪ AMOS 'N' ANDY—Comedy
"The Kingfish Teaches Andy to Fly." Kingfish convinces Andy he needs flying lessons. Spencer Williams, Tim Moore.

⑬ JUNGLE—Documentary
1. "Deep Sea." 2. "The Avocet." This is a long-legged shore bird.

㊸ NEWS—Walter Dibble

6:35 ㊸ **FILM SHORTS**
6:40 ④ **WEATHER AND SPORTS**
    ⑧ **WEATHER—Brace Gilson**
6:45 ④ **NEWS—Huntley, Brinkley**
    ⑧ **SPORTS—Syd Jaffe**
6:55 ⑧ **TOWN CRIER—Carol Hill**
7:00 ② **WORLD NEWS—Robert Trout**
    ③ **NEWS—Bruce Kern**
    ④ ⑧ **HONEYMOONERS**
"Pardon My Glove." Alice tries to surprise Ralph by redecorating the apartment without any cost to him. Jackie Gleason, Art Carney. (Film)

    ⑤ **JUDGE ROY BEAN—Western**
"The Eyes of Texas." Judge Bean uncovers a plot to steal Indian territory when he loses his sight for a short while. Edgar Buchanan, Jack Beutel. (Film)

    ⑦ **SPORTS—Howard Cosell**
    ⑨ **CARTOONS—Claude Kirchner**
    ⑪ **NEWS—Kevin Kennedy**
    ⑬ **SPORTS-O-PHONE—Bert Lee**

THE DODGERS AND GIANTS
this week on channel 9
New York's National League Station

7:05 ② **LOCAL NEWS—Ron Cochran**
7:10 ② **WEATHER—Carol Reed**
    ⑪ **WEATHER—Joe Bolton**
7:15 ② **NEWS—Douglas Edwards**
    ③ **WEATHER**
    ⑦ ㊸ **NEWS—John Daly**
    ⑪ **NEWS—John Tillman**
7:20 ③ **SPORTS—Bob Steele**
7:30 ② **NAME THAT TUNE—Quiz**
    ③ **MOVIE—Drama**
"Lulu Belle." (1948) A Natchez saloon-singer rises to fame and wealth on Broadway. Dorothy Lamour, George Montgomery, Albert Dekker, Otto Kruger.

    ④ **WIN WITH A WINNER—Quiz**
Final show of the series.

    ⑤ **WATERFRONT—Adventure**
"Farnum's Folly." Residents of the Seamen's Home are upset by a shipping tycoon's announcement that he plans to build an office building on the site. Preston Foster, Barry Kelly. (Film)

    ⑦ ⑧ **CHEYENNE—Western**
"The Angry Sky." While searching for an escaped murderer, Cheyenne is shot and left for dead on the trail. He is found by a woman who takes him to her sister's cabin. Then the sister's husband returns from a trip to town. Clint Walker. (Film)

Cast
Granger Ward ....Andrew Duggan
Rose ....Adele Mara
Lilac ....Joan Evans

Clint Walker is going thataway. His replacement as star of this series is Ty Hardin. Read about him on page 24.

    ⑨ **BASEBALL—Phillies**
Los Angeles Dodgers vs. Philadelphia Phillies from Philadelphia. Al Helfer and Rex Barney are the sportscasters. Before the start of tonight's regularly scheduled game, the teams will complete a game begun on July 27.

    ⑪ **SPORTSCHOLAR—Uttal**
Questions on shot putting, motorcycling and women's wrestling are answered. Fred Uttal hosts. (Film)

    ㊸ **FILM SHORTS**
7:45 ⑪ **SPORTS—Bill Stern**
    ⑬ **SPORTS—Bert Lee, Jr.**
Guest is Auggie Leo, Georgetown U. All-American football player.

TUESDAY, OCTOBER 14, 1958

7 8 RIFLEMAN—Western
"End of a Young Gun." Young Mark McCain gets caught on a mountain ledge. A young outlaw risks his life to rescue Mark and breaks his leg doing so. Lucas McCain has to decide whether to care for the injured outlaw or to turn him over to the law. Chuck Connors. (Film)

9 SCIENCE FICTION THEATER
An attorney refuses to believe in extrasensory perception. Donald Curtis. (Film)

11 WRESTLING FILM
Tom Thumb vs. Tiny Tim. Iron Mike and Bulldog Plecher vs. Gene Kelly and Duke Keomuka in a tag match.

13 MOVIE—Romance
Request Performance: "Great Expectations." (English; 1947) A poor young boy of the English marshlands falls in love with his playmate, the young mistress of a nearby estate. John Mills, Valerie Hobson, Jean Simmons, Alec Guinness.

4 FILM SHORTS

9:30 2 RED SKELTON—Comedy
Ralph Story of "The $64,000 Challenge" and blonde actress Barbara Nichols are the guests. Bolivar Shagnasty, played by Red Skelton, somehow ends up as the final contestant on a big TV quiz show called "The Million-Dollar Challenge." Quizmaster: Ralph Story. Gertie: Barbara Nichols. David Rose Orchestra.

3 AFRICAN PATROL—Adventure
DEBUT John Bentley stars as Inspector Derek in this adventure series filmed in East Africa. Tonight: "The Hunt." Insp. Derek uncovers a narcotics ring. (Film)

7 8 NAKED CITY—Police
"Line of Duty." Detective Jim Halloran is forced to kill a young gunman in order to save his own life. Later when he returns the gunman's belongings to his elderly mother, she berates him for the killing. Halloran begins to brood about the tragedy and Detective Lt. Dan Muldoon tries to help him. John McIntire, James Franciscus. (Film)

Cast
Katina	Eugenie Leontovich
Arcaro	Harry Bellaver
Janet	Suzanne Storrs
Yankee	Diane Ladd

The sidewalks of New York are this show's stage. Turn to pages 28 and 29.

9 HARNESS RACE—Yonkers

10:00 2 GARRY MOORE—Variety
Garry's guests are vocalists Tommy Sands and Louise O'Brien, comedienne Audrey Meadows and comic actor Andy Devine. In a Western sketch Devine sings "Dear Horse," and multi-voiced Mel Blanc is the front end of a horse. Marion Lorne, Durward Kirby, Howard Smith's orchestra.

3 DECOY—Police
"The Lost Ones." A teen-aged problem child is visited by Casey Jones. (Film)

4 CALIFORNIANS—Western
Marshal Matt Wayne finds a teen-ager, Samantha Jackson, after her wagon has been plundered by the notorious Bandanna Gang. He takes her to a mining camp where they meet a group of settlers anxious to go to San Francisco but afraid of the Bandanna Gang. Matt agrees to escort them. Richard Coogan. (Film)

7 MYSTERY PLAYHOUSE
"Nineteen Rue Marie." An American in Paris can't remember who he is or why he's there. David Brian. (Film)

8 SEA HUNT—Adventure
Mike investigates the death of a telephone company employe and discovers an underwater cable has been tapped. Lloyd Bridges. (Film)

11 PRO FOOTBALL HIGHLIGHTS
Films show high spots of last Sunday's games: Eagles vs. Steelers, Cards vs. Browns, 49ers vs. Bears, Giants vs. Redskins, Colts vs. Packers, Rams, vs. Lions. Jim Leaming comments.

13 THIS AND THAT—Variety
A fashion show is featured.

10:30 3 NEWS—Bob Ellsworth

TONIGHT at 9:30
HARNESS RACING
FROM YONKERS

channel 9

SATURDAY, OCTOBER 25, 1958

some Frenchman. ZaSu Pitts, Roy Roberts, James Fairfax. Roland: Jacques Bergerac. Gertie: Patricia Michon. (Film)

Producer Alex Gottlieb sprang a practical joke on Jacques. See page 28.

4 STEVE CANYON—Adventure

"Operation Survival." Canyon is accidentally thrown out of a plane while observing a training jump by Air Force "Paramedics." Though Canyon parachutes to safety, the drillmaster bails out in an effort to help Canyon and fractures his leg on landing. The two men find themselves in wild mountain country, miles from civilzation. Dean Fredericks. (Film)

Cast

Sgt. Harry Triver	Eddie Firestone
Gen. Hall	Frank Gerstle
Capt. Milt. Newberry	Fred Eisley
Lt. Kress	Buck Young

5 SEE THE PROS—Glenn Davis

Glenn Davis's guest is Bill Wade, Los Angeles Rams quarterback. (Film)

7 8 DANCING PARTY—Welk

Selections scheduled for tonight include "I Remember It Well" and "A Bushel and a Peck." Alice Lon, the Lennon Sisters, Aladdin, Larry Dean, Buddy Hayes and the Lawrence Welk Orchestra. (Hollywood)

9 MOVIE—Drama

"Walk Softly, Stranger." See 7:30 P.M., Ch. 9, for details.

13 WRESTLING FILMS

9:20 **3 WEATHER**

9:25 **3 POLITICAL PROGRAM**

9:30 **2 HAVE GUN, WILL TRAVEL**

Beaten up and robbed, Paladin finds that his enemy is a crooked sheriff who lets his family run the town. Unable to leave town because of the loss of his horse, money and gun, Paladin is helped to reclaim his possessions by a beautiful dance-hall girl. In return Paladin has to promise to take her with him when he leaves. Richard Boone. (Film)

Cast

Susan	Christine White
Sherriff Jack Goodfellow	Harry Carey, Jr.
Sol Goodfellow	Rayford Burns

3 OFFICIAL DETECTIVE—Police

"Beauty in the Bag." The body of a murder victim is found in a laundry bag. Everett Sloane. (Film)

4 CIMARRON CITY—Western

"To Become a Man." Matt Rockford tries to arrest a thief and in the ensuing struggle the man's gun goes off and he is killed. Matt is shocked when he recognizes the thief as an old friend. He tries to help the dead man's two sons, but one, a rebellious teen-ager, vows to hunt down the man responsible for his father's death. George Montgomery, Audrey Totter. (Film)

Cast

Joey Conway	Robin Riley
Mr. Conway	William Talman
Sheriff Sampson	Stuart Randall
Cal Demming	Jason Robards, Sr.

5 MOVIE—Drama

"The Macomber Affair." (1947) Hemingway's tale of a lion-hunting expedition in Africa, and how the wife of a rich American becomes enamored of their guide is dramatized. It is taken from the story "The Short and Happy Life of Francis Macomber." Gregory Peck, Robert Preston, Joan Bennett.

11 GUY LOMBARDO—Music

Jose Melis and the Funtane Sisters are the guests. (Film)

10:00 **2 GUNSMOKE—Western**

Caught in a stagecoach holdup, Matt and Chester see one of the bandits wounded in the arm during the getaway. Back in Dodge City, Matt recognizes the voice of a man trying to get medicine from Doc for a wounded friend. Matt is certain that the man is one of the bandits. James Arness, Dennis Weaver. (Film)

3 CHAMPIONSHIP BOWLING

7 8 SAMMY KAYE—Music

Regulars include Johnny Amoroso, Ray Michaels, Larry Ellis, Lynn Roberts, Hank Kanui, Harry Reser, Teddy Auletta, Johnny McAfee, Jerry Mercer, J. Blasingame Bond and the Dixieland Quartet.

11 MOVIE—Western

"Red Canyon." (1949) A wild-horse tracker comes to a small western town and tries to capture a beautiful horse. Howard Duff, Ann Blyth.

13 JUNGLE—Adventure

"The Scorpion." Life cycle study of the scorpion. Also "Jungle Ranger," "Underwater Spear Fishing," "The Grebe" and "The Eel Show" are shown. (Film)

MONDAY

the RAF, a pilot must change his tactics for daylight raids when he is transferred to the U.S. Eighth Air Force. (Film)

43 MOVIE—To Be Announced

9:00 2 3 DANNY THOMAS
Uncle Tonoose comes to visit. He tells Danny that he has decided to become a professional playboy and asks Danny to help him. Marjorie Lord, Rusty Hamer, Angela Cartwright. (Film)

4 PETER GUNN—Drama
"The Man with the Scar." Two days before the trial of a notorious racketeer, the district attorney is told to drop the case or else his son will be charged with murder. The district attorney hires Peter Gunn to investigate. Craig Stevens, Lola Albright, Hope Emerson. (Film)

Cast

Yale Lubin	Richard Wessel
Liz Hatton	Joan Taylor
Roy Davidson	Roy Thinnes

The review in next week's TV GUIDE is aimed at Peter Gunn.

5 MOVIE—Comedy
"The Amazing Mr. Williams." (1939) The love-life of a slap-happy detective is continually disrupted by one murder case after another. Melvyn Douglas, Joan Blondell, Clarence Kolb.

7 8 VOICE OF FIRESTONE
"Best of Opera" marks the show's 30th birthday. Details, page A-36.

11 MAN WITHOUT A GUN
"The Kidder." A vicious practical joker accepts money from a poor farmer in exchange for a worthless silver mine. Rex Reason, Mort Mills. (Film)

13 MOVIE—Drama
"The House of Rothschild." Time approximate. See 7:30 P.M., Ch. 13.

9:30 2 3 ANN SOTHERN—Comedy
Katy suspects that young Donald Carpenter has taken up with a gang of hoodlums. But when she learns he is organiz-

ing the boys into a band, she manages to get them a job at Bartley House. Both she and Donald are in trouble when the boys steal the turkeys the chef was to prepare for the hotel staff's Thanksgiving party. Ann Sothern, Ernest Truex, Reta Shaw, Jack Mullaney. (Film)

4 GOODYEAR THEATER—Drama
Richard Kiley in "Guy in Ward 4," by Paul Monash. An overworked psychiatrist continually drives himself to treat the airmen under his care at an air-base hospital in England during World War 2. He becomes interested in a young gunner being treated for wounds in Ward 4 and also needing a psychiatrist. (Film)

Cast

Capt. Josiah Newman	Richard Kiley
Alvarado	Charles Aidman
Jim Tompkins	Edward Ryder
Briscoe	Keith Vincent

7 8 ANYBODY CAN PLAY
The home-viewer prize is worth $5000 and the quiz contest now runs for two weeks instead of the previous four. Contestants tonight play a series of games involving mothers of movie stars, baby pictures of celebrities, anagrams, distorted music and a stunt connected with the sense of touch. George Fenneman, Judy Bamber. (Hollywood)

9 SCIENCE FICTION THEATER
"Barrier of Silence." A U.S. scientist working on top-secret atomic projects, mysteriously disappears. Adolphe Menjou, Warren Stevens, Phyllis Coates. (Film)

"Martin Kane" will not be seen tonight

11 SILENT SERVICE—Drama
"Mine for Keeps." For her third war patrol, the USS Trigger's assignment is to mine the coastal waters of Japan to drive enemy shipping into deep water. Carl Betz, Russ Johnson. (Film)

10:00 2 3 DESILU PLAYHOUSE
William Bendix in "The Time Element,"

What's the best thing on television?

FRIDAY, JANUARY 2, 1959

J A N U A R Y 2
Evening

⑤ WATERFRONT—Adventure
"Warehouse Incident." A young Korean
War veteran leads the captain to shady
dealings by a "respectable" merchant.
Preston Foster, Skip Homeier. (Film)

⑦ ⑧ ⑮ 77 SUNSET STRIP
"Hit and Run." Stuart Bailey's admirer,
Kookie, the parking-lot attendant, is involved in an auto crash. A former film
beauty claims she was disfigured in the
crash, and Kookie, charged with criminal
negligence, appeals to Bailey for help.
Efrem Zimbalist, Jr., Edward Byrnes,
Roger Smith. (Film)

Cast
Liz Murray .... Gloria Robertson
Robert Carter Murray .... Robert H. Harris
Chick Hammons .... Sue Randall

⑪ DAVID NIVEN—Drama
"Finale." An aspiring Broadway actor
finds his career thwarted by his look-
alike cousin. Martha Hyer. (Film)

10:00 ② ③ LINEUP—Police
"The Rorschach Murder Case." A murder
victim is found half-buried in the woods.
Lt. Guthrie and Inspector Greb enlist the
aid of a psychiatrist in finding the killer.
Warner Anderson, Tom Tully, Marshall
Reed. (Film)

Cast
Harry Beasley .... Howard McNear
Frank Gibbs .... Ken Mayer
Tony Armbruster .... Gordon Polk

④ BOXING—New York City
Gaspar Ortega, Tijuana, Mexico, vs. Denny
Moyer, Portland, Ore., welterweights, 10
rounds. Jimmy Powers reports from
Madison Square Garden, New York City.

TV GUIDE RATINGS

	Rating	Bouts	W	L	D	KO's
Ortega	3	59	44	13	2	19
Moyer	Unrated	18.	18	0	0	4

The undefeated Moyer has a dangerous
left hand and he likes to mix. He is
making his TV debut against a top-rated
scrapper in Ortega. Moyer, who comes
from a family of boxers, whipped tough
Tony Dupas and former welterweight
champ Johnny Saxton in his last two
fights. Ortega's two most recent appearances saw him lose split decisions
to Don Jordan, who subsequently won the
welterweight crown. Scoring by rounds.
Compiled by Nat Fleischer (The Ring)

⑤ JIM BOWIE—Adventure
"Ursula." Jim Bowie meets the woman
who will eventually become his wife.
Scott Forbes, Sidney Blackmer. (Film)

⑪ DIVORCE COURT—Drama
"Barnes vs. Barnes." A mother of two
sues her actor-husband, stating she would
prefer he returned to plumbing. (Tape)

10:30 ② ③ PERSON TO PERSON
Edward R. Murrow visits Roy Campanella
and George Jessel. Campanella, former
Brooklyn Dodgers catcher who was critically injured in an auto accident last
year, is seen with his wife and six children in their Glen Cove, N.Y., home.
Jessel, "toastmaster general of the world,"
is seen at home in Santa Monica, Cal.

⑤ OFFICIAL DETECTIVE—Police
"Missing." The skeletons of two women
are found in a condemned area. Robert
Anderson. (Film)

⑦ ⑧ ⑯ NEWS—John Daly

④ MOVIE—Drama
Million Dollar Movie: "The Hunchback
of Notre Dame." See 7:30 P.M., Ch. 9.

KHRUSHCHEV'S VISIT ON TV

A n historic event—the American visit of Soviet Premier Nikita Khrushchev—is about to take place before the eyes of the American people, because television makes it possible. The TV networks will cover some events live, and there will be frequent half-hour evening roundups of the day's activities.

Programs of the coming week are listed below, plus a suggestion of next week's activities, to be covered in greater detail in next week's TV GUIDE. Khrushchev being no rigid respecter of schedules, the networks are alerted for changes in plan.

TUESDAY, SEPTEMBER 15

10 A.M. (2, 3, 4, 7, 8) Arrival in U.S.
Khrushchev's plane is seen landing at an airport near Washington, D.C. After President Eisenhower's greeting, there is a motorcade to Blair House, near the White House, which will be the Premier's Washington residence.

7:30 P.M. (2, 3) Day's Roundup

9 P.M. (4) Day's Roundup

10:30 P.M. (7) Day's Roundup

WEDNESDAY, SEPTEMBER 16

1:30 P.M. (2, 3, 4, 7) and 2 P.M. (8) Press Conference
Premier Khrushchev addresses the National Press Club in Washington and is interrogated by the members in a telecast of approximately two hours. The Premier's remarks will receive a "consecutive" translation for which he will pause from time to time.

8 P.M. (2, 3) Day's Roundup
Khrushchev's second day in Washington is reviewed with excerpts from his press conference at the National Press Club. Commentators are Walter Cronkite, Paul Niven and Whitman Bassow.

9:30 P.M. (7) Day's Roundup

11 P.M. (8) Day's Roundup

THURSDAY, SEPTEMBER 17

7 P.M. (2) Day's Roundup

7:30 P.M. (2, 3, 4) Day's Roundup
Khrushchev's day in New York is recapped.

10 P.M. (7) Day's Roundup

10:45 P.M. (8) Day's Roundup

FRIDAY, SEPTEMBER 18

3 P.M. (2, 3, 4, 7, 8) Talk at UN
Soviet Premier Khrushchev addresses the United Nations General Assembly in New York City.

8:30 P.M. (7, 8) Day's Roundup

SEPTEMBER 19-27

SATURDAY, SEPTEMBER 19
Today Premier Khrushchev flies to Los Angeles, where his arrival may be covered.

SUNDAY, SEPTEMBER 20
Khrushchev flies from Los Angeles to San Francisco, where his arrival at about 8 P.M. may be picked up.

MONDAY, SEPTEMBER 21
The Soviet leader appears before the San Francisco Press Club at a luncheon.

TUESDAY-WEDNESDAY, SEPTEMBER 22-23
Premier Khrushchev is in the Midwest. He plans to visit Iowa State College at Ames, and the Roswell Garst farm at Coon Rapids, Iowa.

THURSDAY, SEPTEMBER 24
After arriving in Washington from a brief stay in Pittsburgh, the Soviet leader is to attend an evening reception at the Russian Embassy.

SUNDAY, SEPTEMBER 27
After TV talk to the American people the Soviet Premier will be seen taking off on his return trip to his own country.

SATURDAY

Barbara Bel Geddes, Wendell Corey and Aline MacMahon in "The Desperate Age," by Abby Mann. It is puzzling to the family and office friends of Letty Greene that she has little interest in men and apparently no concern over the prospect that she may never marry. But Letty has been seeing a co-worker, Will Robbins, who is unhappily married. She tries not to face the future of this romance.

Cast

Letty Greene	Barbara Bel Geddes
Will Robbins	Wendell Corey
Mrs. Greene	Aline MacMahon
Ed Coyne	Martin Balsam

④ BONANZA—Western

[DEBUT] [COLOR] This hour-long filmed Western series, set in the mid-1880's, centers around the activities of three half-brothers and their father, who run a timberland area near Virginia City, Nev. They are portrayed by Lorne Greene, Pernell Roberts, Dan Blocker and Michael Landon. Tonight: Yvonne DeCarlo is a guest star in "A Rose for Lotta." The powerful men of silver-rich Virginia City hire Lotta Crabtree, the famous entertainer, to lure Little Joe, the youngest brother, into the city so that they can hold him prisoner and demand valuable timber from the Cartwrights as ransom.

Cast

Ben Cartwright	Lorne Greene
Adam Cartwright	Pernell Roberts
Hoss Cartwright	Dan Blocker
Little Joe Cartwright	Michael Landon
Lotta Crabtree	Yvonne DeCarlo
Alpheus Troy	George Macready
Aaron Hopper	Barry Kelley
George Garvy	Willis Bouchey

⑤ MR. AND MRS. NORTH

"Climax." A crazed killer, who chooses sailors as his victims, makes a mistake when he attempts to challenge the Norths' detective prowess. Barbara Britton, Richard Denning.

⑦ ⑧ DICK CLARK—Music

Dick returns to New York after three weeks in Hollywood. His guests are Paul Anka, Roberta Shaw and Bobby Rydell, and the Fleetwoods, vocal group.

⑨ MOVIE—Drama

Million Dollar Movie: "Born to be Bad." (1950) An unscrupulous young woman begins her series of romantic conquests by wedding her friend's fiance. Joan Fontaine, Robert Ryan.

8:00 ⑤ ALAN FREED—Music

Guests include singers Lloyd Price, Johnny October; Vocal groups the crests, the Skyliners, Dion and the Belmonts.

Highlights

"I'm Gonna Get Married"	Price
"The Angels Listen"	Crests
"This I Swear"	Skyliners
"My Wish"	Grant

⑦ JUBILEE U.S.A.—Variety

Red Foley's guests are comedian Smiley Burnette and singer Faron Young. Leroy van Dyke, Promenaders. (Live)

⑧ MYSTERY IS MY BUSINESS

"Once a Killer." Ellery has three suspects from which to select the murderer of a young girl. Hugh Marlowe.

8:30 ② ⑥ WANTED—DEAD OR ALIVE—Western

Rejecting orthodox medicine, a desperate father takes his seriously ill son to an old medicine woman. Bounty hunter Josh Randall attempts to take the boy from the raving medicine woman and angry father so that a doctor can operate. Steve McQueen.

Cast

Carey Summers	John Collier
Tom Summers	Mort Mills
Manda	Virginia Gregg

④ MAN AND THE CHALLENGE

[DEBUT] George Nader stars as Glenn Barton in this weekly half-hour adventure series. Barton, athlete, medical expert and ex-Marine, is assigned by the government to find people who are willing to undergo dangerous missions for the purposes of scientific observation. Tonight: "The Sphere of No Return." Barton and two other men are asked to test a balloon at a very high altitude. When one of the men fails to measure up to the required standards, he is replaced by a girl.

Cast

Glenn Barton	George Nader
Corey	Paul Burke
Lynn Allan	Joyce Meadows
Kramer	Raymond Bailey

⑧ JUBILEE U.S.A.—Variety

This program is picked up in progress. See 8 P.M., Ch. 7, for details.

SUNDAY

boasts he owns a gun that has never been seen before and asks sharpshooter Annie to test it out. Gail Davis.

⑧ SHERIFF OF COCHISE

⑨ BUCCANEERS—Adventure
"The Ghost Ship." A drifting ship and a spectral crew plague merchant craft. Robert Shaw.

⑬ MOVIE—Western
"The Marshal's Daughter." Hoot Gibson.

6:30 ② ③ TWENTIETH CENTURY
"Enter with Caution: The Atomic Age," second half of a two-part series. Today's film, narrated by Walter Cronkite, shows the problems which resulted from a 1957 accident in which two men unknowingly carried radioactive particles with them out of a Houston laboratory and into the city. Rep. Chet Holifield of California, chairman of a Congressional subcommittee which has been studying problems produced by atomic radiation, analyzes the perils of the atomic age.

⑦ HAWKEYE—Adventure
"Revenge." A Shawnee war chief kills the parents of a young lieutenant. John Hart.

⑧ RESCUE 8—Drama
"The Ammonia Trap." An earthquake shakes Los Angeles and traps three men in an ice vault. Lang Jeffries, Jim Davis.

⑨ BOSTON BLACKIE—Adventure
A woman believes she has killed the girl her son wants to marry. Kent Taylor.

7:00 ② ⑧ LASSIE—Drama
Although the Martin farm is operated chiefly by mechanical implements, Timmy is sure there is still a way for an old horse to earn his keep. Trying to find a task for the animal, Timmy is trapped in a cave-in in an old mine. Jon Provost, June Lockhart, Hugh Reilly.

Cast
Higgins	Emory Parnell
Old Farmer	Ed Cobb
Carter	Robert Burton

③ DIAL 999—Police
Inspector Maguire risks his life attempting to capture an ex-commando.

④ RIVERBOAT—Adventure
[DEBUT] Darren McGavin stars as Capt Grey Holden, owner of the Mississippi riverboat Enterprise, in this weekly hour-long adventure series set in the 1840's Burt Reynolds is featured as the pilot of the Enterprise. Aldo Ray, Barbara Bel Geddes and Louis Hayward are guest stars in the first episode, "Payment in Full." For a price, simple-minded Hunk Farber is only too happy to reveal the whereabouts of his boss Monte, who accidentally killed a senator's son. The reward in his pocket, Hunk heads for Captain Holden's riverboat. He is eager to impress a particular young lady aboard the Enterprise

Cast
Capt. Grey Holden	Darren McGavin
Ben Frazer	Burt Reynolds
Hunk Farber	Aldo Ray
Missy	Barbara Bel Geddes
Ash Cowan	Louis Hayward
Monte	William Bishop
Sister Angela	Nancy Gates

A number of new shows will sail onto the airwaves this season. TV GUIDE previews them next week.

⑦ YOU ASKED FOR IT—Smith
"You Asked for It" presents a special program of New Orleans jazz. Among the guests are famed jazzman Louis Armstrong and Peter Davis, who taught Armstrong how to play the trumpet years ago Jack Smith is the host.

⑨ TERRYTOONS—Kirchner

⑬ BETWEEN THE LINES

㊸ TO BE ANNOUNCED

7:30 ② THAT'S MY BOY—Comedy
Jarrin' Jack plans for a pre-Christmas party at the office seem fated never to materialize when the boss' wife contemplates calling the party off Eddie Maye

THURSDAY

Highlights

"Four Leaf Clover" ....Fora
"Exotica" ....Top 20
"Old Rugged Cross" ....Ford

7 8 UNTOUCHABLES—Drama
DEBUT This weekly hour-long series, narrated by Walter Winchell, stars Robert Stack as Treasury Agent Eliot Ness, crime fighter of the Prohibition era. Tonight: "The Empty Chair." As underworld boss Al Capone leaves for prison, his chief lieutenants, Jake "Greasy Thumb" Guzik and Frank "The Enforcer" Nitti seek to take over the Capone empire.

Cast

Eliot Ness	Robert Stack
Jake Guzik	Nehemiah Persoff
Martin Flaherty	Jerry Paris
William Youngfellow	Abel Fernandez
Frank Nitti	Bruce Gordon
Enrico Rossi	Nick Georgiade
Brandy La France	Barbara Nichols
Aunt Norman	Betty Garde

11 LOVE STORY—Drama
"The Returning." A professor at a school

in Tokyo falls in love with a Eurasian. Dick Powell, Joan Elan.

10:00 4 GROUCHO MARX—Quiz
German airline Stewardess Margot Rohde and Hollywood Limerick writer Kent W Shelby return to the show.

11 TRACKDOWN—Western
"Law in Lampasas." The citizens of a small town convict an innocent man of murder. Robert Culp.

13 FULL COVERAGE—Barry Gray
43 FILM SHORTS

10:30 4 LAWLESS YEARS—Police
"The Art Harris Story." Hoodlum Art Harris assures district leader Pat Reilly of his support in the upcoming primaries. But the hood stipulates that Reilly must be prepared to grant various favors to Harris and his mob. James Gregory. Art Harris: John Beradino. Pat Reilly: Barry Kelley. Max: Robert Karnes.

7 HARBOR COMMAND—Police
A man kills his brother and then frames an innocent man. Wendell Corey.

8 TO BE ANNOUNCED

NEW KIND OF SERIES!

Beginning the true story of FBI man Eliot Ness' savage war on the notorious mobsters Al Capone left behind!

Robert Stack in **the UNTOUCHABLES**
9:30-10:30 THURSDAYS ABC 7

THURSDAY

⑪ YOU ARE THERE
The Boston Tea Party
⑬ NEWSBEAT—Mike Wallace
7:55 ⑬ **WEATHER—Margo Lee**
8:00 ② ③ **BETTY HUTTON—Comedy**
"Goldie Goes Broke." Goldie decides that her charges, the Strickland youngsters, need a lesson in the value of money. She tricks them into believing that their fortune has disappeared

Cast
Goldie	Betty Hutton
Pat Strickland	Gigi Perreau
Nicky Strickland	Richard Miles
Roy Strickland	Dennis Joel

④ BAT MASTERSON—Western
"Lady Luck." Casino owner Jess Porter plots a blackmail scheme against two sisters. Coming to the aid of the defenseless girls, Bat Masterson assumes the role of a detective.

Cast
Bat Masterson	Gene Barry
Mary Lowery	Diane Cannon
Rachel Lowery	Pamela Duncan

THE
BettyHutton
SHOW

8:00 TONIGHT WCBS ② TV CHANNEL 2

Jess Porter	Don Haggerty
Ron Davis	Charles Maxwell

⑤ BYLINE, STEVE WILSON
Steve Wilson investigates a detective agency. Mark Stevens.
⑦ ⑧ DONNA REED—Comedy
"Nothing Like a Good Book." Alex and Donna Stone reflect on the lack of culture in their family life. They set about bringing their children closer to good music and literature.

Cast
Donna Stone	Donna Reed
Dr. Alex Stone	Carl Betz
Mary	Shelley Fabares
Jeff	Paul Petersen
Lydia Langley	Mary Shipp

⑪ MEET McGRAW—Mystery
See Sunday, 9 P.M., Ch. 11, for details
⑬ PLAY OF THE WEEK—Drama
"Back to Back." See Mon. 8 P.M.. Ch 13, for details.
㊼ TO BE ANNOUNCED
8:30 ② ③ **JOHNNY RINGO—Western**
"Ghost Coach." An old Civil War wagon with a skeleton on board is found near town. After Ringo wires the Army to come and investigate, two mysterious killings take place.

Cast
Johnny Ringo	Don Durant
Laura	Karen Sharpe
Cully	Mark Goddard
Cason	Terence DeMarney
Colonel	Carl Benton Reid
Nelson	Phillip Pine

④ STACCATO—Mystery
"The Wild Reed." Johnny Staccato visits his old friend Frankie Aspen, who plays saxophone in a night club. Although Frankie is sure his playing is as great as ever, Staccato knows he has lost his touch and tries to find out why.

Cast
Johnny Staccato	John Cassavetes
Frankie Aspen	Harry Guardino
Waldo	Eduardo Ciannelli
Sally	Olive Deering
Mike	Joseph Sargent

⑤ DOUGLAS FAIRBANKS
"Ship's Doctor." The mate of the S.S. Romulus dies mysteriously. Miles Malleson.

TV GUIDE

SUNDAY, NOVEMBER 15, 1959

4 POLITICS 1960—News Analysis
SPECIAL David Brinkley moderates a 30- minute preview of the 1960 Presidential nomination and election contests. White House correspondent Ray Scherer speculates on the role President Eisenhower will play in 1960 politics. Richard Harkness analyzes the Republican nomination race and Frank McGee the Democratic race. Brinkley reports the results of a poll of news editors of NBC stations in which they reported on the strength of the leading Presidential candidates in their areas. The program concludes with a round-table discussion of the poll.

7 43 COLT .45—Western
"Yellow Terror." Chris Colt and his prisoner are aboard a river boat when the prisoner is killed. Then a yellow-fever panic seizes the passengers, and Colt hopes that fear will cause the killer to reveal himself. Chris Colt: Wayde Preston. Ed Pike: Richard Devon. Lucie: Kaye Elhardt. Captain Gibbs: Andy Clyde.

9 TERRYTOONS—Kirchner

11 WHIRLYBIRDS—Adventure
"Buy Me a Miracle." A man charters a plane to Mexico. Kenneth Tobey.

13 BETWEEN THE LINES

7:30 2 3 DENNIS THE MENACE
"The New Neighbors." Mr. Burnley offers a commission to anyone who helps sell his house. Preparing to show the Burnley house to prospective buyers, crafty George Wilson gives Dennis and Tommy money and sends them to the movies. Dennis: Jay North. Henry Mitchell: Herbert Anderson. Alice Mitchell: Gloria Henry.

4 HALLMARK HALL OF FAME
SPECIAL **COLOR** Ibsen's "A Doll's House," with Julie Harris, Christopher Plummer, Hume Cronyn, Eileen Heckart and Jason Robards Jr. in the cast. See the Close-up below for details of this 90-minute adaptation by James Costigan.

This is a reunion for Julie Harris, Christopher Plummer, James Costigan and George Schaefer. They combined talents for the award-winning "Hallmark"

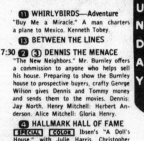

TV GUIDE CLOSE-UP **7:30 4 HALLMARK HALL OF FAME**

Henrik Ibsen's
'a doll's house'

Cronyn Harris Plummer

SPECIAL **COLOR** To Torvald Helmer his wife Nora is a child, a doll-wife. He does not know that his "little sparrow" saved his life some years earlier by borrowing money when he was critically ill. Now, on Christmas Eve, the man who lent the money re-enters Nora's past. He intends to exploit his knowledge of her past.

Henrik Ibsen's 1879 play is set in 19th-century Norway. This 90-minute adaptation was written by James Costigan. The producer-director is George Schaefer. (Live)

Cast
Nora Helmer	Julie Harris
Torvald Helmer	Christopher Plummer
Nils Krogstad	Hume Cronyn
Kristine Linde	Eileen Heckart
Dr. Rank	Jason Robards Jr.
Anne-Marie	Katharine Raht
Helene	Mildred Trares
Emmy	Maggie King
Bobby	Randy Gaynes
Ivor	Richard E. Thomas

SATURDAY, DECEMBER 5, 1959

⓫ BOOTS AND SADDLES
"The Coward." A young soldier must prove himself to his uncle who is a sergeant. Jack Pickard.

⓭ NEWSBEAT—Mike Wallace

8:00 ⑤ BIG BEAT—Richard Hayes

⑦ ⑧ ㊸ HIGH ROAD—Gunther
"Caves and Mountains." John Gunther narrates films of the only successful climb of the Devil's Needles, five Alpine peaks near Mont Blanc, and of explorations of the caves of Pierre St. Martin in the Pyrenees. The extensive preparations of mountain climber Marcel Ichac and four Alpine guides are seen, as well as their ascent of the dangerous peaks. Cave explorer Norbert Casteret leads a team of speleologists in an exploration of the perilous caves and in an attempt to recover the body of an earlier explorer who died trying to map the caves.

⓫ PRO FOOTBALL HIGHLIGHTS
Packers vs. Lions, Bears vs. Cards, Rams vs. Colts, Eagles vs. Steelers, 49ers vs. Browns, Redskins vs. Giants.

⓭ JAI ALAI—Tijuana

8:30 ② ③ WANTED—DEAD OR ALIVE—Western
Jay "Tonto" Silverheels appears in tonight's episode. Charley Red Cloud, an Apache accused of murder, flees into wild Indian country. Josh Randall decides that the high bounty offered for Charley is worth the risk of tracking him down. Josh Randall: Steve McQueen. Charley Red Cloud: Jay Silverheels. Merv Bascomb: Fred Beir. Myra: Jeanne Cooper.

④ MAN AND THE CHALLENGE
An Air Force plane, believed lost at sea, is found in the Amazon jungle. Officials assign Glenn Barton to find out why the crewmen are still missing. Glenn Barton: George Nader; Jim Connor: Dean Harens; Anne Sanders: Marcia Henderson.

⑦ ⑧ LEAVE IT TO BEAVER
"Beaver's Fortune." Armed with a fortune-card prediction that this is his lucky day, Beaver is not afraid to fight the school bully. Beaver: Jerry Mathers. Wally: Tony Dow. June Cleaver: Barbara Billingsley. Ward Cleaver: Hugh Beaumont.

⓫ I SEARCH FOR ADVENTURE
"Call of the Ganges." Host Jack Douglas presents Dr. Michael Hagopian, who traveled through India.

㊸ QUEST FOR ADVENTURE

9:00 ② ③ MR. LUCKY—Adventure
"The Gordon Caper." Maybelle Towers, a former employee of millionaire Walter Gordon, goes into debt while gambling aboard Lucky's ship. Seeing her old boss gambling, Maybelle devises a blackmail scheme to obtain the money she needs. Mr. Lucky: John Vivyan. Andamo: Ross Martin. Maggie: Pippa Scott. Walter Gordon: Berry Kroeger. Joyce: Nora Hayden.

④ DEPUTY—Western
"The Deal." Outlaws kidnap Fran McCord and force her brother Clay to agree to help them rob a mine payroll. Chief Marshal Simon Fry suspects that Clay will go through with the deal to save his sister's life. Simon Fry: Henry Fonda. Clay McCord: Allen Case. Fran McCord: Betty Lou Keim. Hamish: Kelly Thordsen. Jack Usher: Mel Welles.

⑤ ROLLER DERBY—Westerners vs. Chiefs
Chicago Westerners meet the New York

TUESDAY   **DECEMBER 29**
Evening

④ LARAMIE—Western
The Pass." Army troops, preparing an all-out campaign against the Sioux, engage Slim Sherman to block a pass. Plans are delayed when Sherman discovers that the Indians are holding a girl a captive. (60 min.)

Cast
Slim Sherman .... John Smith
Eva .... Madlyn Rhue
Jess Harper .... Robert Fuller
Ben Sears .... Richard Shannon

⑤ BIG STORY—Drama
A cadet, expelled from a maritime academy, claims that the dismissal was unjust. James Stephens, Donald Hastings

⑦ ⑧ ㊸ BRONCO—Western
"Night Train to Denver." Bronco Layne is escorting a body on a train when a large sum of money is stolen from the car carrying the body. Bronco is held for the theft, but his bail is paid by two strangers. The strangers then demand that Bronco give them the stolen money. (60 min.)

Cast
Bronco Layne .... Ty Hardin
Laura Winters .... Jacquelyn McKeever
Al Simon .... Brad Dexter
Matt Larker .... Myron Healey

⑨ MOVIE—Comedy
Million Dollar Movie: "Hold That Ghost." (1941) Two of America's most incredible citizens find themselves in a haunted house. Bud Abbott, Lou Costello, Richard Carlson, Joan Davis.

⑪ FLIGHT—Drama
"Japanese Code." When the Japanese decide to put a new code into operation, an American colonel is assigned to get it.

⑬ NEWSBEAT—Mike Wallace

7:55 ⑬ WEATHER—Margo Lea

8:00 ② ③ DENNIS O'KEEFE
Hal Towne offers his apartment as emergency quarters to the expectant wife of a United Nations delegate. The woman gives birth to quintuplets.

Cast
Hal Towne .... Dennis O'Keefe
Sarge .... Hope Emerson
Mrs. Martine .... Lisa Simone
Dumbroski .... Arthur Keane

⑤ SHERLOCK HOLMES—Mystery
Dr. Watson believes Holmes is losing his senses when he appears to be in league with thieves. Ronald Howard.

⑪ PUBLIC DEFENDER—Drama
A judge's son helps a convict escape from the law. Reed Hadley.

⑬ PLAY OF THE WEEK—Drama
Helen Hayes stars in "The Cherry Orchard." See Close-up on A-35.

8:30 ② ③ DOBIE GILLIS—Comedy
"Couchville, U.S.A." In order to raise money for a date with Thalia Menninger, Dobie is finally forced to go to work in his father's store. But his methods almost put the store out of business, and Mr Gillis turns to psychoanalysis for comfort

Cast
Dobie Gillis .... Dwayne Hickman
Herbert Gillis .... Frank Faylen
Winifred Gillis .... Florida Friebus
Thalia Menninger .... Tuesday Weld
Dr. Kligger .... Harvey Stephens

④ FIBBER McGEE AND MOLLY

⑤ CITY ASSIGNMENT—Drama
Lorelei becomes involved in a mystery while doing a series of stories

FRIDAY, MAY 6, 1960

A **TV** GUIDE CLOSE UP

ROYAL WEDDING

Antony Armstrong-Jones and Princess Margaret

At about 11 o'clock this morning, London time, the Glass Coach will pull away from Clarence House, carrying Great Britain's Princess Margaret Rose and her brother-in-law Prince Philip, Duke of Edinburgh, who will give her hand in marriage to Antony Armstrong-Jones. On its way to Westminster Abbey the carriage, escorted by a troop of the household cavalry, will pass under arches of roses and other spring flowers, along avenues flanked by thousands of cheering spectators.

In the Abbey, an audience of 2000 dignitaries, including Queen Elizabeth II, will view the 50-minute ceremony conducted by Dr. Geoffrey Fisher, Archbishop of Canterbury, assisted by Rev. C. T. H. Dams, precentor of the Abbey, and Dr. Eric Abbott, dean of Westminster.

After the ceremony, the couple will ride in an open carriage to Buckingham Palace for a wedding breakfast with their families and close friends. This will be followed by a large reception attended by hundreds of British and foreign dignitaries.

Elaborate television plans have been made to bring the pageantry to those who are unable to view it in person. Even within Westminster Abbey, many of the guests whose view is obstructed will watch on specially installed TV sets. In America, the first visual reports will be provided by the film-by-wire transmission system. These brief pictures will be followed within a few hours by films and tapes flown across the Atlantic.

TODAY'S TELECASTS

8:30 and 9:30 A.M. 4, EARLY FILMS

Times approximate. Brief film-by-wire reports are shown during the "Today" show and approximately an hour later, breaking into regularly scheduled programs.

1 P.M. 2, 3 Early Tapes

Time approximate. The first of two Royal Air Force bombers will deliver taped reports to Canada. Shortly thereafter Chs. 2, 3 and possibly other stations telecast 30-minutes of these tapes.

3 P.M. 4 Royal Wedding—Part 1

Time approximate. Joseph C. Harsch, John Chancellor and Merrill Mueller narrate a 30-minute taped report.

5 P.M. 7 Princess's Wedding

Robert Sturdevant and Yale Newman narrate a half-hour taped report.

7:30 P.M. 4 Royal Wedding—Part 2

Joseph C. Harsch, John Chancellor and Merrill Mueller narrates a 60-minute taped report of the wedding, including the colorful processions to and from Westminster Abbey, and the ceremony itself.

11:15 P.M. 2, 3 Wedding of Princess

Eric Sevareid, Alexander Kendrick and Walter Cronkite, with BBC commentator Richard Dimbleby, narrate a 45-minute report.

FRIDAY, AUGUST 5, 1960

AUGUST 5
Evening

⑨ TERRYTOONS—Kirchner
⑪ NEWS—Kevin Kennedy
7:05 **② LOCAL NEWS**—Richard Bate
7:10 **② WEATHER**—Carol Reed
⑪ NEWS—John Tillman
7:15 **⑪ NEWS**—Walter Cronkite
7:25 **⑪ WEATHER**—Gloria Okon
7:30 **② ③ RAWHIDE**—Western

Vera Miles in "Incident at the Buffalo Smoke House." Looking for a good place for the herd to cross the river, Gil Favor stops to get information from Mr. and Mrs. Jeremiah Walsh. Then outlaw Wes Thomas and his gang ride up and take the three of them hostage. Favor: Eric Fleming. Yates: Clint Eastwood. Nolan: Sheb Wooley. (60 min.)

Guest Cast

Helen Walsh	Vera Miles
Wes Thomas	Gene Evans
Jeremiah Walsh	Leif Erickson
Rose Morton	Allison Hayes
Lon Grant	John Agar

④ CIMARRON CITY—Western

Dorothy Malone and Glenda Farrell in "A Respectable Girl." Nora Atkins and her mother settle in Cimarron City with the hope of starting a new life. But this may be a difficult task—Deputy Lane Temple questions Nora about her past. Temple: John Smith. Beth: Audrey Totter. George Montgomery narrates. (60 min.)

Guest Cast

Nora Atkins	Dorothy Malone
Maggie Atkins	Glenda Farrell
Fred Barker	Harold J. Stone
Sam Jethro	John Beradino

⑤ CANNONBALL—Adventure

"Sights on Safety." A careless driver "borrows" Mike and Jerry's truck. Paul Birch, William Campbell.

⑦ ⑧ WALT DISNEY—Adventure

"Geronimo's Revenge." Texas John Slaughter challenges the Apache renegade Geronimo to single combat—to the death. Slaughter: Tom Tryon. Ashley Carstairs: Darryl Hickman.

Guest Cast

Viola	Betty Lynn
Willy	Brian Corcoran
Addie	Annette Gorman
Geronimo	Pat Hogan
Chief Natchez	Jay Silverheels

⑨ MOVIE—Musical Comedy

Million Dollar Movie: "George White's Scandals." (1945) A young man's stern sister interferes with his romance. Joan Davis, Jack Haley, Philip Terry.

⑪ BASEBALL—Little League

SPECIAL Top teams in the city vie for the New York City championship at Hy Turkin field, Staten Island.

⑬ HIGHWAY PATROL—Police

A young man whose efforts at bank robberies are unsuccessful always makes a clean getaway. Broderick Crawford.

8:00 **⑤ NIGHT COURT**—Drama

The court roster includes a man accused of selling baseball tickets illegally.

⑬ TO BE ANNOUNCED

8:30 **② ③ CALIFORNIA RODEO**

SPECIAL For details of this rodeo, see the Close-up on page A-66. (60 min.)

"Hotel De Paree" and "Video Village" will not be seen tonight.

⑦ WICHITA TOWN—Western

"Out of the Past." Bounty hunter Murdock rides into town looking for a man who escaped from prison some years before. The man being sought turns out to be Gus Ritter, a German immigrant who is a skilled gunsmith. Mike Dunbar: Joel McCrea. Ben Matheson: Jody McCrea.

Guest Cast

Murdock	Skip Homeier
Gus Ritter	Robert H. Harris
Gus Ritter Jr.	Jan Stine

⑤ TOMBSTONE TERRITORY

Hollister kills a man during a gun fight. He's tricked into riding into the man's home town, where the citizens put him on trial for murder. Pat Conway, James Seay, Robert J. Wilke.

⑦ ⑧ MAN FROM BLACKHAWK

"Death Is the Best Policy." Ambushed by two of the Schuler brothers, Martin Harris kills one of them in self-defense. Realizing that the remaining Schulers will be gunning for him, Harris asks Sam Logan to protect his young son Paul. Logan: Robert Rockwell.

Guest Cast

Tyce	Walter Burke
Mary Schuler	Virginia Christine
Early Schuler	Ted Markland

MONDAY

TV CLOSE UP GUIDE

9:30 NIXON-KENNEDY DEBATE

Vice President Richard M. Nixon

SPECIAL In an unprecedented event the citizen in his living room witnesses a campaign debate staged specifically for him. Vice President Richard M. Nixon, Republican candidate for President of the United States, meets Sen. John F Kennedy, the Democratic candidate, in face-to-face discussion. Originating live from Chicago, this nationally televised program is seen locally over Chs. 2 3 4 7 and 8

This is the first of four joint telecasts by the two candidates. Topic of this one is domestic policy. Howard K Smith of CBS is the moderator and here is the order of business

Sen. John F. Kennedy

Opening Statements. These are limited to eight minutes each, with Senator Kennedy leading off.

Questions. There should be time for two from each newsman: Robert Fleming of ABC, Stuart Novins of CBS, Charles Warren of Mutual Radio and Sander Vanocur of NBC. The questions are addressed to the candidates in turn, with Senator Kennedy getting the first Answers are limited to two and a half minutes.

Comments. Each candidate may comment on the other's answer, and has a minute and a half to do it.

Summations. For their closing statements the two men divide the remaining time. The Vice President is first

Quotes from the Candidates

Nixon: "We believe that the Republican program is based on a sounder understanding of the action and scope of government There are many things a free government cannot do for its people as well as they can do them for themselves There are some things no governments should promise . . . ."

Kennedy: "My chief argument with the Republican Party has been that they have not had faith in the free system Where we would set before the American people the unfinished business of our society, this Administration has set ceilings and limitations

SUNDAY, OCTOBER 9, 1960

The money was left to the town of Laramie. Troop: John Russell. McKay: Peter Brown. Lily: Peggie Castle.
Guest Cast
Bess Harper ....Lee Patrick
Jason McQuade ....Arch Johnson
Oren Slauson ....Vinton Hayworth

⓫ LAWBREAKERS—Police
"Martos Case."

9:00 ❷ ❸ JACKIE GLEASON
[SPECIAL] "The Big Sell" is Jackie's first special of the season. Details are contained in the Close-up on page A-22. (60 min.)

❹ DINAH SHORE—Variety
[RETURN] [COLOR] Dinah's first show of the new season is built around her theme song, "See the U.S.A." Red Skelton, Nat King Cole and Tuesday Weld join in to review such serious matters as love and religion in America, and other assorted items like clothing styles, suburbia —and TV. (60 min.)
Highlights
"Home" ....Shore
Sketch: "Back Yard Barbecue"
.................................Skelton, Shore
"Mr. Cole Won't Rock 'n' Roll" ......Cole
"Teen-Ager and the Telephone" ....Weld
Musical Comedy Satire ....Shore
"Swing Low" ....Shore

❺ I LED THREE LIVES—Drama
Herb Philbrick is ordered by the Reds to aid in a campaign. Richard Carlson.

❼ ❽ REBEL—Western
"The Waiting." Yuma enters a saloon and finds three people inside: Mike the bartender, a bounty hunter named Tom Hall, and Cassie, a barmaid whose fugitive husband is the man Hall's looking for. Yuma: Nick Adams.
Guest Cast
Tom Hall ....Claude Akins
Cassie ....Joan Evans
Mike ....Harry Whisner

❾ MOVIE—Drama
"Lola Le Piconera." (Spanish) French forces besieging the Spanish city of Zaragoza attempt to contact a traitor within the city's walls. Juania Reina.

⓫ NAVY LOG—Drama
"A Guy Called Mickey." An American pilot's plane is shot down on the coast of Japan. Lane Nakano.

9:30 ❺ MEDIC—Drama
"My Child's Keeper." Two children find a cigaret lighter and accidentally set fire to the kitchen curtains. Richard Boone.

❼ ❽ ISLANDERS—Adventure
"Flight from Terror." A small band of Buddhists have hidden the young Padhama of the Roshadis from the Chinese Communists. Willie agrees to have her partners fly the god-king to safety in Amboina. Sandy: William Reynolds. Zack: James Philbrook. Willie: Diane Brewster. (60 min.)
Guest Cast
Padhama of Roshadis ....Warren Hsieh
Julie Staunton ....Gigi Perreau
Mrs. Staunton ....Fay Wray

⓫ WORLD CRIME HUNT—Police
"Dressed to Kill." A fashion designer suspects that someone is out to copy his original dress designs. Charles Korvin.

⓭ PLAYBACK—Music
Guest is conductor Leonard Bernstein.

9:35 ⓭ NEWSPROBE

10:00 ❷ ❸ CANDID CAMERA
Yankee sluggers Mickey Mantle and Yogi Berra and comedian Jonathan Winters are guests. The ballplayers are seen on a golf course and Winters plays a toll collector. Dorothy Collins, Allen Funt and host Arthur Godfrey chat about the results.

In the article beginning on page 17, Godfrey discusses the days when he was a center of controversy.

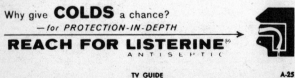

MONDAY

9:30 **2** **3** **ANDY GRIFFITH**—Comedy
"Ellie Comes to Town." She's the niece of the local druggist and comes to Mayberry to help him out. The trouble is, her modern ways don't quite sit with the local folk. Andy: Andy Griffith. Barney: Don Knotts. Auntie Bee: Frances Bavier. Opie: Ronny Howard. Ellie: Elinor Donahue.

4 **HALLMARK HALL OF FAME**
SPECIAL **COLOR** Claude Rains, Richard Basehart and Marisa Pavan in "Shangri-La." Details are contained in the Close-up on page A-39. (Live; 90 min.)

7 **8** **ADVENTURES IN PARA-DISE**—Drama
"Away from It All." Prof. Flanders enjoys life in the South Sea islands, but his wife doesn't like it a bit. She keeps nagging him, until he books passage on the Tiki—for one. Troy: Gardner McKay. Clay: James Holden. Penrose: George Tobias. Sondi: Sondi Sodsai. (60 min.)

Guest Cast
Prof. Flanders ....Henry Jones
Harriet Flanders ....Maxine Stuart
Tapou ....Barbara Luna
Queen Atea ....Betty Garde
Native Girl ....Vivianne Cervanees
Young Girl ....Luana Campos

9 **KINGDOM OF THE SEA**
COLOR "Man Beneath the Sea." A history of underwater diving.

11 **THIS MAN DAWSON**—Police
A payroll robbery amounting to nearly a million dollars is committed.

10:00 **2** **3** **HENNESEY**—Comedy
Harvey Spencer Blair III is back on the base—and in his usual hot water. This time a gal named Consuelo Maddox claims to be his wife. Chick: Jackie Cooper. Martha: Abby Dalton. Shafer: Roscoe Karns. Bronsky: Henry Kulky.

Guest Cast
Harvey Spencer Blair III ..James Komack
Consuela Maddox ....Joan Marshall

5 **WALTER WINCHELL FILE**
"Exclusive Story." An actress seeks columnist Winchell's help for her boy friend who has gotten mixed up with a pair of murderers. John Larch.

9 **SCIENCE FICTION THEATER**
"Dead Storage." A baby mammoth comes to life after being frozen for thousands of years. Virginia Bruce.

11 **STATE TROOPER**—Police
"The Last War Party." Convicted of cattle-rustling, a bewildered Piute Indian finds himself serving a jail sentence. Trooper Blake: Rod Cameron.

Last show at this time. Next week, a new series of boxing matches from St. Nick's Arena.

10:30 **2** **3** **PRESIDENTIAL COUNT-DOWN**—News Analysis

5 **BIG STORY**—Drama
"Thanksgiving for Dr. Joe." Because of his refugee status, a young doctor is barred from medical practice in Minnesota. Victor Cohn: Al Markim.

7 **8** **PETER GUNN**—Mystery
"The Candidate" for governor, Adrian Grimmett, runs on a no-taxes platform. Who could possibly want to assassinate him? Gunn: Craig Stevens. Edie: Lola Albright. Jacoby: Herschel Bernardi. Leslie: James Lamphier. Emmett: Bill Chadney.

Guest Cast
Adrian Grimmett ....Lloyd Corrigan
Harold Canfield ....Alexander Lockwood
Yuki ....Yuki Shimoda
Jim Oakland ....Ken Mayer
Fenton R. Warwick ..Raymond Greenleaf
Clerk ....Robert Ball
Blonde ....Vikki Dougan

9 **MOVIE**—Drama
See 7:30 P.M. Ch. 9 for details.

11 **SILENT SERVICE**—Drama
"Royal Submarines." An embittered young

WEDNESDAY, NOVEMBER 9, 1960

WEDNESDAY

⓫ SAN FRANCISCO BEAT
"The Big Score Case." San Francisco police investigate the robbery of a taxicab office. Warner Anderson.

⓭ PLAY OF THE WEEK—Drama
Two William Saroyan plays are featured tonight: "My Heart's in the Highlands," starring Eddie Hodges, Myron McCormick and Walter Matthau and "Once Around the Block." See the Close-up on page A-34 for details. (Two hours)

9:00 ❷ ❸ MY SISTER EILEEN
"Eileen's Big Chance." Playwright Doug Cartwright hears Eileen read and gives her a part in his new play. Ruth, who doesn't know that Eileen has been given the role persuades Cartwright to rewrite the part for an older woman. Ruth: Elaine Stritch. Eileen: Shirley Bonne. Beaumont: Raymond Bailey.

Guest Cast
Randy Connerly .... Bert Convy
Doug Cartwright .... John Shay

❹ PERRY COMO—Variety
[COLOR] Perry's guests tonight are Ginger Rogers, comedian Alan King and songstress Della Reese. Renne Taylor, Ray Charles singers, Hugh Lambert dancers. (60 min.)

❺ MOVIE—Comedy
"Heaven Can Wait." (1943) A man who has led a life of wine, women and song, must tell his life story to the guardian of the gates of Hades. Gene Tierney, Don Ameche. Charles Coburn.

❼ ⑧ HAWAIIAN EYE—Mystery
"The Kahuna Curtain." Wealthy Jennifer Morgan is grief-stricken over the death of her father, and clings to a toy teddy bear as the one link to her happy life with him. Then the teddy disappears, and a strange young man named Mark Wallace suggests holding a few seances to help find the toy. Tom: Robert Conrad. Tracy: Anthony Eisley. (60 min.)

Guest Cast
Jennifer Morgan .... Shirley Knight
Mark Wallace .... Chad Everett
George Wallace .... Lyle Talbot

❾ FAVORITE STORY—Drama
"The Man Who Sold His Shadow," adapted from a story by Adelbert von Chamisso. A young man is refused the hand of the girl he loves. DeForest Kelley

⓫ TRACKDOWN—Western
"The Gang." Led by Texas Ranger Hoby Gilman, the people of Porter, Texas, try to get rid of an outlaw gang that has taken over the town. Robert Culp.

9:30 ❷ ❸ I'VE GOT A SECRET
❾ HARNESS RACING
⓫ CALIFORNIANS—Western
"Corpus Delicti." Marshal Matt Wayne tries to convict a man of murder on circumstantial evidence. Richard Coogan.

10:00 ❷ ❸ ARMSTRONG CIRCLE THEATER—Drama
"The Antique Swindle," tonight's play by Howard Gast, points up some unscrupulous practices of the trade in antiques and the pitfalls for unwary customers. Ralph and Meg Pearson, a young couple who are just beginning their antique collection, and Mrs. Crandall, an old and experienced hand, test their eye for a buy against the craftiness of Mr. Lambeth, an unethical antique dealer. Narrator: Douglas Edwards. (60 min.)

Cast
Oliver Lambeth .... Harry Townes
Ralph Pearson .... Robert Gerringer
Meg Pearson .... Kathleen Murray
Mrs. Crandall .... Joan Copeland

❹ PETER LOVES MARY
"Life with Father-in-Law." Mary's parents, the Gibneys, have come for a visit. Mr. Gibney finds out that a couple of Peter's bookings have been canceled. So naturally he assumes that Peter and Mary must be broke. Peter: Peter Lind Hayes. Mary: Mary Healy. Leslie: Merry Martin.

Guest Cast
Horace Gibney .... Howard Smith
Mrs. Gibney .... Harriet MacGibbon

❼ ⑧ NAKED CITY—Police
Claude Rains in "To Walk in Silence." Financial expert John Weston likes to play the horses. One day while visiting his favorite bookie, he's accidentally shot. Afraid of the notoriety, he doesn't want to have the bullet extracted. Flint: Paul Burke. Parker: Horace McMahon. Arcaro. Harry Bellaver. (60 min.)

Guest Cast
John Weston .... Claude Rains
Gabe Hody .... Telly Savalas

⓫ DECOY—Police
"Reasonable Doubt." A web of guilt

them. Bat: Gene Barry. Angie Pierce: Paula Raymond.

9 MOVIE—Musical

COLOR Million Dollar Movie: "Bundle of Joy." (1956) When a young woman picks up a baby on the steps of a foundling home she is tabbed as its unwed mother. Eddie Fisher, Debbie Reynolds, Adolphe Menjou.

11 U.S. BORDER PATROL

"The Quota Case." Swindlers attempt to con foreign-born Americans into paying for priority listing for their relatives on the immigration list. Richard Webb.

8:00 5 NIGHT COURT—Drama

Up for deliberation are the cases of a musician accused of disturbing the peace by holding a midnight jam session and an old man charged with molesting young girls in a theater.

7 8 HARRIGAN AND SON

"Miss Claridge Finds Romance," and her daydreaming is cutting down her typing speed too much to suit Harrigan Sr. Things don't get any better when her dream man, Tracy Oakhurst is suspected of being a swindler. Harrigan Sr.: Pat O'Brien. Harrigan Jr.: Roger Perry.

Guest Cast

Tracy Oakhurst ....Gerald Mohr
Larry ....Paul Kent

11 HIGH ROAD—John Gunther

"Canadian Profile." John Gunther narrates the story of how the young people of Canada are participating in the rapid economic expansion of their huge underpopulated country.

13 MIKE WALLACE—Interview

French novelist Andre Maurois discusses his philosophy of love, marriage, old age and death.

8:20 13 DIALING THE NEWS

8:30 2 3 ROUTE 66—Adventure

"Layout at Glen Canyon." A group of models are sent to a construction camp at Glen Canyon to do some fashion shots for a magazine. Jeff Grady, head of the crew working at the dam, appoints Tod and Buz as bodyguards. Tod: Martin Milner. Buz: George Maharis. (60 min.)

Guest Cast

Jeff Grady ....Charles McGraw
Jo ....Bethel Leslie

4 WESTERNER—Drama

5 TOMBSTONE TERRITORY

The wounded leader of a gang of outlaws is pursued by Sheriff Clay Hollister. Pat Conway, Richard Eastham.

7 8 FLINTSTONES—Cartoons

"Hollyrock, Here I Come." Wilma wins a trip for two to Hollyrock, film capital of the world. She and Betty are trailed to a TV studio by Fred and Barney, who have grown suspicious. Characters' voices . . . Fred: Alan Reed. Barney: Mel Blanc. Wilma: Jean Vander Pyl.

11 MEET McGRAW—Mystery

"The Lie That Came True." A man invents a fake robbery. Frank Lovejoy.

13 PLAY OF THE WEEK—Variety

"Highlights of New Faces." See Mon. 8:30 P.M. Ch. 13 for details.

9:00 4 OUR AMERICAN HERITAGE

SPECIAL Bill Travers, Barbara Rush and Farley Granger in "Born a Giant," a play about Andrew Jackson. Details are in the Close-up on the opposite page. (60 min.)

5 AWARD THEATER—Drama

"Dr. Mike." The husband of a woman on the operating table threatens to kill the doctor if the surgery is not successful. Keith Andes.

7 8 77 SUNSET STRIP

"The Antwerp Caper." Carl Neuman and his wife thought that their daughter Gabriella had been killed by the Nazis, but now they've received word that she is alive in Belgium. They ask Bailey to meet with magician Alexis Manet, who claims to know the girl's whereabouts. Bailey: Efrem Zimbalist Jr. Spencer: Roger Smith. (60 min.)

Continued on next page

9 PM
BOLD JOURNEY
WPIX 11

MONDAY

⑨ MOVIE—Mystery
Million Dollar Movie: "Ride the Pink Horse." (1947) NY TV Debut. An embittered war veteran goes to a New Mexico town during fiesta time to wreak vengeance on a master crook.

Cast

Gagin	Robert Montgomery
Pila	Wanda Hendrix
Marjorie	Andrea King
Pancho	Thomas Gomez
Hugo	Fred Clark

⑪ INVISIBLE MAN—Drama
"Jailbreak." Convinced that prisoner Joe Green is innocent, Brady plots his escape. Green: Dermot Walsh.

8:00 ② PETE AND GLADYS—Comedy
It's the Coltons' anniversary, and Pete and Gladys are trying to think of an appropriate gift. Gladys can't remember just what kind of gift they got from the Coltons, but figures it must have been a cheap one—so she and Pete reciprocate with a worthless doodad. Pete: Harry Morgan. Gladys: Cara Williams.

TONIGHT: 8:30 P.M. CHANNEL 2

BRINGING UP BUDDY
STARRING: DORO MERANDE
ENID MARKEY · FRANK ALETTER

Guest Cast

George Colton	Peter Leeds
Janet Colton	Shirley Mitchell
Claire	Muriel Landers

⑤ DIAL 999—Police
To rob warehouses, a gang of robbers uses a plan inspired by the story of "Ali Baba and the 40 Thieves." Robert Beatty.

⑪ THIS MAN DAWSON

⑬ MIKE WALLACE—Interview
Connie Francis tells Mike about her career as a singer.

8:20 ⑬ DIALING THE NEWS

8:30 ② BRINGING UP BUDDY
Violet and Iris have always had a special birthday dinner for Buddy. But this year Buddy's friend Jay Fuller has planned a big surprise party with the office staff. Buddy: Frank Aletter.

Guest Cast

Penny	Yvonne Lime
Jay Fuller	George Neise

④ WELLS FARGO—Western
"Jeff Davis' Treasure." Some time ago a Wells Fargo stage was robbed by two bandits, one of whom disappeared with a fortune in gold. When Birely, the bandit who was captured, is released, Hardie trails him. Hardie: Dale Robertson.

Guest Cast

Wade Cather	John Dehner
Henry Moore	John McLiam
Adam Kemper	Leo Gordon
Amos Birely	Lennie Geer

⑤ DIVORCE HEARING
A woman sues for divorce claiming that she and her husband have serious religious differences and that he is now in trouble with the law.

⑦ ⑧ SURFSIDE 6—Adventure
"The International Net." Producer Mike Hogan is in Miami with the idea of getting a backer for a show. But that might not be his only idea. His wife gets a note saying that he's a murderer—and that she's slated to be his next victim. Ken: Van Williams. Sandy: Troy Donahue. (60 min.)

Guest Cast

Mike Hogan	Claude Akins
Alixe Hogan	Anna-Lisa
Frederik Lundstrom	John Van Dreelen
Ann Trevor	Myrna Fahey

Postponed from last week

SATURDAY

DECEMBER 24
Evening

consequences to archeologist Professor Jody's announcement that he's discovered an ancient necklace that keeps its wearer from harm. George Reeves, Noel Neill.

13 MOVIE—Drama
Picture of the Week: "God Is My Partner." (1957) A doctor has a habit of giving his money away. Walter Brennan, Jesse White.

7:10 4 WEATHER—Frank Field

7:15 4 NEWS

7:30 2 3 PERRY MASON—Mystery
"The Case of the Corresponding Corpse" is rerun. Perry gets a phone call from George Beaumont. He finds this a bit odd, since he attended Beaumont's funeral three years before. Mason: Raymond Burr. Della: Barbara Hale. Drake: William Hopper. Burger: William Talman. Tragg: Ray Collins. (60 min.)

Guest Cast

George Baumont	Ross Elliott
Ruth Whittaker	Joan Camden
Harry Folsom	Vaughn Taylor

The producer of this series is a woman—but she avoids taking undue advantage of it. See next week's TV GUIDE.

4 BONANZA—Western
[COLOR] "El Toro Grande" is rerun. Ben sends Hoss and Little Joe to California to buy a prize bull. The first night on the trail, the pair are ambushed by gunmen. Little Joe: Michael Landon. Hoss: Dan Blocker. Ben: Lorne Greene. Adam: Pernell Roberts. (60 min.)

Guest Cast

Don Xavier	Ricardo Cortez
Cayetena	Barbara Luna
Eduardo	Armand Alzamora

5 JUDGE ROY BEAN—Western
"Desperate Journey." An unscrupulous father plots his son's kidnaping. Edgar Buchanan, Jack Beutel.

7 8 ROARING 20'S—Drama
"Burnett's Woman" is rerun. Julie Fiore, Big Lou Burnett's moll, is tagged for the murder of her boy friend's underworld rival. But Garrison uncovers some information, and writes a story which gives Julie an alibi. Garrison: Donald May. Norris: Rex Reason. Pinky: Dorothy Provine. Chris: Gary Vinson. (60 min.)

Guest Cast

Julie Fiore	Madlyn Rhue

Big Lou Burnett ....Lawrence Dobkin
Mrs. Fiore ....Penny Santon
Read about Dorothy Provine in next week's TV GUIDE.

9 MOVIE—Drama
"The Voice of Silence." Rosanna Podesta, Aldo Fabrizi.

11 AQUA-LUNG ADVENTURE
A demonstration of the use of fins in underwater diving.

8:00 5 BIG BEAT—Richard Hayes
Richard Hayes guests are singers Frank Gari and Dodie Stevens and the Catholic Youth Organization Boys Choir. (60 min.)

Highlights

The Power of Prayer	Hayes
"Utopia"	Gari
"Merry Christmas, Baby," "Jingle Bells" and "Yes, I'm Lonesome Tonight"	Stevens
Christmas Medley	Choir

11 I SEARCH FOR ADVENTURE
"Case of the Happy Dragon." Blake goes to San Francisco's Chinatown to capture a man who is holding a dead body for ransom. Rod Cameron, Frances Fong.

8:30 2 3 CHECKMATE—Mystery
Joseph Cotten in the repeat episode, "Face in the Window." Archeologist George Mallinson and his fiancee Janet Evans are shopping in a store when Mallinson thinks he sees a man he knows at the window. This upsets him, but he refuses to tell Janet why. Corey: Anthony George. Jed: Doug McClure. (60 min.)

Guest Cast

Dr. George Mallinson	Joseph Cotten
Janet Evans	Julie Adams
Louis Roche	John Hoyt

4 TALL MAN—Western
"Billy's Baby." Billy and his girl Rita are whooping it up in the local tavern, and along comes a tough guy named Hartman who insults Rita. Billy shoots Hartman and takes a quick powder. Billy: Clu Gulager. Garrett: Barry Sullivan.

Guest Cast

Rita	Marianna Hill
Hartman	K. L. Smith
Bartender	Pedro Regas

7 8 LEAVE IT TO BEAVER
"Beaver's Accordion" is a big problem. He sent away for the instrument and neglected to tell his parents. Now, who's

Guest Cast

Barbara Wentworth ....Patrice Wymore
Wilmer Zaleski ....Jay Novello
Schoolboy Slidell ....Gordon Jones
Harvey Stokes ....Robert Hutton
Flip ....William Kendis

⑨ JEAN SHEPHERD—Comment

9:30 ⑤ NIGHT COURT—Drama
Standing at the bench are an alcoholic;
a girl accused of being a Peeping Tom
when she spies on her boy friend; and a
poverty-stricken man who stole some
food for his children. Jay Jostyn.

⑨ PLAYBOY'S PENTHOUSE
Hugh Hefner sings and plays host to
the Buddy Rich sextet, Jazz combo;
singers Roberta Sherwood, Don Lanning
(her son) and Duke Hazlett; singer-
pianist Dorothy Donegan and her trio;
and ventriloquist Wayne Roland. Marty
Ruberstein trio. (60 min.)

10:00 ② ③ TWILIGHT ZONE—Drama
"Rip Van Winkle Caper." Four men place
themselves in coffins, using gas that
will keep them alive inside for 100 years.
Rod Serling is host.

Cast

Farwell ....Oscar Beregi
DeCruz ....Simon Oakland
Brooks ....Lew Gallo
Erbe ....John Mitchum

④ MICHAEL SHAYNE—Mystery
Comedian Dick Shawn in "The Trouble
with Ernie." Billy Jack comes looking for
aquacade star Ernie Trask, and a clown
jokingly says that he is Trask—so Billy
shoots him. Shayne: Richard Denning.
Lucy: Margie Regan. Joe: Meade Martin.
McCord: Richard Banke (60 min.)

Guest Cast

Ernie Trask ....Dick Shawn
Billy Jack ....Ron Nicholas
Ross Colby ....Harold J. Stone
Marcella Colby ....Sue Randall

⑤ AWARD THEATER—Drama
"Boyden vs Bunty." A freshman with an
inquisitive mind and an eye for mischief
arrives at Boyden University. Robert
Trumbull, Kathy Red.

⑦ ⑧ DETECTIVES—Taylor
"The Short Way Home." Even though their
late father was a policeman, Nikki and
Frankie Williams aren't convinced that
crime does not pay. Holbrook Rob-

ert Taylor. Russo: Tige Andrews. Ballard:
Mark Goddard. Lindstrom: Russell Thor-
son. Lisa: Ursula Thiess.

Guest Cast

Frankie ....Don Quine
Nikki ....Enid Janes
Eddie Benson ....Johnny Seven

10 30 ② ③ EYEWITNESS TO HISTORY
Walter Cronkite is anchor man for an
in-depth study of the week's top news
events.

⑤ MANHUNT—Police
An exclusive jewelry shop is robbed of a
fortune in diamonds. Victor Jory.

⑦ ⑧ LAW AND MR. JONES
"A Fool for a Client." Emily Barron
wants Jones to handle a petition that will
commit her father to a hospital for alcoholics. Jones: James Whitmore. Marsha:
Janet DeGore. Carruthers: Conlan Carter.

Guest Cast

Franklyn Malleson Ghent .....Otto Kruger
Emily Barron ....Stanja Lowe
Howard Barron ....Whit Bissell
Thorn ....Milton Parsons

SUNDAY, SEPTEMBER 17, 1961

Arthur Space. Edwina: Carole Wells. Donald: Joey Scott.

Guest Cast

Bradley Walton	Emory Parnell
Bradley Walton III	Roger Mobley
Marilyn	Beverly Lunsford
Forsythe	Richard Deacon

Last show for this series in this time spot. "National Velvet" begins its new season tomorrow at 8:00 P.M.

⑪ CHARLES FARRELL—Comedy

8:30 **④ CAR 54**—Comedy

DEBUT In "Car 54, Where Are You?" Fred Gwynne and Joe E. Ross play Francis Muldoon and Gunther Toody, a pair of slightly wacky big-city cops. Beatrice Pons plays Gunther's wife Lucille, Paul Reed plays Police Captain Block, and seen from time to time as patrolmen on this taped, half-hour show are Al Henderson (O'Hara), Nipsey Russell (Anderson), Hank Garrett (Nicholson), Jerry Guardino (Antonnucci), Joe Warren (Steinmetz), Fred O'Neal (Wallace), Duke Farley (Reilly), Shelley Burton (Murdock) and Jim Gormley (Nelson). Tonight, in "Who's for Swordfish?" Toody and Muldoon have been invited on a fishing trip—and they'd rather catch fish than traffic violators.

Guest Cast

Conroy	Ralph Stantley
Helen O'Hara	Sybil Lamb
Inspector	Milo Boulton
Fishing Captain	Jack McGraw

This series was created by Nat Hiken, who was also responsible for the "Sergeant Bilko" series in which Gwynne and Ross had featured roles.

⑤ TO BE ANNOUNCED

⑦ ⑧ LAWMAN—Western

"Trapped" begins this series' fourth season. Outlaw leader Hale Connors arrives in Laramie to notify Troop that the stage won't be arriving on schedule—Connors is holding the passengers for ransom. Troop: John Russell. McKay: Peter Brown. Lily: Peggie Castle.

Guest Cast

Hale Connors	Peter Breck
Oren Slausen	Vinton Hayworth
Joe Poole	House Peters Jr.
Ben Toomey	Grady Sutton

⑪ ADVENTURE THEATER

Tonight's features are: 1. A ranger hunts down a "Rogue Lion. 2. "Biography of a Fish" shows the habits of the stickleback.

⑬ MOVIE—Mystery

"Murder Without Crime." (English; 1951) After a fight with his wife, a drunken man becomes involved in murder. Dennis Price, Derek Farr, Patricia Plunkett.

9:00 **② ③ G.E. THEATER**—Drama

"The Red Balloon," a prizewinning French film, is shown. The story, which is without dialog, is about a boy who finds a red balloon. But instead of floating about like any normal balloon, this one follows the boy wherever he goes. Albert Lamorisse wrote, produced and directed the film. His son Pascal plays the boy, his daughter Sabine, the girl. Ronald Reagan introduces the film.

④ SUNDAY MYSTERY—Drama

COLOR "Murder by the Book." Mystery writer John Clayton hasn't produced anything for two years, and his publisher Arthur Chandler is getting tired of supporting him. Clayton gets a manuscript

SEPTEMBER 23

Evening

Captain Kennedy .... Grant Richards
Sergeant McKnight .... Lee Torrance

④ BONANZA—Western

COLOR "The Infernal Machine." Hoss and his friend Daniel Pettibone offer the citizens of Virginia City a chance to get the inside track on the Automobile Age—by financing Pettibone's version of the horseless carriage. The two visionaries are roundly rejected, however, until a con man named Throckmorton offers to help them out. Hoss: Dan Blocker. Ben: Lorne Greene. Adam: Pernell Roberts. Little Joe: Michael Landon. (60 min.)

Guest Cast

Daniel Pettibone ....Eddie Ryder
Throckmorton ....Willard Waterman
Robin ....June Kenney
Big Red ....Nora Hayden

Last show in this time period for "Bonanza," which begins its new season tomorrow night at 9 P.M. Next week, an hour-long version of "Wells Fargo" will be seen at this time.

⑤ JUNGLE JIM—Adventure

Jim finds danger awaiting him when he accompanies two archeologists to a lost Inca city. Johnny Weissmuller.

⑦ ⑧ ASSASSINATION PLOT AT TEHERAN—Drama

SPECIAL First of a two-part drama about a plot to kill the Allied Big Three during World War II. Details in the Close-up on page A-10. (60 min.)

"Roaring 20's" will not be seen tonight.

⑨ GREAT MUSIC—Chicago

RETURN The Chicago Symphony Orchestra begins a weekly series of hour-long taped concerts. Guest conductors and soloists are featured and the selections range from pops to the classics. Today: Andre Kostelanetz conducts. Concertmaster Sidney Harth and first cellist Frank Miller are soloists.

⑪ NFL HIGHLIGHTS—Leaming

RETURN Jim Leaming narrates this 30-minute, filmed show, covering the highlights of the previous weekend's games.

8:00 **⑤ JIM BOWIE—Adventure**

⑪ YOU ARE THERE—History

8:30 **② ③ DEFENDERS—Drama**

"Killer Instinct." When mild-mannered Jim

McCleery kills a bully in a fight it seems a clear case of self-defense. But the district attorney brings out an interesting point during the trial: During the war McCleery was trained to kill any attacker on sight. Preston: E.G. Marshall. Kenneth: Robert Reed. Helen: Polly Rowles. Joan: Joan Hackett. (60 min.)

Guest Cast

Jim McCleery ....William Shatner
Peg McCleery ....Joanne Linville
Mrs. Kane ....Edith Meiser

④ TALL MAN—Western

"Where Is Sylvia?" Garrett and Billy take in a poker session in El Paso, and take their opponents to the cleaners. Then a woman named Sylvia approaches and pleads with Pat and the Kid to return her husband's losses. Garrett: Barry Sullivan. Billy: Clu Gulager.

Guest Cast

Sylvia ....Patricia Barry
Gerber ....Stanley Kohn
Marshal ....Jeff DeBenning
Jake ....Tom Monroe

⑤ DIVORCE HEARING—Drama

⑦ ⑧ LEAVE IT TO BEAVER

Curiosity gets Beaver "In the Soup." He climbs up a billboard to take a close look at its three-dimensional ad—a huge, steaming soup bowl. Naturally, he falls in. Beaver: Jerry Mathers. Wally: Tony Dow. Ward: Hugh Beaumont. June: Barbara Billingsley. Whitey: Stanley Fafara.

⑨ MOVIE—Musical

COLOR "Great Vaudeville." (Italian) Three episodes make up this film. 1 Vittorio De Sica stars as "The Fading Actor," a vaudevillian neglecting his fellow performers. 2. "Birth of a Star." A poor country girl becomes a star overnight. Maria Fiore. 3. "Military Affair" tells the story of a performer drafted into the army. Carl Croccolo.

⑪ VICTORY AT SEA

9:00 **④ MOVIE—Comedy**

DEBUT COLOR "How to Marry a Millionaire" is the first of a series of feature films to be seen on the network Details in the Close-up on the opposite page. (Two hours)

⑤ WRESTLING—Bridgeport

⑦ ⑧ LAWRENCE WELK

"Shine On, Harvest Moon," tonight's theme is a salute to the arrival of aut

MONDAY, OCTOBER 2, 1961

Ansel Mowrer, author of "An End to Make-Believe"; Dr. Ronald Lamont-Havers, medical director of the Arthritis and Rheumatism Foundation; and Herbert Anderson, an authority on Wall Street. (Live; 90 min.)

8:30 ② WINDOW ON MAIN STREET

[DEBUT] Robert Young stars as Cameron Brooks in this half-hour film series about a novelist who returns to his home town to write about the people he knows best. Other regulars are Ford Rainey as newspaper editor Lloyd Ramsey, and Constance Moore as widow Chris Logan, Ramsey's assistant. Also seen in this debut, and scheduled for some later episodes, are Brad Berwick as Chris Logan's teen-age son Arny, and Warner Jones as Harry McGil, desk clerk in the local hotel. Tonight: "The Return." Cameron is intent on recapturing memories of his youth in Millburg—but his old friend Ramsey tells him the town he knew no longer exists.

Guest Cast
Miss Kelly .... Erin O'Brien-Moore
Young Cameron .... Rick Bache
Tina .... Joy Lane

④ [COLOR] PRICE IS RIGHT

⑦ ⑧ RIFLEMAN—Western

"The Vaqueros" begins this series' fourth season. Lucas and Mark go to Mexico to buy a bull for the ranch. Before they reach their destination, they're captured by a bandit named Miguel. Lucas: Chuck Connors Mark: Johnny Crawford.

Guest Cast
Maria .... Ziva Rodann
Miguel .... Martin Landau
Ramos .... Than Wyenn

⑪ I SEARCH FOR ADVENTURE

Adventurers John Daggett and Bill Beer attempt a daring feat when they try to swim the violent waters of the Grand Canyon rapids. Jack Douglas hosts.

9:00 ② ③ DANNY THOMAS—Comedy

[RETURN] Danny Thomas launches his fifth season as Danny Williams, husband, father and night-club performer. Marjorie Lord is seen as his wife Kathy, and Rusty Hamer and Angela Cartwright appear as his children, Rusty and Linda. Tonight, comedian Bill Dana appears as his own character creation, Jose Jimenez. Jose, the elevator operator in Danny's building,

is afraid to talk to the girl of his dreams. He asks Danny to write her a love letter —seems he writes with an accent too. Hilda: Lisa Kraal.

④ 87TH PRECINCT—Police

"Lady in Waiting." A woman dressed in black walks into the precinct, levels a gun at Kling and Meyer, and announces that she's going to wait for Carella to arrive— so she can kill him. Carella: Robert Lansing. Havilland: Gregory Walcott. Meyer: Norman Fell. Kling: Ron Harper. Teddy: Gena Rowlands. (60 min.)

Guest Cast
Virginia Colt .... Constance Ford
Angelica Galindo .... Margarita Cordova
Captain Howard .... Emile Meyer

⑤ FOUR JUST MEN—Drama

"The Moment of Truth." Tim Collier travels to Spain to witness the debut of a young bullfighter. Dan Dailey.

⑦ ⑧ SURFSIDE 6—Mystery

"One for the Road." Ernie Jordan wanders into Daphne Dutton's life, and captures her heart. Sandy, Dave and Ken warn her to go slow since the mysterious Ernie won't talk about his past or his plans for the future—but their advice falls on love-deafened ears. Sandy: Troy Donahue. Dave: Lee Patterson. Ken: Van Williams. Daphne: Diane McBain. Cha Cha: Margarita Sierra. (60 min.)

Guest Cast
Ernie Jordan .... James Best
Margia Knight .... Elizabeth MacRae
Lt. Gene Plehn .... Richard Crane

⑪ MAN AND THE CHALLENGE

A pilot who was forced to parachute is trapped on a cliff above the Colorado River. George Nader, John Archer.

9:20 ⑨ PLAYBACK—Music

The guest for this week will be jazz musician Olatunji.

9:25 ⑨ NEWS—John Wingate

9:30 ② ③ ANDY GRIFFITH—Comedy

[RETURN] Andy Griffith resumes his role as small-town Sheriff Andy Taylor, as this half-hour filmed series begins its second season. Other regulars are Don Knotts as Andy's deputy Barney Fife; Ronny Howard as Opie, Andy's son; and Frances Bavier as the Taylor housekeeper, Aunt Bee. Tonight: Andy's baffled when he learns that Opie has asked for milk money from both Aunt Bee and himself. But he's not as baffled as Opie, who's trying to figure a way to stop paying protection to a bully named Sheldon. Sheldon: Terry Dickinson.

5 MIAMI UNDERCOVER
The co-owner of a sweepstakes ticket meets an unlucky end. Lee Bowman.

9 KINGDOM OF THE SEA
COLOR "Marineland." A visit to the world's largest oceanarium in Palos Verdes, Cal. Bob Stevenson is host.

11 MEN INTO SPACE—Adventure
The first newsmen are taken to the moon. William Lundigan, Harry Lauter.

13 MOVIE—To Be Announced

10:00 2 3 HENNESEY—Comedy
"The Holdout" is a young patient named Nicky Rocco, who refuses to join in a birthday party at the base hospital children's ward. Hennesey sets out to learn the reason for the boy's refusal. Hennesey: Jackie Cooper. Martha: Abby Dalton. Shafer: Roscoe Karns. Bronsky: Henry Kulky.

Guest Cast
Nicky Rocco .... Gregory Irvin
Lou Rocco .... Arch Johnson

4 THRILLER—Mystery
Host Boris Karloff in "The Premature Burial." The wedding bells in young Victorine Lafourcade's life are replaced by a dirge when her fiance Edward Stapleton suddenly dies. Dr. Thorne, who can't believe that his friend died of natural causes, exhumes the body to perform a post-mortem—and discovers that the "corpse" is still alive. (60 min.)

Cast
Dr. Thorne .... Boris Karloff
Edward Stapleton .... Sidney Blackmer
Julian Boucher .... Scott Marlowe
Victorine Lafourcade ..... Patricia Medina

This story is based on research once done by Edgar Allan Poe on the subject of premature burial.

5 MANHUNT—Police
"The Accidental Truth." A teen-age boy finds a stolen deposit bag. Victor Jory.

7 8 9 BEN CASEY—Drama
DEBUT This hour-long film series stars Vincent Edwards as Dr. Ben Casey, resident neurosurgeon at a metropolitan hospital, and Sam Jaffe as Dr David

Zorba, his friend and mentor. Other regulars are Harry Landers and Bettye Ackerman as Casey's associates, Drs. Ted Hoffman and Maggie Graham. Nick Dennis plays orderly Nick Kanavaras, and Jeanne Bates appears as Nurse Wills. Tonight: "59 the Pure." Casey's diagnosis of young Pete Salazar indicates that brain surgery is necessary—but Dr. Harold Jensen, the head of the hospital, disagrees. (60 min.)

Guest Cast
Dr. Paul Cain .... Barton Heyman
Pete Salazar .... Rafael Lopez
Mrs. Salazar .... Angela Clarke
Dr. Harold Jensen .... Maurice Manson

Casey is only one of "TV's Big Operators." See pages 8-11.

9 TREASURE—Documentary
COLOR "Strange Case of Sir Harry Oakes." "Treasure" producer Gene McCabe travels to the island of Nassau.

11 GRAND JURY—Police
An agent has been planted in a home for retired ladies. Lyle Bettger.

8:00 ② ③ DICK VAN DYKE—Comedy
[DEBUT] Comedian and musical-comedy performer Dick Van Dyke portrays comedy writer Rob Petrie in this half-hour film series. Mary Tyler Moore plays his wife Laura, and Larry Mathews plays Ritchie, their six-year-old son. Other regulars are Rose Marie and comedian Morey Amsterdam as Sally Rogers and Buddy Sorrell, Rob's co-writers of a TV venture called "The Alan Brady Show." Richard Deacon plays Melvin Cooley, the show's producer. Tonight, Ritchie is feverish, but Bob insists on taking Laura to a party given by the show's sponsor.

Guest Cast
Dr. Miller ....Stacy Keach
Sam ....Michael Keith
Dotty ....Barbara Eiler

Carl Reiner, a former regular on the old "Sid Ceasar Show," is the producer of this series and author of tonight's script.

⑤ WALTER WINCHELL—Police
Two hoodlums are captured after robbing and shooting a man. Pat Donohue.

⑦ ⑧ BACHELOR FATHER
"The King's English" launches this series' new season. Bentley can't seem to get a hot evening meal any more. Peter's taking one night off a week for night school—and the rest of his evenings are devoted to helping fellow student Rosie Sue Ming. Bentley: John Forsythe. Kelly: Noreen Corcoran. Peter: Sammee Tong.

Guest Cast
Suzanne Collins ....Jeanne Bal
Gino ....Joey Faye
Rosie Sue Ming ....Anna Shin

⑪ PASSING PARADE
[DEBUT] A series of half-hour vignettes concerning personalities who have made an important contribution to history, medicine, science, art or literature.

⑬ BETTY FURNESS
Among the guests tonight are Dr. John McClenahan, radiologist, and Don Schiffer, world series expert. (Live; 90 min.)

8:30 ② ③ DOBIE GILLIS—Comedy
"Like Mother, Like Daughter, Like Wow." When Dobie mentions his name to pretty

DECEMBER 28

Evening

❹ DR. KILDARE—Drama
Young Johnny Temple is slashed in a teen-age street brawl and brought to Blair Hospital. Kildare tries to administer emergency treatment, but the boy pulls a switchblade on him. Kildare: Richard Chamberlain. Gillespie: Raymond Massey. Susan: Joan Patrick. Nurse Johnson: Della Sharman. (60 min.)

Guest Cast

Johnny Temple	Doug Lambert
Carl Temple	Peter Whitney
Grace Temple	Virginia Gregg
Dr. Golden	Karl Weber
Dr. Galmeir	Laurence Haddon
Markle	Martin Garralaga
Policeman	Ray Montgomery

❺ TRANSPORT WORKERS' UNION—Talk

❼ ❽ REAL McCOYS—Comedy
"The Handsome Salesman." Just as Kate is beginning to feel self-conscious about her bedraggled appearance, Fate comes knocking on the door. Fate, in this case, takes the form of a Mr. Lomax, who represents the Princess Regina Cosmetics Company. Grampa: Walter Brennan. Luke: Richard Crenna. Kate: Kathy Nolan.

Guest Cast

Florrie MacMichael	Madge Blake
Mr. Lomax	Grant Richards
Mrs. Lomax	Grace Lee Whitney

Madge Blake regularly portrays Mrs. Barnes on "The Joey Bishop Show."

9:00 ❷ ❸ INVESTIGATORS—Drama
Henry Jones and Pat Carroll in "The Dead End Men." Margaret Ransom is about to have her missing husband Henry declared legally dead, and then collect on his life-insurance policy. But a derelict known as Gov suddenly appears in the office of the insurance company and says he wants her name removed as the beneficiary. Andrews: James Franciscus. Banks: James Philbrook. Maggie: Mary Murphy. (60 min.)

Guest Cast

Gov	Henry Jones
Blossom Taylor	Pat Carroll
Margaret Ransom	Dorothy Green
Charley Brent	Philip Ober
Poncho	William Fawcett
Raven	Ted Newton
Eelie	William Challee

Derelict	Clegg Hoyt
Mr. Pinkham	James Millhollin
Stanley	Dick Wessel
Judge	Vaughn Taylor

In next week's TV GUIDE, Mary Murphy models a basic sheath—and accessories.

Last show of the series.

❺ WRESTLING—Washington

❼ ❽ MY THREE SONS
"Damon and Pythias." Robbie is tired of living in the shadow of his older brother Mike. So when he gets an invitation to join the Chieftains, the one school club Mike never belonged to, he jumps at the chance. Steve: Fred MacMurray. Bub: William Frawley. Robbie: Don Grady. Mike: Tim Considine. Chip: Stanley Livingston.

Guest Cast

Don	John Rockwell
Hank Ferguson	Peter Brooks
Sudsy Pfeiffer	Ricky Allen
Hal	Buddy Hart
Tony	Bruce Baxter
Mr. Cronkite	Fred Kruger
Miss Elliot	Jeanne Dante

⓫ TO BE ANNOUNCED

9:25 ❾ MAHALIA JACKSON—Music

9:30 ❹ HAZEL—Comedy
"Dorothy's Obsession." George gets nervous when he hears that Dorothy plans to attend an auction—he knows how she likes to spend money. George decides to send Hazel along to guard his wife's purse strings. Hazel: Shirley Booth. George: Don DeFore. Dorothy: Whitney Blake. Harold: Bobby Buntrock. Peggy Baldwin: Frances Helm. Phil Baldwin: Lauren Gilbert.

close-up view of underwater divers hunting trophies such as giant sea bass with spears and cameras. Bob Stevenson hosts.

⓫ M SQUAD—Police

"Shakedown." Ballinger goes after racketeers selling "protection" to cleaning establishments. Ballinger: Lee Marvin. Louise: Katherine Bard.

10:00 ❷ ❸ ❹ WHITE HOUSE TOUR

SPECIAL Mrs. John F. Kennedy is our guide for this tour of the White House. For details see the Close-up below. (60 min.)

"Armstrong Circle Theater," "The Bob Newhart Show" and "David Brinkley's Journal" will not be seen tonight.

On pages 22-25, read about Mrs. Kennedy's role in this tour.

❼ ❽ NAKED CITY—Police

Jack Klugman in "Let Me Die Before I Wake," by Abram S. Ginnes. Joe Calageras is a successful businessman, generous, thoughtful and well-liked. So why does somebody try to run him down with a truck? Flint: Paul Burke. Parker:

Horace McMahon. Arcaro: Harry Bellaver. (60 min.)

Guest Cast

Joe	Jack Klugman
Rosie	Joanne Linville
Vito	Michael Constantine
Nick	Paul Stevens
Phil	Al Viola
Angie	Louis Guss

❾ TREASURE—Documentary

COLOR "Mystery of Brewster Island." Host Bill Burrud talks to Edward Rowe Snow, who found a pirate treasure by means of a code hidden in an ancient book. This lead him to Cape Cod.

⓫ HIGH ROAD—Documentary

"Canadian Profile." John Gunther narrates the story of how the young people of Canada are participating in the rapid economic expansion of their country.

10:30 ❾ MOVIE—Drama

Million Dollar Movie: "The Wild North." See 7:30 P.M. Ch. 9 for details.

⓫ WILD CARGO—Documentary

"Siam." Films of two snakes in combat.

10:00 ❷ ❸ ❹ WHITE HOUSE TOUR

SPECIAL Mrs. John F. Kennedy has spent a good part of her first year in the White House restoring many rooms of the executive mansion. Tonight, she welcomes newsman Charles Collingwood for this taped, hour-long tour.

Points of interest include the Diplomatic Reception Room; the original kitchen, later used by President Roosevelt for his "fireside chats"; the East Room; the State Dining Room, where the table is set for an official function; the Red Room; the Blue Room, furnished in the style of the Monroe period; and the Green Room, which in style represents the earliest period of the White House.

We also visit, for the first time on television, Lincoln's cabinet room, now used as a bedroom for distinguished visitors, and the Monroe Room, where President Kennedy joins the tour briefly.

Mrs. John F. Kennedy in the Red Room of the White House.

his father, an Air Force officer stationed in Paris. His strategy: a campaign designed to get himself dismissed from Westfield. The colonel obliges by heaping demerits on the young cadet—until he learns that Washington is impressed by McKeever's score on a leadership test.

Guest Cast

Major McKeever	Peter Hansen
Lieutenant Postelwaite	Jim Houghton
Mrs. Davenport	Shirley Mitchell

(9) MERRYTOON CIRCUS

(11) SCOTT ISLAND—Adventure
A man, believed dead, returns to Scott Island. Scott: Barry Sullivan.

7:00 (2) LASSIE—Drama
Timmy is keeping a careful eye on Lassie and making sure that she doesn't become overexcited—because she is soon going to have puppies. Timmy: Jon Provost. Ruth: June Lockhart. Paul: Hugh Reilly.

Guest Cast

Deputy Sheriff	Lane Bradford
Dr. Porter	James Seay

(3) ZOORAMA—San Diego
Bob Dale visits the polar bears, seals and walruses at the San Diego Zoo.

(4) ENSIGN O'TOOLE—Comedy
DEBUT This half-hour comedy series involves the zany crew of the destroyer USS Appleby. Dean Jones stars as Ensign O'Toole, an expert on almost every subject—but a hard man to find when there's work to be done. Jack Mullaney is featured regularly as Lt. (j.g.) Rex St. John, Jay C. Flippen as Chief Petty Officer Nelson, and Jack Albertson as Lt. Cdr. Virgil Stoner. Other regulars include Harvey Lembeck as Seaman Gabby Di Julio, Beau Bridges and Bob Sorrells as Seamen Spicer and White. Tonight: "Operation Kowana." The Appleby puts into the Japanese port of Kowana. The sailors are given shore leave with a stern warning to mind their manners—the last crew flooded the town with play money.

(9) TIMES SQUARE PLAYHOUSE
"Slide Rule Blonde." A beautiful math-

NEW SEASON: WALT DISNEY'S WONDERFUL WORLD OF COLOR! 7:30 CHANNEL 4

NEW SEASON: BONANZA! IN COLOR NBC TONIGHT AT 9:00 PM ON CHANNEL 4

〜〜 Of Unusual Interest

> At press time, Ch. 13 was off the air because of a strike by the American Federation of Television and Radio Artists. When the strike is settled, the station will resume broadcasting.

⑬ BOOKS FOR OUR TIME

8:00 ② I'VE GOT A SECRET—Panel
Andy Griffith brings along a secret for panelists Henry Morgan, Bess Myerson, Bill Cullen and Betsy Palmer.

⑤ TIGHTROPE!—Police
"The Perfect Circle." Barney MacCready puts his train-robbery plan into action.
MacCready: Dennis Patrick

8:30 ② LUCILLE BALL—Comedy
[DEBUT] Back with a half-hour situation comedy series, Lucille Ball stars as Lucy Carmichael, a widow with two children (Candy Moore as daughter Chris and Jimmy Garrett as Jerry), who lives in suburban Westchester County, New York. Sharing her home—and problems—are divorcee Vivian Bagley, played by Vivian Vance, and Vivian's son Sherman (Ralph Hart). Tonight: "Lucy Waits Up for Chris." Chris goes out on a date, and the rest of the household settles down for an evening of TV—and waiting for Chris.

Off-camera, Lucy's anything but feather headed. See the article on page 6

④ SAINTS AND SINNERS
[COLOR] Joseph Cotten and Kathy Nolan (formerly of "The Real McCoys") in "The Man on the Rim." Though crippled by the labor racketeers he's striving to expose, reporter Preston Cooper hasn't stopped fighting. He persuades Nick to act as his "legs" to get evidence on racket boss Willy Arnold. Nick: Nick Adams. Grainger: John Larkin. Klugie: Richard Erdman. Tabak: Robert F Simon. Charlie: Nicky Blair. (60 min.)

Guest Cast

Preston Cooper	Joseph Cotten
Summer Day	Kathy Nolan
Willy Arnold	Alan Baxter
Alice Cooper	Coleen Gray

⑤ PETER GUNN—Mystery
"The Missing Night Watchman." Art dealer Charles Quimby is robbed of a valuable collection of jewels. Gunn: Craig Stevens Quimby Howard McNear

CBS ② NEW SERIES
8:30-9:00PM WCBS-TV
THE LUCY SHOW
Odd-Ball antics abound when Lucy and Vivian Vance team up just for the fun of it!

CBS ②
9:00-9:30PM WCBS-TV
DANNY THOMAS SHOW
In tonight's comedy unpredictable Danny keeps his family in turmoil—and yours in stitches!

MONDAY, OCTOBER 1, 1962

7 8 RIFLEMAN—Western
RETURN "Waste," the first episode of a two-part drama by actor-writer Robert Culp, opens this series' fifth season. After a cattle sale, Marshal Torrance, Lucas and Mark are traveling home through desolate border country. When the old lawman disappears, Lucas and his son pick up his trail in a nearby Mexican ghost town. Lucas: Chuck Connors. Mark: Johnny Crawford. Torrance: Paul Fix.

Guest Cast
Alphonso ....Vito Scotti
Horse Teeth Looker ....Tony Rosa

11 ONE STEP BEYOND—Drama
'Doomsday.' During the 1680's, the Earl of Donamoor forbids a lady to continue seeing his son. Torin Thatcher.

13 ART OF SEEING—Ernst Haas

8:45 **3 POLITICAL TALK**—Republican
Republican candidates in Connecticut.

9:00 **2 3 DANNY THOMAS**
RETURN This series begins its sixth season tonight. Charley Halper's baby son laughs happily even when complete

CBS 2
9:30-10:00PM WCBS-TV
ANDY GRIFFITH SHOW
The slow-talking sheriff with the hair-trigger wit, aided and abetted by deputy Don Knotts.

strangers play with him. But whenever Charley comes near—the baby screams. Danny: Danny Thomas. Kathy: Marjorie Lord. Rusty: Rusty Hamer. Linda: Angela Cartwright. Charley: Sid Melton. Bunny: Pat Carroll.

Guest Cast
Tramp ....Benny Rubin
Policeman ....Joe Devlin

Danny filmed eight episodes in Europe this summer. See next week's TV GUIDE.

5 CAIN'S HUNDRED—Drama
"Blue Water, White Beach." There's an accident at Ed Hoagley's bootleg distillery. Cain: Mark Richman. Ed Hoagley: Ed Begley. (60 min.)

7 8 STONEY BURKE—Drama
DEBUT In this film Jack Lord plays a rodeo rider who's out to win the Gold Buckle, the trophy given to the world's champion saddle bronc rider. Other regulars include Warren Oates as Ves Painter, Robert Dowdell as Cody Bristol and Bruce Dern as E.J. Stocker, rodeo performers and Stoney's pals. Tonight: "The Contender." Stoney's ambitions suffer a setback when he draws Megaton, the meanest bronc in the rodeo. The Gold Buckle seems a long way off after the horse throws and injures him before he can be dragged to safety. (60 min.)

Guest Cast
Cleo Ann ....Ruby Lee
Erlie Bristol ....Kate Manx
Royce Hamilton ....Philip Abbott
Clay Bristol ....Carl Benton Reid

11 I SEARCH FOR ADVENTURE
Adventurer Sylvia Christian presents her story of Ethiopia and its king.

STARTING TONIGHT AT 9:30
MAVERICK
JAMES GARNER
CHANNEL 9

MONDAY, OCTOBER 1, 1962

④ TONIGHT—Johnny Carson
✔✔ COLOR Johnny Carson takes over as permanent host. Details are in the Close-up below. (One hour, 45 min.)

⑦ MOVIE—Comedy
Night Show: "Father Was a Fullback." (1949) A football coach is harassed by domestic problems. Fred MacMurray, Maureen O'Hara. (One hour, 45 min.)

⑧ LOCAL NEWS—Dull, Thompson

11:25 **⑧ SPORTS VIEW**—Carl Grande

11:30 **⑧ STEVE ALLEN**—Variety
See 11 P.M. Ch. 11 for details.

12:15 **② MOVIE**—Drama
Time approximate. Late Late Show: "Rumba." (1935) NY TV Debut. A Broadway dancer becomes famous by reviving rumba dancing. George Raft, Carole Lombard, Margo. (85 min.)

12:30 **⑪ NEWS**

12:50 **③ NEWS AND WEATHER**
⑤ NEWS

12:55 **③ MOMENT OF MEDITATION**

1:00 **④ NEWS**—Bill Rippe

⑤ MOVIE—Drama
"The Reckless Moment." (1949) A woman's daughter is involved with a man of questionable character. James Mason, Joan Bennett. (90 min.)

⑦ MOVIE—Comedy
"In the Meantime, Darling." (1944) A girl, used to luxury, weds an Army officer. Jeanne Crain, Frank Latimer. (75 min.)

⑧ MOVIE—Mystery
"The Curse of the Cat People." (1944) Simone Simon, Kent Smith. (90 min.)

⑨ PLAYBACK—Music
Guests are Lester Flatt and Earl Scruggs, performing country music

1:05 **④ MAYOR OF THE TOWN**
⑨ ALMANAC NEWSREEL
See 8:55 A.M. Ch. 9 for detail.

1:10 **④ NEWS AND WEATHER**

1:30 **⑤ NEWS**

1:35 **④ SERMONETTE**—Religion

1:40 **② NEWS**

1:45 **② GIVE US THIS DAY**—Religion

2:15 **⑦ EVENING PRAYER**—Religion

TV CLOSE-UP GUIDE

11:15 ④ **TONIGHT**—Johnny Carson

COLOR It's not really a debut if you come right down to it. Steve Allen once ruled this particular roost, and Jack Paar also made the feathers fly. But tonight's the night Johnny Carson takes over, after a brief interregnum following Paar's abdication.

Johnny doesn't face the music (provided by Skitch Henderson and crew) alone. He has brought along Joan Crawford, who acts; Tony Bennett and Rudy Vallee, who sing; and Mel Brooks, who writes comedy lines and delivers them too. Keep an ear peeled for references to Joan's current film-making—she's co-starred with another great lady of the cinema, Bette Davis.

Otherwise, the "Tonight" show will follow a format not unlike Paar's. There will be plenty of Carson's satiric shafts, plenty of guests paying encore visits, plenty of off-the-cuff conversation and, they hope, plenty of commercials. Ed McMahon handles the announcing.

✔✔ Of Unusual Interest

Guest Cast

Lafferty ....Dennis Patrick
Gogarty ....Mousie Garner

⑤ CALL MR. D.—Mystery
A woman is threatened because of snapshots in her possession. David Janssen. Nancy: Judith Braun. Adams: George Neise. Carlin: James Nolan.

⑦ ⑧ GOING MY WAY—Drama
DEBUT In this 60-minute series, Gene Kelly plays Father Chuck O'Malley, a young priest with progressive ideas, who is sent to a New York parish to aid crusty old Father Fitzgibbons, portrayed by Leo G. Carroll. Dick York appears regularly as Tom Colwell, director of the community center and close friend of Father O'Malley. Nydia Westman is seen as Mrs. Featherstone. Tonight: "Back to Ballymora." Father Fitzgibbons is preparing for a visit home to Ballymora, Ireland, but Father O'Malley has heard some disturbing news: Ballymora is no longer a romantic village, but a thriving industrial center (60 min.)

⑪ SILENTS PLEASE
"Three Musketeers," conclusion. As D'Artagnan, the young soldier who saves the queen from her enemies, and the country from a probable revolution, Douglas Fairbanks Sr. leads the cast in this swashbuckling film version of 'Alexandre Dumas' novel.

Cast

D'Artagnan ....Douglas Fairbanks Sr.
Athos ....Leon Barry
Porthos ....George Seigmann
Aramis ....Eugene Pallette

⑬ COURT OF REASON—Debate

9:00 ② ③ BEVERLY HILLBILLIES
Jed Clampett is now the largest depositor in Milburn Drysdale's Beverly Hills bank. Drysdale wants to make sure the Clampetts are kept happy, so he assigns his secretary Jane Hathaway to help them get settled in their luxurious mansion. Jed: Buddy Ebsen. Granny: Irene Ryan. Elly May: Donna Douglas. Jethro: Max Baer. Drysdale: Raymond Bailey. Jane Hathaway: Nancy Kulp.

TV GUIDE   A-79

WEDNESDAY, OCTOBER 3, 1962

EVENING

6:00 ⑦ **NEWS**—Ron Cochran
⑧ **ROCKY AND HIS FRIENDS**
⑪ **THREE STOOGES**—Russell
⑬ **WHAT'S NEW**—Children
6:15 ⑪ **LOCAL NEWS**—Scott Vincent
6:20 ⑪ **WEATHER**—Rosemary Haley
6:25 ③ **WEATHER**
⑦ **SPORTS**—Howard Cosell
6:30 ③ **NEWS**—Bruce Kern
⑤ **MICKEY MOUSE CLUB**
"Anything Can Happen Day" features "Spin and Marty," Part 13.
⑦ **RESCUE 8**—Drama
A juvenile gang attacks a high school principal. Jim Davis, Jan Jeffries.
⑧ **DEPUTY**—Western
⑨ **ZOORAMA**—San Diego
⑪ **BRAVE STALLION**—Adventure
See Sunday 6 P.M. Ch. 11 for details.
⑬ **PROFILE: NEW JERSEY**
6:40 ③ **SPORTS**—Bob Steele
6:45 ③ **NEWS**—Walter Cronkite
④ **NEWS**—Huntley, Brinkley
7:00 ② **NEWS**—Douglas Edwards
③ **PROBE**—Albert E. Burke
④ **DEATH VALLEY DAYS**
RETURN This is a series of half-hour dramas set in old California. "Hangtown Fry" is an elaborate omelette which Paul Duval ordered for his last meal—to delay his execution for murder. Paul Duval: Fabrizio Mioni. Ann: Nancy Rennick.
⑤ **AQUANAUTS**—Adventure
A department store has sold five defective diving tanks. Larry: Jeremy Slate. Mike: Ron Ely. (60 min.)
⑦ **REBEL**—Western
When townspeople accuse Johnny of robbery, he proves his innocence. Yuma: Nick Adams. Cannon: Walter Sande.
⑧ **NEWS**—Ron Cochran
⑪ **MERRYTOON CIRCUS**
⑪ **NEWS**—Kevin Kennedy
⑬ **RUSSIAN FOR BEGINNERS**
7:10 ② **WEATHER**—Carol Reed
⑪ **LOCAL NEWS**—John Tillman
7:15 ② **NEWS**—Walter Cronkite
⑧ **LOCAL NEWS**—Dull, Thompson

7:25 ⑧ **WEATHER**—Jim Burnes
⑪ **WEATHER**—Gloria Okon
7:30 ② ③ **CBS REPORTS**
"Showdown in the Congo," a look behind the "suspended animation" of the current Congo situation, is reported by newsman Richard C. Hottelet. An appraisal of the work done in the young republic by UN Special Forces is offered. Interviewed are Premier Cyrille Adoula of the Central Congolese Government, President Moise Tshombe of secessionist Katanga Province, acting UN Secretary-General U Thant and Edmund Guillion, U.S. ambassador to the Congo. (60 min.)
④ **VIRGINIAN**—Western
COLOR Jack Warden in "Throw a Long Rope." Judge Garth and the other cattlemen are threatened by homesteaders moving in on their land and rustling their steers. Despite the Virginian's disapproval and his own misgivings, Garth joins the violent Major Cass in a "war of extermination"—with settler Jubal Tatum as the first victim. Garth: Lee J. Cobb. Virginian: James Drury. Trampas: Doug McClure. Steve: Gary Clarke. Betsy: Roberta Shore. Script by Harold Swanton. (90 min.)
Guest Cast
Jubal Tatum ....Jack Warden
Major Cass ....John Anderson
Melissa Tatum ....Jacqueline Scott
⑦ ⑧ **WAGON TRAIN**—Western
Thelma Ritter in "The Madame Sagittarius Story." Madame Sagittarius, a con woman, is snubbed by the other passengers. But Charley takes pity on her and the two soon develop a very close relationship. Hale: John McIntire. Charley: Frank McGrath. Hawks: Terry Wilson. Duke: Scott Miller. (60 min.)
Guest Cast
Madame Sagittarius ....Thelma Ritter
Dennie ....Doug Lambert
⑨ **MOVIE**—Mystery
Million Dollar Movie: "Strangers on a Train." (1951) Bruno Anthony suggests a diabolical scheme to tennis star Guy Haines. Farley Granger, Robert Walker, Ruth Roman, Leo G. Carroll. (Two hours)
⑪ **HONEYMOONERS**—Comedy
Alice tries to surprise Ralph by redecorating the apartment. Audrey Meadows

OCTOBER 25

Evening

🕊️ Of Unusual Interest

Guest Cast

Ted Galahad ....Jack Carter
Debby Lawton ....Georgann Johnson
Dr. Atkinson ....James Callahan
Terry Galahad ....Mark Murray

⑤ MANHUNT—Police

Finucane receives his first clue about an elusive burglar from Mrs. Barlow—she says her husband is the robber. Victor Jory. Mrs. Barlow: Doris Kemper.

⑦ ⑧ LEAVE IT TO BEAVER

"Double Date." Carolyn Stewart cancels her movie date with Wally because she has to baby-sit with her younger sister Susan. Wally suggests they take Susan along—if they can talk Beaver into escorting her. Wally: Tony Dow. Beaver: Jerry Mathers. Ward: Hugh Beaumont. June: Barbara Billingsley.

Guest Cast

Carolyn Stewart ....Vicki Albright
Susan ....Diane Mountford

9:00 ② ③ NURSES—Drama

"Dr. Lillian." Surgeon Lillian Bauer's fellow staff members have been harassing her since she arrived. They call her Lily the Knife behind her back. But Dr. Bauer is haunted by the memory of a child's death during one of her operations. Liz Thorpe: Shirl Conway. Gail Lucas: Zina Bethune. (60 min.)

Guest Cast

Dr. Lillian Bauer ....Virginia Gilmore
Dr. Mehli ....Lonny Chapman
Dr. Campbell ....Roy Poole

⑤ WRESTLING—Washington

⑦ ⑧ MY THREE SONS

"The Ghost Next Door." Chip and his pal Sudsy are out trick-or-treating on Halloween when a sinister old ragpicker scares them back to Chip's house. But there's a bigger fright in store—a ghostly figure carrying a candle appears in the upstairs window of the vacant house next door. Steve: Fred MacMurray. Bub: William Frawley. Chip: Stanley Livingston. Sudsy: Ricky Allen. Mike: Tim Considine.

⑪ TRUE ADVENTURE

"Bottom of the World." A look at the scientific work being done at a U.S. Navy station in Antarctica.

⑬ TO BE ANNOUNCED

9:30 ④ HAZEL—Comedy

`COLOR` "A Four-Bit Word to Chew On." Hazel's self-improvement program includes learning a new word each day—a practice that is annoying George. Hazel: Shirley Booth. George: Don DeFore. Dorothy: Whitney Blake. Harold: Bobby Buntrock.

Guest Cast

Mr. Griffin ....Howard Smith
Bob ....Bert Whaley

⑦ ⑧ McHALE'S NAVY

"McHale and His Seven Cupids." Ensign Parker has a crush on Lt. Casey Brown, a Navy nurse. But Parker is too shy to do anything about it, so McHale and his crew decide that it's up to them to bring them together. McHale: Ernest Borgnine. Binghamton: Joe Flynn. Parker: Tim Conway. Gruber: Carl Ballantine. Christy: Gary Vinson. Tinker: Billy Sands. Virgil: Edson Stroll. Happy: Gavin MacLeod. Lt. Casey Brown: Betsy Jones-Moreland.

⑨ ON STAGE—Drama

"Kiss Mama Goodbye." Mama has plans for her son Ralphie. She wants him to be a doctor—but Ralphie has a great aver-

TUESDAY, DECEMBER 18, 1962

5 LOCK UP—Drama
A fancy suit of clothes aids lawyer Maris
in clearing a GI of a murder charge.

7 COMBAT—Drama
"A Day in June" follows Sergeant Saunders' infantry squad through the D-Day landing at Omaha Beach. Private Braddock thinks that he's pretty lucky when he wins a cash pool by picking June 6 as the day for the Normandy invasion. Saunders: Vic Morrow. Hanley: Rick Jason. Braddock. Shecky Greene. (60 min.)

Guest Cast
Theo .... Max Dommar
Marcelle .... Lisa Montell
Gardello .... Frankie Ray
Beecham .... Dean Stanton

8 MOVIE—War Drama
"The Scarlet Coat." (1955) New Haven TV Debut. A spy for the colonists attempts to uncover a plot. Cornel Wilde. Michael Wilding. (Two hours)

9 MOVIE—Drama
Million Dollar Movie: "The Angry Silence." (English 1960) A labor organizer in-

TUESDAY NIGHT AT 9:30
CHANNEL 9

The pageant of British monarchy from Victoria to Elizabeth

CROWN and CRISIS
PERSPECTIVE ON GREATNESS

TIMEX®
presents

"MR. MAGOO'S CHRISTMAS CAROL"

TONIGHT IN COLOR

Bumbling, loveable Mr. Magoo plays Ebenezer Scrooge in the most original and enchanting version of Charles Dickens' "A Christmas Carol." Specially created for Timex, with a memorable score by Jule Styne and Bob Merrill.

Channel 4
7:30 p.m.

NBC

a crooked gambler named Lucky Leo (Gordon). The MacRaes will do some of their imitations of well-known entertainers. Modernaires, Skelton dancers, David Rose orchestra. (60 min.)

Highlights

"I'll Try," "Salute to the Stool"
....Gordon, Sheila
"So In Love" ....Gordon

④ EMPIRE—Drama

[COLOR] "When the Gods Laugh." Sharecropper Theron Haskell doesn't want his son Kieran to get an education, and one morning he tries to stop him from going to school. The boy manages to reach his classroom, but Haskell follows him and threatens to beat him—until Connie intervenes. Redigo: Richard Egan. Connie: Terry Moore. Lucia: Anne Seymour. Tal: Ryan O'Neal. (60 min.)

Guest Cast

Theron Haskell ....James Gregory
Kieran Haskell ....Roger Mobley
Collin Haskell ....Jenny Maxwell
Arnie Haskel ....Dexter Dupont

⑦ HAWAIIAN EYE—Mystery

Miss "Shannon Malloy" receives a rather valuable anonymous gift—a half-million dollars worth of pearls. Cricket feels that such generosity is a little suspicious, so she asks Shannon to stay with her while Tom checks up on the origin of the gems. Tom: Robert Conrad. Cricket: Connie Stevens. Phil: Troy Donahue. Kim: Poncie Ponce. (60 min.)

Guest Cast

Shannon Malloy ....Susan Silo
Mavis Sloan ....Virginia Gregg
Cozy Neal ....H. M. Wynant

⑪ YOU ARE THERE—History

"The Fall of Fort Sumter." Southern guns fire on Fort Sumter, S.C., in April, 1861. Walter Cronkite narrates.

9:00 ⑤ PLAY OF THE WEEK—Drama

Dane Clark and Kim Hunter in "The Closing Door," by Alexander Knox. Vail Traherne's wife is frightened by his behavior. She had hoped that what the doctors told her wasn't true, but now she agrees that he is ill. (Two hours)

Cast

Vail Traherne ....Dane Clark
Norma Traherne ....Kim Hunter
David Traherne ....Kevin Coughlin

⑪ LOCAL ISSUE—John Tillman

⑬ JAZZ INVENTIONS: THREE HORNS—Music

[SPECIAL] An explanation of jazz is offered by author and jazz expert Martin Williams, and six musicians, consisting of a rhythm section (piano, bass and drums) and three trumpet players.

9:30 ② ③ JACK BENNY—Comedy

Jack's guest is Louis Nye. Jack and Don Wilson take a taxi to the airport, but the cab driver becomes emotional when they arrive—he just hates to say goodbye. Don Wilson, Eddie "Rochester" Anderson.

Cast

Taxi Driver ....Louis Nye
Clerk ....Frank Nelson
Mexican and P.A. Voice ....Mel Blanc

④ DICK POWELL—Drama

Dana Andrews, Barry Sullivan, Rip Torn and Vera Miles in F. Scott Fitzgerald's "Crazy Sunday," adapted by James Poe. Although Joel Coles wants to make it big as a Hollywood writer, he ignores the

THE MIRACLE OF OUR LADY OF FATIMA
SATURDAY, DEC. 22, ON THE LATE SHOW
●2

MONDAY, APRIL 15, 1963

⑦ ⑧ GENERAL HOSPITAL
Dr. Hardy keeps a date with Peggy.

⑪ RAMAR OF THE JUNGLE
"The Burning Barrier." Ramar sets out to find an old witch doctor. Jon Hall.

1:25 ⑤ NEWS

1:30 ② ③ AS THE WORLD TURNS

⑤ MOVIE—Science Fiction
"Immediate Disaster." See 10 A.M. Ch. 5 for details.

⑦ GIRL TALK—Virginia Graham
Panelists are singer Beverly Aadland, author Eleanor Harris and TV publicist Meredith Anderson.

⑧ MOVIE—Comedy
"Four Girls in White." (1939) One of four nurses in training is out to bag herself a rich husband. Florence Rice, Una Merkel, Ann Rutherford. (90 min.)

⑨ STAR AND STORY—Drama
"Black Savannah." A wealthy and ambitious mine-owner in the Dutch East Indies dies under mysterious circumstances. Lester Mathers, Joan Leslie.

⑪ GLOBAL ZOBEL—Travel
"Child Bride of Nepal." Myron Zobel visits Nepal, the buffer state between India and Communist Tibet.

1:55 ④ NEWS—Bob Wilson

2:00 ② ③ PASSWORD—Allen Ludden
Susan Strasberg and Orson Bean are this week's guest celebrities.

④ BEN JERROD—Serial
[COLOR] Ben learns about the "other men" in the life of murder suspect Janet Donelli.

⑦ DAY IN COURT—Drama
Today's re-enacted case: A widow accuses a printer of taking her life savings.

⑨ INQUIRING MIND—Education
"Reform in Mathematics." Prof. Charles Brumfiel discusses new theories for teaching mathematics.

⑪ DIVORCE COURT—Drama
A husband tries to block his wife's motion for a final divorce decree. (60 min.)

2:25 ④ NEWS—Floyd Kalber
⑦ NEWS—Alex Dreier

2:30 ② ③ HOUSE PARTY—Linkletter
Art Linkletter interviews Dr. S. Mark Doran and a mental patient about the problems of returning home.

④ DOCTORS—Drama
A student nurse is threatened with dismissal two weeks before graduation. Dr. Chandler: Richard Roat.

Soap operas are reviewed on page 23.

⑦ JANE WYMAN—Drama
The life of a storekeeper is disrupted by the return of David, the black sheep of the family. David: Robert Horton.

⑨ MOVIE—Science Fiction
"The Man from Planet X." See 10 A.M. Ch. 9 for details.

2:55 ⑤ NEWS

3:00 ② TO TELL THE TRUTH—Collyer
Carol Channing, Joan Fontaine, Skitch Henderson and Henry Morgan begin a week on the panel. Bud Collyer is host.

③ EDGE OF NIGHT—Serial

④ LORETTA YOUNG—Drama
"Royal Partners" begins a two-part story. Dancer Edie Royal, blinded in an accident, may never dance again. Edie: Loretta Young. Willie Royal: Robert Fortier.

⑤ DOORWAY TO DESTINY
"My Brother Joe." A 10-year-old boy is injured in an automobile accident.

⑦ ⑧ QUEEN FOR A DAY

⑪ HOW TO MARRY A MILLION-AIRE—Comedy
A wealthy Texan takes Greta out.

⑬ MUSIC INTERLUDE

3:25 ② NEWS—Douglas Edwards

3:30 ② MILLIONAIRE—Drama
Fred Graham, editor of a country newspaper, has been waging a one-man war against corruption. Graham: Jack Kelly.

③ TO TELL THE TRUTH—Collyer
Darren McGavin, Florence Henderson, Sam Levenson and Phyllis Newman are this week's panelists. Bud Collyer is host.

④ YOU DON'T SAY!—Kennedy
[COLOR] This week's guest celebrities are Lee Marvin and Beverly Garland.

⑤ TEXAN—Western
Longley has just delivered some cattle and collected several thousand dollars for his work. Longley: Rory Calhoun.

⑦ WHO DO YOU TRUST?

⑧ GALE STORM—Comedy

⑪ BEST OF GROUCHO—Quiz

⑬ MUSIC FOR YOUNG PEOPLE
"The Personality of Music" is defined by

FRIDAY

⑬ COMPLEAT GARDENER

7:45 ⑬ **JOY OF ANTIQUES**

[DEBUT] This fifteen-minute series is designed to show the progression and evolution of styles through objects used by Americans in the 17th and 18th Centuries. Lorraine Pearce, former curator of the White House Fine Arts Advisory Committee, is the lecturer.

8:00 ⑨ **SPORTS—Ralph Branca**

⑪ **SAN FRANCISCO BEAT**

"The Changeable Blonde Case." When an aging dancer attempts suicide, police suspect attempted murder. Tom Tully.

⑬ **YEARS WITHOUT HARVEST**

[SPECIAL] See Tuesday 8 P.M. Ch. 13.

8:10 ⑨ **BASEBALL—Mets**

The New York Mets play the Pittsburgh Pirates at Forbes Field, Pittsburgh. Lindsey Nelson, Bob Murphy and Ralph Kiner are the sportscasters. (Live)

8:30 ② ③ **ROUTE 66—Drama**

"What a Shining Young Man Was Our Gallant Lieutenant." On impulse, Tod and Linc leave Tampa, Fla., to visit Linc's idol, an officer he knew in Vietnam—but the lieutenant isn't the man he used to be. Tob: Martin Milner. Linc: Glenn Corbett. (60 min.)

Guest Cast

Lieutenant School ....Dick York
Mrs. School ....Jane Rose
Mr. School ....John Litel
Brothers ....James Brown, James Olson
Beth ....Dianne Ramey
Dock Foreman ....Arnold Soboloff

④ **SING ALONG WITH MITCH**

[COLOR] Mitch Miller welcomes Sandy Stewart, Gloria Lambert and Leslie Uggams. Sandy sings "Blue Room." Bill Ventura joins her for "Getting to Know You." Gloria does "Diamonds Are a Girl's Best Friend." Leslie solos "Smoke Gets in Your Eyes" and "The Man I Love." Family photos segment: "I Wanna go Back to Michigan," "The Victors." (60 min.)

⑤ **TEXAN—Western**

"No Way Out." Seeking protection from a storm, Longley stops at a way station. Longley: Rory Calhoun.

⑦ ⑧ **FLINTSTONES—Cartoon**

[COLOR] "Hawaiian Escapade." Wilma and Betty enter a TV contest offering a chance to meet torrid TV star Larry Lava, and a part in Lava's series "Hawaiian Spy." Voices . . . Fred: Alan Reed. Wilma: Jean Vander Pyl. Barney: Mel Blanc. Betty: Bea Benaderet. Larry Lava and Goldrock: John Stephenson.

Ch. 8 will not colorcast this program.

⑪ **WAR THAT CREEPS**

[SPECIAL] Dane Clark narrates this documentary on guerrilla warfare in Vietnam. Filmed on location with a guerrilla band by M. Fujita Associates, a Japanese company. (60 min.)

⑬ **AGE OF KINGS—Shakespeare**

"The Sun in Splendor," Acts 4 and 5 of Shakespeare's "Henry VI, Part 3." Frank Windsor, John Greenwood, Mary Morris, Terry Scully. (75 min.)

9:00 ⑤ **BRONCO—Western**

"One Came Back." A group of investors, seeking to buy up Mexican ranchland, ask Bronco to guide them across the border. Bronco: Ty Hardin. Jeremy: Robert McQueeney. (60 min.)

⑦ ⑧ **DICKENS . . . FENSTER**

"Big Opening at the Hospital." Kate is going to sing in a benefit show and Harry's afraid that a talent scout will spot Kate and make her into a big star. Harry: John Astin. Arch: Marty Ingels. Kate: Emmaline Henry. Mel: Dave Ketchum. Benson: Jesse White.

9:30 ② ③ **ALFRED HITCHCOCK**

John Forsythe in a rerun of "I Saw the Whole Thing," directed by Alfred Hitchcock. After a hit-and-run accident, mystery writer Mike Barnes gives himself up and decides to handle his own defense—but five eyewitnesses swear that he ran a stop sign just before the collision. The script by Henry Slesar and Henry Cecil derives its suspense from the question of reliability of these eyewitnesses. (60 min.)

Cast

Mike Barnes ....John Forsythe
Jerry O'Hara ....Kent Smith
Penny Sanford ....Evans Evans

④ [COLOR] **PRICE IS RIGHT**

⑦ ⑧ **77 SUNSET STRIP**

"The Left Field Caper." Little Leaguer Danny Saunders yearns for a dad to root for him, little dreaming that the father he has never known is very much alive—

SATURDAY, MAY 18, 1963

⑪ ADVENTURES IN JAPAN
Raw octopus and sea weed are among the delicacies served at a formal Japanese dinner to which Eunice and Skipper are invited. Rev. Everett C. Parker.

10:00 ② ALVIN—Cartoon

③ DEPUTY DAWG—Cartoon

④ SHARI LEWIS—Children
COLOR Margaret Hamilton guests. Lamb Chop has joined a secret club.

⑦ COURAGEOUS CAT—Cartoons

⑧ CARTOONIES—Paul Winchell

⑪ CHRISTOPHER PROGRAM
"Be Hopeful, Not Cynical." Father James G. Keller tells how a writer's negative attitudes are reflected in his stories.

10:15 ⑪ LIVING WORD—Religion
"Hughie." A teacher faces a difficult decision when one of her students persists in taking things. Gertrude Tyas, Ivor Barry, Ken Stevenson.

10:30 ② ③ MIGHTY MOUSE

④ COLOR KING LEONARDO

⑤ LITTLE RASCALS—Cartoons

⑧ MR. GOOBER—Children

⑨ COOKING—Bontemps
The menu features a macaroni picnic loaf, lamb-kabobs, veal steak in mushroom sauce and raisin pie. Guest: sculptor Anthony Cipriano. (90 min.)

⑪ THIS IS THE LIFE—Religion
"Out of the Mouths of Babes." A 12-year-old boy is hit by a truck while practicing baseball with his father.

11:00 ② ③ RIN TIN TIN—Western
Mrs. O'Hara plans to be married. Rusty: Lee Aaker. Mrs. O'Hara: Connie Gilchrist. Masters: James Brown.

④ FURY—Drama
Joey tries to help a young crippled boy regain his self-confidence. Joey: Bobby Diamond. Tim: Peter Votrian.

⑦ CARTOONIES—Paul Winchell

⑪ RELIGIOUS LEADERS
"The Church's Attitude on Segregation." Rev. Eugene Callandar of The Church of the Master in Harlem speaks.

11:30 ② ③ ROY ROGERS—Western
Gunmen attempt to murder a prospector. Roy Rogers, Dale Evans.

④ MAKE ROOM FOR DADDY
Danny and Margaret are considering

sending Rusty to military school. Danny: Danny Thomas. Rusty: Rusty Hamer. Margaret: Jean Hagen.

⑤ JUST FOR FUN—Cartoons

⑦ ⑧ BEANY AND CECIL

⑪ YOUR RIGHT TO SAY IT
"Does Transportation Cost Too Much?" Panelists are executives William Quinn and J.R. Staley, and Donald Rogers, business editor of the New York Herald Tribune. James H. McBurney.

12:00 ② ③ SKY KING—Adventure
At the rodeo, bandits plan to turn a mock holdup into the real thing. Sky: Kirby Grant. Penny: Gloria Winters.

④ MR. WIZARD—Science
Mr. Wizard and Rita investigate "The Science of Orbiting." (Live)

⑦ ⑧ BUGS BUNNY—Cartoons

⑨ PRINCESS AND THE SOLDIER
SPECIAL A royal princess locked in a tower, a soldier in search of his fortune, and magic dogs with eyes big as saucers and grindstones—these are some of the elements that go into this fairytale told through cartoons. (60 min.)

⑪ MR. ADAMS AND EVE
"The Social Crowd." Howard and Eve spend a weekend at the home of a socialite. Ida Lupino, Howard Duff, Olive Carey, Hayden Rorke, Norma Varden.

12:30 ② SPACE: NEW OCEAN
"The Man and His Ships" shows how astronauts are trained and how space capsules are designed. Herb Clarke is host.

③ NEWS—Robert Trout

④ EXPLORING—Children
COLOR Actor Brock Peters narrates the legend of John Henry. "Toccata for Toy Trains" (an experimental film) has music by Elmer Bernstein. The Ritts Puppets learn about chain-reaction reasoning. Albert Hibbs narrates a film about cattle drives and outlines the principles of thermodynamics. (60 min.)

⑤ YANCY DERRINGER
"Two of a Kind." Yancy and Pahoo, blamed for a local crime wave, are arrested and sentenced to death. Yancy: Jock Mahoney. Pahoo: X Brands.

Tuesday September 17, 1963

Anny Williams	Susan Oliver
Sheriff Baird	Dewey Martin
Ad Wiley	Robert Emhardt
Kenneth Morgan	David Janssen

Last show of the series. Next week the new "Richard Boone Show" debuts in the 9-10 P.M. time period.

10:00 ② ③ KEEFE BRASSELLE
Keefe welcomes singer Marguerite Piazza, jazzman Lionel Hampton and the Gospel Jazz Singers. Ricky Graziano, Noelle Adam, Bill Foster dancers, Charles Sanford orchestra. (60 min.)

Highlights

'A Lot of Living to Do," "What Kind of Fool Am I?", "Smile"	Piazza
"Broadway," "Vibes Boogie," "Flying Home"	Hampton
"Let Me Entertain You," Dixieland Medley, "Robert E. Lee"	Brasselle
'Down By the Riverside," "Cotton Fields"	Singers

Last show of the series. Garry Moore returns in this time spot next week.

Premiere on
Best of Broadway

NIGHT OF THE QUARTER MOON
with Julie London

11:20 PM on the new ⑦

⑤ DETECTIVES—Police
'The Queen of Craven Point." The Bohemian crowd has invaded the resort town of Craven Point Beach and Edna Craven doesn't seem to mind. Holbrook: Robert Taylor. Edna Craven: Lola Albright. Russo: Tige Andrews. (60 min.)

⑦ ⑧ FUGITIVE—Drama
[DEBUT] Created by Roy "Maverick" Huggins, this series stars David Janssen as Richard Kimble, a fugitive wrongly convicted of murdering his wife, and Barry Morse as Philip Gerard, the police lieutenant dedicated to capturing Kimble. Tonight: "Fear in a Desert City." Gerard is taking Kimble to prison when their train is derailed. Execution plans get off the track, too: Kimble escapes, and gets a job as a bartender in Tucson. (60 min.)

Guest Cast

Monica Welles	Vera Miles
Ed Welles	Brian Keith
Sergeant Burden	Harry Townes
Mark Welles	Donald Losby
Detective Fairfield	Dabbs Greer
Cleve Brown	Barney Phillips
Captain Carpenter	Paul Birch
Evelyn	Abbagail Shelton

10:30 ④ CHET HUNTLEY
[SPECIAL] "A Chance to Achieve." New York City's Mobilization for Youth project is helping residents of a Lower East Side slum cope with unemployment, language difficulties, juvenile delinquency and narcotics addiction. Films of the various programs under way are shown. Project director James E. McCarthy is interviewed. Chet Huntley narrates.

10:45 ⑪ SPORTS—Jerry Coleman
11:00 ② NEWS—Douglas Edwards
③ NEWS AND SPORTS
④ NEWS—Frank McGee
⑤ NEWS
⑦ ⑧ NEWS—Murphy Martin
⑪ NEWS—John K.M. McCaffery
11:10 ③ WEATHER
④ WEATHER—Tex Antoine
⑤ MOVIE—Drama
"The Unsuspected." (1947) After surviving a shipwreck a girl returns home to her guardian. Joan Caulfield, Claude Rains, Audrey Totter, Constance Bennett, Hurd Hatfield. (Two hours, 15 min.)

7:10 ⑪ LOCAL NEWS—John Tillman

7:25 ⑪ WEATHER—Gloria Okon

7:30 ② MARSHAL DILLON—Western

Belle Ainsley turns up in Dodge City and gets a cool reception. The last time she was seen, she was with a wanted criminal. Matt: James Arness. Chester: Dennis Weaver. Belle: Nina Talbot. Phyllis: Nancy Rennick.

　③ JOHN DANDO—Quiz

"Belgium" is the subject.

　④ MR. NOVAK—Drama

[DEBUT] The setting: a high school. The hero: John Novak, young, good-natured teacher of English. Hero's patron: Albert Vane, the understanding principal. First episode: "First Year, First Day." A brilliant student challenges Novak to give him one reason for continuing his education. Novak: James Franciscus. Vane: Dean Jagger. Jean Pagano: Jeanne Bal. Jerry Allen: Stephen Franken. Marilyn Scott: Marian Collier. Miss Harvey: Gloria Talbott. Evelyn Rose: Ann Shoemaker. (60 min.)

Guest Cast

Paul Christopher	Lee Kinsolving
Harmon Stern	Edward Asner
Anthony Gallo	Donald Barry
Mr. Christopher	Paul Genge
Mr. Clyde	Shirley O'Hara

　⑦ ⑧ COMBAT—Drama

"Bridgehead." Sergeant Saunders' squad is given an extremely hazardous assignment—and Pvt. Mick Heller isn't too happy about going along. Saunders: Vic Morrow. Hanley: Rick Jason. Kirby: Jack Hogan. Nelson: Tom Lowell. Littlejohn: Dick Peabody. Doc: Conlan Carter. (60 min.)

Guest Cast

Pvt. Mick Heller	Nick Adams
Private Scott	Noam Pitlik
German Sergeant	Paul Bush
Private Shrope	Richard Jury
Cole	Fred Harris

　⑨ MOVIE—Comedy

Million Dollar Movie: "The Pure Hell of St. Trinian's." See Monday 7:30 P.M. Ch. 9 for details.

　⑪ SPORTSMAN'S CLUB

7:40 ⑪ SPORTS—Jerry Coleman

7:55 ⑪ BASEBALL—Yankees

The Los Angeles Angels vs. the New

Wednesday October 2, 1963

Butterball .... Barbara Harris
Goriot, Dr. Rieux ....Sorrell Booke
Count de Breville ....Alfred Ryder
Emily, Mme. Loiseau ....Marian Seldes

④ VIRGINIAN—Western

[COLOR] "No Tears for Savannah." The Virginian, trailing a swindler named Madden, runs into his old flame Savannah—who's Madden's girl friend. Virginian: James Drury. Garth: Lee J. Cobb. (90 min.)

Guest Cast

Savannah ....Gena Rowlands
Henry T. Madden ....Everett Sloane
Gordon Madden ....Robert Coleman
Sheriff Avedon ....Stephen McNally
Fitz Warren ....Arthur Franz

⑦ ⑧ OZZIE AND HARRIET

"Ozzie, Joe and the Fashion Models." To keep out of trouble with their wives, Ozzie and Joe take off for a secluded mountain trout stream. The Nelsons portray themselves. Joe: Lyle Talbot. Clara: Mary Croft.

⑨ MOVIE—Adventure

Million Dollar Movie: "The Big Sky." See Monday 7:30 P.M. Ch. 9.

espionage

A unique series. Authentic stories from the shadow world of the undercover agents. Filmed in Europe with top acting talent. Tonight: "Covenant with Death"!

PREMIERE TONIGHT 9:00
▼ CHANNEL 4

⑪ NAKED CITY—Drama

"Murder Is a Face I Know." Joey Ross can't believe it when his father Nickolas is picked up by the police and charged with multiple murder. Nickolas: Theodore Bikel. Joey: Keir Dullea. Flint: Paul Burke. Solly Dillman: David J. Stewart. Tolya: Peg Feury. (60 min.)

⑬ MASK, MYTH AND DREAM

[DEBUT] Joseph Campbell, professor of literature at Sarah Lawrence College, conducts this course examining Oriental and Occidental primitive mythologies.

8:00 ⑤ UNTOUCHABLES—Drama

"You Can't Pick the Number." Ness and Flaherty attempt to persuade a father-and-son team to get out of the numbers racket. They refuse, and later the old man is killed. Ness: Robert Stack. Phil Morrisey: Darryl Hickman. Joe Morrisey: Jay C. Flippen. Flaherty: Jerry Paris. Mrs. Pollock: Betty Lou Gerson. (60 min.)

⑦ ⑧ PATTY DUKE—Comedy

"The Elopement." Martin's boss J. R. Castle spots Patty and Richard at the license bureau—and thinks they're planning to elope. But they're actually getting a fishing license. Patty and Cathy: Patty Duke. Natalie: Jean Byron. Martin: William Schallert. Richard: Eddie Applegate. J. R. Castle: John McGiven.

⑬ IMPORTANCE OF BEING WILD

[SPECIAL] Irma Jurist, composer, writer, pianist and singer, satirizes, among other "fads" of the 20th Century, the New Wave in French films; Brecht; and "momism."

8:30 ② ③ GLYNIS—Comedy

Glynis is writing a story about a taxi dance hall and becomes a taxi dancer to learn more about the subject. Glynis: Glynis Johns. Keith: Keith Andes. Chick: George Mathews.

⑦ ⑧ PRICE IS RIGHT—Cullen

Joe E. Ross is the guest celebrity.

⑪ MIKE HAMMER—Mystery

A robbery is committed in a room equipped with every burglar-proof device possible. Darren McGavin.

⑬ COURT OF REASON—Debate

[RETURN] In this debate of a current issue, two "advocates" each take a different side and their arguments are later

Monday October 7, 1963

⓭ QUANDARY OF INDIA

[SPECIAL] India's attitude toward China, internal politics, and industrial emergence into modern technology, is discussed. (60 min.)

9:30 **❷ ❸ ANDY GRIFFITH—Comedy**
Opie and his friend John hit a baseball through the window of the old Remshaw place and are scared away by "ghosts." Andy: Andy Griffith. Opie: Ronny Howard. Barney: Don Knotts. Gomer: Jim Nabors. Otis: Hal Smith. John: Ronnie Dapo.

❹ HOLLYWOOD AND THE STARS
"Sirens, Symbols and Glamor Girls," first of two parts to be concluded next week, is a parade of love goddesses who have dominated the Hollywood scene for six decades: Mary Pickford, America's Sweetheart; Theda Bara, the original vamp; Gloria Swanson, the sophisticate; flapper Clara Bow; Greta Garbo, the enigma; Jean Harlow, the Blonde Bombshell; seductive Mae West; emancipate Bette Davis; comedienne Carole Lombard; soap-opera heroine Loretta Young; rich girl Katharine Hepburn; Sweater Girl Lana Turner; pin ups Ann Sheridan, Betty Grable, Dorothy Lamour, Veronica Lake, Jane Russell, Rita Hayworth; and their post-war replacements: Kim Novak, Elizabeth Taylor, Brigitte Bardot and Marilyn Monroe.

Joseph Cotten and family found an unusual solution to a leaky-ceiling problem. See next week's TV GUIDE.

❾ WRESTLING—Sunnyside

10:00 **❷ EAST SIDE/WEST SIDE**
"You Can't Beat the System." While working on a case, Neil runs into an emotionally disturbed veteran who has lived like a hermit for a decade. Neil: George C. Scott. Frieda: Elizabeth Wilson. Jane: Cicely Tyson. (60 min.)

Guest Cast
Richard Bailey .... Joseph Turkel
Doris Arno .... Janet Margolin

❸ DETECTIVES—Police
1. "Little Girl Lost." Her kidnaper has been caught but the child is still missing. Holbrook: Robert Taylor. Conway: Lee Farr. 2. "House Call." Gunman Rod Halleck is wounded escaping from the police. Mary: Fay Spain. Halleck: Pernell Roberts. (60 min.)

❹ SING ALONG WITH MITCH

[COLOR] Mitch shows high school pictures of himself and members of the Singalong gang, and recalls predictions made in their school yearbooks. Soloists are Sandy Stewart, Leslie Uggams, Bob McGrath, Bill Ventura, Hubie Hendrie and Stan Carlson. Songs: "Luck Be a Lady," "Hard-Hearted Hannah," "Animal Crackers," "Little Old Lady" and "Bim Bam Boom." (60 min.)

❺ JOAN SUTHERLAND

[SPECIAL] See Sun. 8 P.M. Ch. 5.

❼ ❽ BREAKING POINT—Drama
"Bird and Snake." Roger Morton finds Doc Mac's group-therapy sessions an ideal exercise ground for his sadistic tendencies. Thompson: Paul Richards. Raymer: Eduard Franz. (60 min.)

Guest Cast
Roger Morton .... Robert Redford
Norma Rossiter .... Marisa Pavan
Sam Keller .... Jack Weston
Mrs. Levinson .... Connie Sawyer
Judy Lawrence .... Mimi Dillard

Listings continue on page A 45

Friday November 1, 1963

Evening

8:30 ❷ ❸ ROUTE 66—Drama
"And Make Thunder His Tribute." Life isn't a bowl of cherries for Minnesota raspberry farmer Mike Donato: He's set in his eccentric ways and his son is determined not to go along with them. Tod: Martin Milner. Linc: Glenn Corbett. Script by Lewis John Carlino. (60 min.)

Guest Cast
Mike Donato	J. Carrol Naish
Joe Sky	Alfred Ryder
Tony	Lou Antonio

Postponed from last week.

❹ BOB HOPE—Drama
[COLOR] "Four Kings." During World War II, Army Intelligence pulls four "lifers" out of prison and sends them into Germany to steal top-secret rocket plans. Script by Mark Rodgers from a story by Clifford Irving. Bob Hope is host. (60 min.)

Cast
Bert Graumann	Peter Falk
Gabriella	Susan Strasberg
Dr. Krug	Paul Lukas
Erwin	Robert Strauss
Harry	Vito Scotti
Leonard	Than Wynn
Major Stern	Simon Oakland
Colonel Nauman	John VanDreelan

❼ ❽ BURKE'S LAW—Drama
Carolyn Jones plays four sisters—Betsy, Jane, Olivia and Meredith—in "Who Killed Sweet Betsy?" Betsy was well-liked by almost everybody, but what killed her wasn't any love pat, and her three sisters are the prime suspects. Burke: Gene Barry. Tim: Gary Conway. Les: Regis Toomey. (60 min.)

Guest Cast
Nels	Richard Carlson
Steiner	Michael Wilding
Aunt Harriet	Gladys Cooper
Gil	John Ericson

⓫ YOU ASKED FOR IT—Smith
Two elephants enact the parts of a barber and a customer.

⓭ FESTIVAL OF THE ARTS
Guest conductor Thomas Schippers directs the Detroit Symphony Orchestra in a performance of "Variations on a Theme by Frank Bridge" by Benjamin Britten and the Symphony No. 2 by Sibelius. (80 min.)

9:00 ❺ BRONCO—Western
"Borrowed Glory." Working as a deputy for Sheriff Lloyd Stover, Bronco learns that the sheriff lied about winning the Congressional Medal of Honor. Ty Hardin, Robert Vaughn. (60 min.)

⓫ GUEST SHOT—Interview
Fabian goes on a mountain-lion hunt in the Sierra Nevada Mountains.

9:30 ❷ ❸ TWILIGHT ZONE—Drama
"The Living Doll." Erich Streator is being tormented by a new member of his household—his stepdaughter's talking doll. Script by Charles Beaumont. Rod Serling is host.

Cast
Erich Streator	Telly Savalas
Annabelle Streator	Mary LaRoche
Christie Streator	Tracy Stratford

❹ HARRY'S GIRLS—Comedy
In desperation, producer Lester Bailey decides to make Harry "The Star" his next film. After all, a producer who's broke can't be too fussy about casting. Harry: Larry Blyden. Lois: Dawn Nickerson. Rusty: Susan Silo. Terry: Diahn Williams.

Guest Cast
Lester	Lionel Murton
Tony	Jared Allen
Secretary	June Monkhouse

❼ ❽ FARMER'S DAUGHTER
At the doctor's office, Katy is "The Stand-In" for an expectant congresswoman who's so busy with her career that she hasn't any time for her approaching motherhood. Katy: Inger Stevens. Glen: William Windom. Congresswoman Jory: Beverly Garland. Walter Jory: Russell Johnson. Dr. Mays: Justin Smith. Edna: Jeanne Arnold.

❾ MOVIE—Drama
"The Gold of Naples." (Italian; 1955) Four stories dealing with life in the city of Naples. 1. "The Racketeer." Toto, Pasquale Cennamo. 2. "Pizza on Credit." Sophia Loren, Giacomo Furia. 3. "The Gambler." Vittorio De Sica, Piero Bilancioni. 4. "Theresa." Silvana Mangano, Ubaldo Maestri. (90 min.)

⓫ ALLIE SHERMAN—Sports

9:50 ⓭ TO BE ANNOUNCED

Listings continue on page A-85

November 10, 1963 **Sunday**

Evening

9 MOVIE—Adventure

[COLOR] Big Preview: "Last of the Vikings." (Italian-French; 1960) NY TV Debut. Harald, son of the Viking king Sigurd, swears vengeance when he returns from sea to find his homeland devastated, his people scattered and his father killed. Cameron Mitchell, Edmund Purdom, Isabelle Corey, Helene Remy. (Two hours)

11 GREAT MUSIC—Concert

Morton Gould conducts the Chicago Symphony Orchestra in a special Christmas program including some of his own arrangements. (60 min.)

13 AMERICAN NATION

The subject is "Jonathan Edwards: 1691-1754." (60 min.)

8:30 4 GRINDL—Imogene Coca

Grindl inherits a British estate and a title: "Lady Grindl." She arrives at Dundstetter Moors to claim her inheritance, but three bent twigs on the family tree try to frighten her away. Grindl: Imogene Coca. Reggie: Arthur Malet. Grimes: Kendrick Huxham. Bertram: Richard Peel.

5 COMMUNITY DIALOGUE

"The Problem in School District # 4." The growing resistance to the spiraling education costs in the suburbs. Allyn Edwards is host.

7 8 ARREST AND TRIAL

"Inquest into a Bleeding Heart." When a doctor refuses to do more than give first aid to an accident victim, former surgeon Alexander Safford performs an emergency operation himself—and faces manslaughter charges when the little girl dies. Egan: Chuck Connors. Anderson: Ben Gazzara. Miller: John Larch. (90 min.)

Guest Cast

Alexander Safford ....Richard Basehart
Eleanor ....Julie Adams
Dr. Ferguson ....Kent Smith

9:00 2 3 JUDY GARLAND—Variety

Judy's guests are Count Basie and his band, Mel Tormé and folksinger Judy Henske. In a parody of folksinging, Jerry Van Dyke joins Mel and Judy Henske as "Peter, Paul and Irving"; then Jerry and his banjo accompany Basie's Kansas City Seven in "One Night Samba." Nick Castle dancers. (60 min.)

4 BONANZA—Western

[COLOR] "Journey Remembered." In a flashback, Ben recalls his journey West by wagon train with his second wife, Hoss's mother Inger. Ben: Lorne Greene. Hoss: Dan Blocker. (60 min.)

Guest Cast

Inger Cartwright ....Inga Swenson
Lucas ....Gene Evans
Simon ....Kevin Hagen
Little Adam ....Johnny Stephens

5 UNDER DISCUSSION—Panel

"A Special Report on the Peace Corps" is offered by its director, R. Sargent Shriver; Bill Moyers, deputy director; Warren Wiggins, associate director; chief psychiatrist Joseph English; other Peace Corps officials; and several volunteers who have completed two years of service in Sierra Leone, Ghana, Tanganyika, Pakistan and the Philippines. (Two hours)

11 LOCAL ISSUE—Discussion

13 ART OF FILM—Kauffmann

Host Stanley Kauffmann talks with Sol Kaplan, film critic for "The New Repub

Friday November 22, 1963

Afternoon

12:55 [4] **NEWS—Ray Scherer**

1:00 [2] **BURNS AND ALLEN—Comedy**
Gracie doesn't know that Harry Von Zell has borrowed George's topcoat.

[3] **MOVIE—Drama**
"Clash by Night." Part 5. See Monday 1 P.M. Ch. 3 for details.

[4] **TELL US MORE—Conrad Nagel**
The lives of Fred Allen and Jack Benny are featured. (Live)

[5] **CARTOONS—Ed Ladd**

[7] [8] **GENERAL HOSPITAL**

[11] **SEVEN LEAGUE BOOTS**
"The Strongmen of Persia." In Teheran, the capital of Iran, we witness a spectacle put on by some traveling strong men who boast that they are the most powerful men in the kingdom.

[13] **FUN AT ONE—Children**
Miss Mary Ellen and Miss Eppie hold an Indian pow-wow.

1:25 [5] **NEWS**

1:30 [2] [3] **AS THE WORLD TURNS**

[4] **BACHELOR FATHER—Comedy**
Kelly is fascinated by a young man who spends all his time at the beach John Forsythe, Noreen Corcoran.

[5] **MOVIE—Mystery**
"The Man in the Trunk." See 10 A.M. Ch. 5 for details.

[7] **ANN SOTHERN—Comedy**
Katy finds herself in a jam when it's discovered that she's the one responsible for giving Donald Carpenter a puppy.

[8] **GALE STORM—Comedy**
Singer Pat Boone books passage aboard the S.S. Ocean Queen, hoping to enjoy a a short vacation. Susanna: Gale Storm.

[9] **MOVIE—Comedy**
"Private Lives." See Wednesday 9:30 A.M Ch. 9 for details.

[11] **STAR FOR TODAY—Drama**
"The Joyful Lunatic." Alexander Scourby.

[13] **PLANET EARTH—Education**
"Seas of Grass: the Plains."

1:50 [13] **PARLONS FRANCAIS II**
Dialog is set in a Paris toy shop.

2:00 [2] [3] **PASSWORD—Allen Ludden**
Lena Horne and Douglas Fairbanks wind up a week as guest celebrities.

[4] [COLOR] **PEOPLE WILL TALK**

[7] **DECEMBER BRIDE—Comedy**
Ruth brags to her friend Phyllis that she got an expensive watch. Lily: Spring Byington. Ruth: Frances Rafferty.

[8] **QUEEN FOR A DAY—Bailey**

[11] **PEOPLE ARE FUNNY**

2:05 [13] **TELL ME A STORY—Children**
Leah Brittman reads "Otto in Africa" by William Pene Du Bois.

2:25 [4] **NEWS—Floyd Kalber**

[13] **ISSUE AND THE CHALLENGE**
"West Germany: the Continuing Crisis."

2:30 [2] [3] **HOUSE PARTY—Linkletter**
Art Linkletter's guest is Georgia White, a grandmother who guides tourists down the Colorado River rapids on rafts.

[4] **DOCTORS—Drama**
Laura breaks her engagement.

[7] [8] **DAY IN COURT—Drama**
Today's re-enacted case: In the first of a two-part court session, a man is charged with murdering his wife for her insurance.

[11] **MARRY A MILLIONAIRE**
Loco is chosen queen of the rodeo. Lori Nelson, Merry Anders.

2:45 [13] **MAGIC OF WORDS—Education**
John N. Robbins traces the growth of a book from the creative idea to the bound volume.

2:50 [5] **METROPOLITAN MEMO**

2:55 [5] **NEWS**

[7] [8] **NEWS—Lisa Howard**

3:00 [2] **TO TELL THE TRUTH—Panel**
Orson Bean, Joan Fontaine, Chester Morris and Phyllis Newman wrap up their two-week stint on the panel.

[3] **EDGE OF NIGHT—Serial**

[4] **LORETTA YOUNG—Drama**
Widow Cora Skinner was forced to leave her son Jimmy at Chicago's Boys Brotherhood Republic. Cora: Loretta Young Jimmy: Bobby Driscoll.

[5] **DOORWAY TO DESTINY**
Gangster Nick Pompey discovers that his wife and daughter have mysteriously disappeared in Italy. Richard Conte.

[7] **QUEEN FOR A DAY—Bailey**

[8] **TRAILMASTER—Western**
Benjamin Burns leads a group in search of a mountain spring. Burns: J. Carrol Naish. John Colter: James Franciscus Adams: Ward Bond. (60 min.)

A permanent record
of what we watched on television
from Nov. 22 to 25, 1963

Walter Cronkite, the anchor man of the CBS team, was the first on the air with the bulletin. At 1:30 (EST) when the soap opera, "As the World Turns," went on live, Cronkite was preparing his regular evening news show, and in every sense the day was an ordinary one, at least judging by the trials and tribulations of the characters in the soap opera. In retrospect, the hero's sudsy dilemma as to whether or not he should remarry his divorced wife, and his mother's subsequent conversation with his grandfather about it, seems about as eerily remote as another galaxy. Actress Helen Wagner was just saying, "I gave it a great deal of thought, Grandpa," when the program was interrupted.

Cronkite's voice came through, dolorous but contained, as a bulletin slide was displayed on the screen.

"Bulletin . . . In Dallas, Texas, three shots were fired at President Kennedy's motorcade. The first reports say the President was seriously wounded, that he slumped over in Mrs. Kennedy's lap, she cried out, 'Oh, no!' and the motorcade went on . . . The wounds perhaps could be fatal . . .'"

Viewers tuned to ABC and NBC at the moment heard similar bulletins. At that point CBS switched back to the soap opera. The actors, unaware, continued their performance, but the show was cut off at the second commercial. ABC and NBC blacked out a variety of local and regional shows. Bulletin:

"Further details . . . The President was shot as he drove from the Dallas airport to downtown, where he was scheduled to speak at a political luncheon in the Dallas Trade Mart . . . Three shots were heard . . . a Secret Service man was heard to shout, 'He's dead!' . . . The President and Mrs. Kennedy were riding with Gov. [John] Connally of Texas and his wife . . ."

It was shortly after this that the video portions of the broadcasts came

The CBS bulletin (right) broke into As the World Turns as Nancy (Helen Wagner was telling Grandpa (Santos Ortega) about the marital problems of son, Bob.

23

February 24, 1964 **Monday**
Evening

annual choral concert, and the chorus director is preparing to replace him. Andy: Andy Griffith. Barney: Don Knotts. Gomer: Jim Nabors. Eleanor Poultice: Reta Shaw. John Masters: Olan Soulé.

Andy's wife Barbara appears as a singer in the chorus.

④ HOLLYWOOD AND THE STARS
Film clips show "The Swashbucklers," such as Douglas Fairbanks and Errol Flynn, who provided flamboyant heroics on the screen—and led almost as colorful private lives. Joseph Cotten narrates.

⑨ SURFSIDE 6—Mystery
"A Piece of Tommy Minor." Gambling czar Vic Tatum has a pocketful of IOU's from Tommy Minor, a rising young singer. Tommy Minor: Kenny Roberts. Vic Tatum: Paul Dubov. Sandy: Troy Donahue. (60 min.)

⑬ PLEASED TO MEET YOU
9:45 **⑬ BRITISH CALENDAR**
10:00 **② EAST SIDE/WEST SIDE**
"The Street," by Millard Lampell, intro-

duces a new regular: Congressman Charles Hanson, played by Linden Chiles. Living on the Lower East Side has had its effect on Angela, a teen-ager who has decided to sleep in cars on the street until her mother gets rid of a man Angela hates. Brock: George C. Scott. (60 min.)

Guest Cast

Bianca	Louise Troy
Angela	Candace Culkin
Joanna	Barbara Feldon

③ DETECTIVES—Police
"The Walls Have Eyes." Though Nora Carver's father has passed away, his works live on—in the form of some counterfeiting plates being used by a ring for making bogus bills. Holbrook: Robert Taylor. Nora: Ellen MacRae. Burt: John Larkin. (60 min.)

④ SING ALONG WITH MITCH
[COLOR] In this repeat show, Mitch and the gang salute the world of entertainment. Featured performers: singers Leslie Uggams and Phil Olson, and dancers Victor Griffin and Diane Davis. Musical-

---

10:00 ⑤ PRE-FIGHT SPECIAL

Cassius Clay   **Sonny Liston**

[SPECIAL] Tomorrow night, Sonny Liston defends his world heavyweight title against Cassius Marcellus Clay in Miami Beach.

How does the stoic champion compare with his brash, poetizing challenger? Former heavyweight champion Joe Louis analyzes their fighting styles, strengths and weaknesses.

Both men are seen in training sessions—Clay in Miami and Liston in Las Vegas—and we follow their boxing careers through films of some of their im-

portant matches. Clay is shown winning his Gold Medal in the 1960 Rome Olympics and in the pro ring with George Logan and Archie Moore, two of his 19-straight professional victims.

Liston is seen in his first title match with Floyd Patterson, and also in the ring with Wayne Bethea, Roy Harris and Cleveland Williams—who Liston says is the hardest puncher he ever fought.

Each fighter picks himself as winner of tomorrow's bout and helps Jim Jacobs narrate (60 min.)

Monday September 14, 1964

Evening

7 8 VOYAGE—Adventure

DEBUT Stars of this adventure series are the Seaview, a glass-nosed, atomic-powered submarine of the 1970's; Richard Basehart as Adm. Harriman Nelson, director of Nelson Oceanographic Research Institute; and David Hedison as Cdr. Lee Crane, the sub's captain. Tonight: "Eleven Days to Zero." Seismologist Fred Wilson and Nelson's crew are headed for the Arctic, where they hope to forestall a predicted quake. But they're attacked en route by an unidentified foe. Other regulars: Bob Dowdell as Lt. Cdr. Chip Morgan, Henry Kulky as Chief Petty Officer Curley Jones and, on the enemy side, Theodore Marcuse as sinister Dr. Gamma. Dr. Fred Wilson: Eddie Albert. Kowalski: Del Moore. (60 min.)

9 MOVIE—Biography

Million Dollar Movie: "The James Dean Story." (1957) NY TV Debut. Martin Gabel narrates this account of the life and legend of the haunted young actor. Using stills by Dennis Stock and others.

Thrill!

Voyage to the Bottom of the Sea

Premiere tonight

abc 7:30 pm 7, 8

it features Dean in his screen test for "East of Eden" and, ironically, a highway safety film he made shortly before his fatal car crash. (90 min.)

11 HONEYMOONERS—Comedy

"The Sleepwalker." Ralph takes Norton with him on a business trip. Jackie Gleason, Art Carney.

13 PROFILE: NEW JERSEY

Members of the Newark Area Redevelopment Corporation discuss how small businesses employ low-skill workers.

8:00 2 I'VE GOT A SECRET—Panel

Garry Moore winds up his last show, handing the reins to co-host Steve Allen. Magician Milbourne Christopher is the celebrity guest. Panelists are Bess Myerson, Henry Morgan, Betsy Palmer and Bill Cullen

5 NEW BREED—Police

"Care Is No Cure." Barney Talltree, a known carrier of typhoid fever, heads for Los Angeles. Barney: Mario Alcalde. Dr. Eric Thor: Leif Erickson. (60 min.)

11 NAKED CITY—Drama

Karen Gunnarson, governess for young Danny Cameron comes to police headquarters with "A Wednesday Night Story" of marital distress in the Cameron home. Karen: Ulla Jacobsson. Blair Cameron: David Janssen. Danny: Tommy Battrell (60 min.)

13 ANTIQUES—George Michael

New Hampshire's collection of bells cast by Paul Revere is displayed

8:30 2 VACATION PLAYHOUSE

"Ivy League." After 30 years in the Marine Corps, Bull Mitchell retires and enrolls as a college freshman. Bull: William Bendix. Timmy: Tim Hovey.

Last show of the series. Andy Griffith takes over this time spot next week.

7 8 NO TIME FOR SERGEANTS—Comedy

DEBUT Based on the best-selling novel, Broadway play and movie of the same title, this half-hour comedy series stars Sammy Jackson as Will Stockdale, a gullible farm boy who joins the Air Force. Other regulars include Hirkox as the harried Sergeant King, Kevin O'Neal as Will's buddy Ben Whitledge, and Laurie Sibbald as Will's girl friend Milly An

September 17, 1964 **Thursday**

Evening

5 WRESTLING—Washington

7 8 BEWITCHED—Comedy

[DEBUT] In this half-hour comedy series, Elizabeth Montgomery is Samantha, a young witch—yes, witch—who marries an advertising man. Dick York is her husband Darrin, who must adjust to his wife's blackmagical ways, and also mother-in-law (Agnes Moorehead) who pops into the picture whenever the spirit moves her. Tonight: "I, Darrin, Take This Witch Samantha." The couple meet, marry and face their first problem: Darrin's former gir friend Sheila. Sheila: Nancy Kovack. Dave: Gene Blakely. Doctor: Lindsey Workman.

There's more information about this show —and other new Thursday-night fall entries in next week's TV GUIDE.

9 JAZZ SCENE, U.S.A.—Music

Host Oscar Brown Jr. and Blues singer Big Miller discuss Miller's early education and experience.

11 MOVIE—Comedy

"Topper." (1937) Cosmo Topper hasn't a ghost of a chance against a couple of dead young people who return to make mischief with the living. Cary Grant, Constance Bennett. (Two hours)

13 METROPOLITAN WONDERLAND—Tour

The Museum of the City of New York, Stamford and Newark Museums are visited by host John Tobias.

9:30 **4** HAZEL—Comedy

[RETURN] [COLOR] George's cousin Fred makes a favorable impression on Hazel, but George considers him a sponge and refuses to lend him the money he wants to borrow. Hazel: Shirley Booth. George: Don DeFore. Cousin Fred: Frederic Downs.

7 8 PEYTON PLACE—Serial

Constance seems strangely apprehensive at meeting Dr. Rossi. Constance: Dorothy Malone. Rossi: Ed Nelson. Allison: Mia Farrow. Rodney: Ryan O'Neal. Leslie: Paul Langton. Norman: Christopher Connelly. Catherine: Mary Anderson. Swain: Warner Anderson.

9 WORLD THEATRE

[SPECIAL] See Monday 9:30 P.M.

13 PERSONAL REPORT

Anita Loos, author of "Gentlemen Prefer Blondes," is interviewed.

10:00 **2 3** NURSES—Drama

"The Seeing Heart." A blind priest, hospitalized with an ulcer, is asked to help a blind Jewish boy prepare for his bar mitzvah. Script by Alvin Boretz. Mrs. Thorpe: Shirl Conway. Miss Luca: Zina Bethune. Dr. Lowry: Stephen Brooks. Dr. King: Edward Binns. Father Wickford: Fritz Weaver. David Kaplan: Paul Mace. David's Father: Sam Gray. (60 min.)

"The Defenders" will be seen in this time period next week. "The Nurses" moves to Tuesday, 10-11 P.M.

4 KRAFT SUSPENSE THEATRE

[COLOR] Conclusion of "The Case Against Paul Ryker." Sergeant Ryker is sentenced to death for treason, but prosecuting officer David Young asks for another hearing on the grounds that Ryker wasn't properly defended. The request is granted, but Young is ordered to handle the defense this time. Capt. David Young: Bradford Dillman. Ann Ry

Sunday

September 20, 1964

Evening

able Londoner. Spencer Tracy, Deborah Kerr. Ian Hunter. (Two hours)

⑪ STATE TROOPER—Police
Love on the Rocks." A domineering woman, hated not only by her family but by the townspeople as well, is murdered. Rod Cameron. Danny: Mason Alan Dinehart. Pat: Nancy Kilgas.

6:30 ② ③ MISTER ED—Comedy
Identifying himself as Wilbur, Mister Ed telephones Los Angeles Dodger coach Leo Durocher with helpful advice for several slumping batters. Dodgers pitcher Sandy Koufax, catcher Johnny Roseboro and outfielder Willie Davis appear as themselves, along with Durocher Wilbur: Alan Young. Carol: Connie Hines. Addison: Larry Keating. Kay: Edna Skinner. Announcer: Vin Scully.

④ NEW YORK ILLUSTRATED
"A Joyful Noise unto God." The art and meaning of gospel singing.

⑧ HENNESEY—Comedy

⑪ PIONEERS—Drama
"The Hangman Waits." More than 15

years go by before a man is accused and brought to trial for a murder. Percy Helton, Clark Howat, Claire Weeks.

6:45 ⑦ ALL-PRO SCOREBOARD

7:00 ② ⑤ LASSIE—Drama
In the conclusion of a three-part story, Forest Ranger Corey Stuart, an old friend of Lassie's, sets out to search for the missing collie. Stuart: Robert Bray. Cully: Andy Clyde. Doc Weaver: Arthur Space. Bill: Richard Tretter.

③ ZOORAMA—San Diego
Bob Dale shows different types of birds' beaks and how they're used

④ FILM COMEDY
An unconventional and nomadic Midwesterner arrives in the city and, almost before he knows it, manages to cause a traffic jam, lose all his money and get "married" to a girl he met for the first time only a few hours before.

Cast
R. B. ....Aldo Ray

Continued on page A-30

TV CLOSE-UP GUIDE **8:00 ② ③ ED SULLIVAN—Variety**

"YEAH, YEAH, YEAH!"

Sullivan and the Beatles

This program received the highest rating of any entertainment show on TV last season when it was originally aired in February.

The Beatles—Ringo Starr, John Lennon, Paul McCartney and George Harrison—were responsible for the peak tune-in and an estimated 50,000 ticket requests. Since the Sullivan theater has only 728 seats, 30 policemen were on hand outside the theater.

The shaggy foursome sang "I Want to Hold Your Hand," "She Loves You," "All My Loving," "Till There Was You" and "I Saw Her Standing There." Despite

an audience reaction that bordered on hysteria, the Beatles reportedly weren't too happy with the engagement because microphone levels occasionally allowed audience screams to drown out their voices.

Also on the show: Georgia Brown and the "Oliver!" youngsters doing "As Long as He Needs Me" and "I'd Do Anything"; impressionist Frank Gorshin; magician Fred Kaps; the comedy team of Mitzie McCall and Charlie Brill; Wells and Four Fays, acrobats; and Tessie O'Shea, who sings an English music hall medley. Ray Bloch orchestra

Tuesday September 22, 1964

Evening

7:55 [9] BASEBALL—Mets

[COLOR] The St. Louis Cardinals vs. the Mets at Shea Stadium. Lindsey Nelson, Bob Murphy and Ralph Kiner report the action. (Live)

8:00 [2] [3] WORLD WAR I

[DEBUT] This half-hour documentary series uses film from news, official and private sources to explore World War I. Tonight: "The Summer of Sarajevo." We see a complacent, confident Europe in 1914 shaken by the assassination of Austria's Archduke Ferdinand, an event seized by German and Austrian leaders as a motive for unleashing the war they had plotted for years. We see war reprisals against Serbia; the crowned heads of Europe mobilizing their armies; and people in Britain and France marching jauntily off to war. In Germany, there was cheering and waving as the Kaiser told his people they would "go to war with God on their side." Script by John Sharnik, who co-produced with Isaac

Kleinerman. Robert Ryan narrates. Music by Morton Gould.

Other new series for Tuesday nights this fall are described on page 16.

[5] WIDE COUNTRY—Drama

"Step over the Sky." Oldtime rodeo rider Johnny Prewitt has only one ambition in life—to capture and conquer the wild stallion that crippled him years ago. Prewitt: Victor Jory. Mitch: Earl Holliman. Andy: Andrew Prine. (60 min.)

[13] FRENCH CHEF—Cooking

Julia Child shows how to make "rognons (kidneys) sautés et flambés."

8:30 [2] [3] RED SKELTON—Comedy

[RETURN] Seen a half-hour later than last season, Red starts his 14th year on television by welcoming as guests Shakespearean actor Maurice Evans and the Greenwood County Singers. Sketch: "The Taming of the Schmo." Shakespearean actor Sir Neville (Evans) develops car trouble while crossing the Kadiddlehopper Farm. In the Silent Spot, Red and Chanin Hale play newlyweds. The Green-

---

<table>
<tr><td>TV CLOSE-UP GUIDE</td><td>8:30 [4] MAN FROM U.N.C.L.E.—Drama</td></tr>
</table>

'THE VULCAN AFFAIR'

[DEBUT] Robert Vaughn portrays Napoleon Solo, a James Bond-type agent of U.N.C.L.E. (United Network Command for Law Enforcement), an organization dedicated to fighting world-wide crime. David McCallum plays Solo's assistant Illya Kuryakin, and Leo G. Carroll is Alexander Waverly, head of U.N.C.L.E. In each episode, an average citizen is yanked out of his everyday life and swept into the action.

Tonight, U.N.C.L.E.'s New York headquarters is attacked by an assault force from THRUSH, a secret international group bent on dominating the world. They intend to kill Waverly, who has learned of his plans to assassinate the visiting premier of a newly independent African nation. Written by producer Sam Rolfe and directed by Don Medford. (60 min.)

Robert Vaughn as Napoleon Solo

Guest Cast

Elaine May Donaldson	Patricia Crowley
Andrew Vulcan	Fritz Weaver
Ashumen	William Marshall
Soumarun	Ivan Dixon
Alfred Ghist	Eric Berry
Gracie Ladovan	Victoria Shaw
Nobuk	Rupert Crosse
Del Floria	Mario Siletti

Thursday September 24. 1964

Evening

⑬ INGLES PARA TODOS

6:45 ⑦ NEWS—Ron Cochran

6:55 ④ WEATHER—Pat Hernon

7:00 ② NEWS—Walter Cronkite

③ WYATT EARP—Western
"The Frameup." When Wyatt's saddlebags are found loaded with money, the town is convinced he has been taking bribes. Wyatt: Hugh O'Brian.

④ NEWS—Huntley, Brinkley

⑤ DAKOTAS—Western
There's "Trouble at French Creek" when a woman mine owner hires gunslingers to bring her striking employees back to work. Jay French: Mercedes McCambridge. Regan: Larry Ward. J. D.: Jack Elam. Del: Chad Everett. (60 min.)

⑦ DICKENS . . . FENSTER
"Kick Me Kate." Harry is sure that Kate doesn't love him any more, because she never shows any signs of jealousy. Harry: John Astin. Kate: Emmaline Henry.

⑪ YOGI BEAR—Cartoon

PREMIERE! Ⓦ2

Cross their welcome mat and you're apt to die laughing! Monstrous fun with Fred Gwynne, Yvonne de Carlo, Al Lewis.

7:30 TONIGHT ON CBS

THE MUNSTERS

⑬ COLUMBIA SEMINARS
"Communications in the Modern World." Guest: Joseph Rothschild, associate professor of government, Columbia U.

7:30 ② ③ MUNSTERS—Comedy
[DEBUT] That creeper-covered plottage at 43 Mockingbird Lane is the residence of the Munster family: Herman (played by Fred "Car 54" Gwynne), Lily (Yvonne DeCarlo), Grandpa (Al Lewis), their son Eddie (Butch Patrick) and niece Marilyn (Beverley Owen). Marilyn's a blonde beauty—the others bear a startling resemblance to m-m-m-monsters. Tonight: "Munster Masquerade." Marilyn has been dating Tom Daly, who's a bit reticent about meeting her folks. Tom Daly: Linden Chiles. Agnes Daly: Mabel Albertson.

The new entries for Thursday nights this fall: read about them on page 22.

④ DANIEL BOONE—Adventure
[DEBUT] Still wearing his Davy Crockett coonskin cap, Fess Parker ventures onto the Colonial frontier as Dan'l Boone in this adventure series. Tonight: George Washington sends Dan'l and his side-kick into "Ken-tuck-e," the "dark and bloody" hunting ground of four Indian nations, to find a site for a fort. Additional series regulars are Albert Salmi as Daniel's side-kick Yadkin; Ed Ames as his Oxford-educated, half-breed friend Mingo; Patricia Blair as his wife Rebecca; Veronica Cartwright as daughter Jemima; and Darby Hinton as his son Israel. (60 min.)

Guest Cast
Chief Blackfish .... Robert Simon
George Washington ....Stephen Courtleigh
Wigeon ....George Lindsey

⑦ ⑧ FLINTSTONES—Cartoon
[COLOR] "Monster Fred." Fred goes to see Dr. Frankenstone after a head injury leaves him with the mind of a child. Voices . . . Fred: Alan Reed. Dr. Frankenstone: Allan Melvin Wilma: Jean Vander Pyl. Dino: Chip Spam.

⑨ MOVIE—Drama
Million Dollar Movie: "While the City Sleeps." See Monday 7:30 P.M. Ch 9

⑪ HAWAIIAN EYE—Mystery
Magazine writer Gloria Matthews asks Steele to help her find Micho "Robinson

Friday September 25, 1964
Evening

🄌 NEW ORLEANS JAZZ
"Papa Jack, the Patriarch." Papa Jack Laine, who calls himself the "inventor" of jazz, talks about the history of jazz from the vantage point of his 80-year career. Vernon Cook is the host.

9:00 🄌 EAST SIDE/WEST SIDE
"The Five Ninety-Eight Dress." With three children to support, Mrs. Stuart is collecting relief checks and working too—so she's thrown in jail. Brock: George C. Scott. Mrs. Stuart: Kathleen Maguire. Hank Stone: George Mathews. Mr. Stuart: Tim O'Connor. Hecky: Elizabeth Wilson. (60 min.)

🄌 VALENTINE'S DAY
"The Life You Save Is Yours." A black-veiled "lady" in the waiting room turns out to be Valentine's old pal Rocky. Rocky: Jack Soo.

Guest Cast
Lola	Pat Priest
Buster	Ricky Layne

🄌 INTERNATIONAL FESTIVAL
SPECIAL Tonight's program is a television documentary from Italy: "Story of the Atom Bomb," about the decisions involved in constructing the atom bomb. Ralph Bunche is host. (One hour, 5 min.)

9:30 🄌 GOMER PYLE, USMC
DEBUT In this half-hour comedy series, Jim Nabors stars as the character he portrayed on "The Andy Griffith" series—naive, easygoing Gomer Pyle. Only now Gomer is a recruit in the Marine Corps. Frank Sutton is seen as Sergeant Carter, a tough drill instructor continually plagued by the harebrained Gomer. In tonight's episode, it's Gomer versus the obstacle course. Duke: Ronnie Schell.

Guest Cast
Sergeant Whipple	Buck Young
Corporal Johnson	Jerry Dexter

Lt. Col. Van Pelt	Peter Hansen
Private Swanson	Mark Slade

"Andy Griffith" producer Aaron Ruben is creator of this series; he also wrote tonight's episode.

Nabors needed a special-type haircut for this role. See next week's TV GUIDE.

🄌 JACK BENNY—Comedy
To begin his 15th year, Jack (still 39) returns to the network where he began on radio 32 years ago. Tonight: Jack visits his new bosses, the NBC vice presidents and sits in on a "panel show" with the Marquis Chimps.

🄌 🄌 12 O'CLOCK HIGH—Drama
"Follow the Leader." To improve bombing effectiveness, General Savage tries a new technique: The whole group is to depend on the accuracy of the bombardier in the lead plane. Savage: Robert Lansing. Crowe: John Larkin. (60 min.)

Guest Cast
Lt. Robert Mellon	Andrew Prine
Lt. Harold Zimmerman	Jud Taylor
Lt. Jesse Bishop	Paul Carr

10:00 🄌 🄌 REPORTER—Drama
DEBUT Novelist-playwright Jerome Weidman is the creator of this series about New York reporter Danny Taylor (Harry Guardino) who works on open assignment under city editor Lou Sheldon (Gary Merrill). Weidman describes his character: "Danny Taylor cares about people. He is not a man standing about with a notebook and jotting down facts. He is a man standing up to his armpits in the facts." Standing by to whisk Danny to the action is cab driver Artie Burns (George O'Hanlon), and standing behind the newsman's favorite bar is Ike Dawson (Remo Pisani). (60 min.)

🄌 JACK PAAR—Variety
COLOR Musical-comedy star Mary Martin and the comedy team of Mike Nichols and Elaine May are Jack's

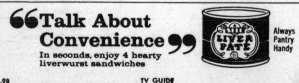

Tuesday September 29, 1964
Evening

7 8 McHALE'S NAVY
While carrying the base payroll, bumbling Ensign Parker stumbles and knocks himself unconscious. When he awakens, the money is gone. McHale: Ernest Borgnine. Binghamton: Joe Flynn. Parker: Tim Conway. Carpenter: Bob Hastings. Betrand: Hoke Howell. Yvette: Jeanne Rainier.

13 OF PEOPLE AND POLITICS
The methods of public-opinion polls and the pollsters' effects on today's politicians are examined in filmed interviews. Sens. William Proxmire (D., Wis.) and Thurston Morton (R., Ky.); pollsters George Gallup, Louis Harris, Elmo Roper and John Kraft; and political commentators Samuel Grafton and Samuel Lubell discuss various aspects of the polls.

9:00 5 MOVIE—Mystery
"The Accused." (1948) A psychology teacher tries to conceal the accidental death of a pupil. Loretta Young, Robert Cummings. (Two hours, 5 min.)

7 8 TYCOON—Comedy
At one of his subsidiary companies, Walter is mistaken for an employee, so he decides to stay on the job—incognito—to find out why plant production has fallen off. Walter: Walter Brennan. Pat: Van Williams. Wilson: Jerome Cowan. Ross Murray: Allen Case. Thompson: Jan Arran.

9 CHAMPIONSHIP BOWLING
Jim Schroeder vs. Pat Patterson. Fred Wolf reports. (60 min.)

13 INTERNATIONAL FESTIVAL
SPECIAL "El Mexicano y la Muerte," televised in Mexico, is a music and dance presentation of folk attitudes towards death; and "The First Pavilion," a science-fiction plot produced in Poland. Ralph Bunche is host. (60 min.)

9:30 2 3 PETTICOAT JUNCTION
One of Kate's neighbors comes to stay at Shady Rest while waiting to have her baby, and a nervous Uncle Joe immediately works out a master plan to get a doctor to Shady Rest when the crucial time arrives. Kate: Bea Benaderet. Uncle Joe: Edgar Buchanan. Billie Jo: Jeannine Riley. Bobbie Jo: Pat Woodell. Betty Jo: Linda Kaye. Elsie: Olive Sturgess. Doc: Frank Ferguson. Henry: Robert Easton.

4 THAT WAS THE WEEK THAT WAS—Satire
RETURN COLOR The topical humor troupe moves to a new time this season with three returning regulars: "TW3 Girl" Nancy Ames, British humorist David Frost and comedienne Phyllis Newman. The roster of semi-regulars includes Bob Dishy, pantomimist-puppeteer Burr Tillstrom and actress Pat Englund. Buck Henry, star of the satirical movie "The Troublemaker," will handle both performing and writing chores. Other writers include Gerald Gardner, author of several political humor books, and Dee Caruso, co-editor of Sick Magazine. (Live)

Postponed from last week.

7 8 PEYTON PLACE—Serial
After Julie has a mysterious accident, Dr. Rossi becomes involved in the internal conflicts of the Anderson family. Rossi: Ed Nelson. Julie: Kasey Rogers. George: Henry Beckman. Betty: Barbara Parkins. Constance: Dorothy Malone. Allison: Mia Farrow. Rodney: Ryan O'Neal. Norman: Christopher Connelly.

Series regular Mia Farrow: Is she a typical Hollywood ingenue? Find out in the article in next week's TV GUIDE.

10:00 2 3 NURSES/DOCTORS
"The Suspect," first of a two-part episode. Dr. Tazinski is on the spot: an old girl friend asks him for help and is later found dead, the victim of an illegal abortion. Tazinski: Michael Tolan. Steffen: Joseph Campanella. Liz: Shirl Conway. (60 min.)

Guest Cast
Detective Sgt. Sam Crowell ..Bert Freed
Edith Robertson ....Jessica Walter
Rachael Mead ....Fran Sharon

Fred J. Scollay appears regularly on another MD series—"The Doctors."

4 CAMPAIGN AND THE CANDIDATES—News Analysis
SPECIAL Frank McGee anchors this coverage of the 1964 campaign. (60 min.)

7 8 FUGITIVE—Drama
"Man on a String." When a philandering husband is found murdered, the chief suspect is his girl friend Lucey Russell. But Lucey has an alibi: She was with Kimble. Kimble: David Janssen. (60 min.)

Saturday October 3, 1964

Evening

⑤ BRONCO—Western

"The Last Letter." Mexican patriots are trying to oust French Emperor Maximilian from their country. Bronco· Ty Hardin. Solado: Ernest Sarracino. Rob Bucklin: Evan McCord. (60 min.)

⑨ MOVIE—Drama

[COLOR] "Land of the Pharaohs." (1955) A great Pharaoh drives his people for 30 years to build a pyramid that will house his body and treasure. Jack Hawkins, Joan Collins. (Two hours)

7:30 ② ③ JACKIE GLEASON

The Great One comments on topless bathing suits, vitamins, quiz shows and school dropouts. Barbara Heller introduces the sketches: Rum Dum's all-out war with a ketchup bottle; a child psychologist's solution to a familiar problem; and the Poor Soul's difficulties in hiding the combination to his safe. Reggie Van Gleason his version of how the West was won and sings "If I Had My Way, Dear." June Taylor dancers. (60 min.)

④ FLIPPER—Drama

[COLOR] "S.O.S. Dolphin." Dr. Rothwell, a marine biologist, discovers a deadly scorpion fish in Coral Key Park and orders Porter to close the area. Porter: Brian Kelly. Sandy: Luke Halpin. Bud: Tommy Norden.

Guest Cast

Dr. Rothwell ....John Lasell
Susan Hadley ....Linda Bennett

⑦ ⑧ OUTER LIMITS

"Behold Eck!" An optics expert designs unique quartz eyeglasses to correct double vision, but the spectacles have another interesting property: They render two-dimensional creatures visible. Script by John Mantley. (60 min.)

Cast

Dr. James Stone ....Pete Lind Hayes
Elizabeth Dunn ....Joan Freeman
Dr. James Stone ....Peter Lind Hayes
Detective Runyan ....Douglas Henderson

⑪ MOVIE—Melodrama

"Return of the Ape Man." (1944) A scientist discovers a way to preserve animals and humans by freezing them. Bela Lugosi, John Carradine. (90 min.)

8:00 ④ MR. MAGOO—Cartoon

[COLOR] In the conclusion of "Treasure Island," Magoo is pirate Long John Silver. Voices . . . Magoo: Jim Backus. Jim Hawkins: Dennis King.

⑤ ROOM FOR ONE MORE

"Girl from Sweden." The Rose's new Swedish maid can't even operate the toaster. Lisa: Sue Ane Langdon. George: Andrew Duggan. Anna: Peggy McCay.

8:30 ② ③ GILLIGAN'S ISLAND

The marooned inhabitants start thinking about some permanent form of shelter, and they set out to build a communal hut to serve as living quarters. Gilligan: Bob Denver. Skipper: Alan Hale. Ginger: Tina Louise. Howell: Jim Backus. Mrs. Howell: Natalie Schafer. Hinkley: Russell Johnson. Mary: Dawn Wells.

④ KENTUCKY JONES—Drama

Ike secretly admires a girl in his Sunday-school class—the same girl whose dad's laundry is putting too much starch in Kentucky's shirts. Kentucky: Dennis Weaver. Ike: Rickey Der. Annie Ng: Cherylene Lee. Judge Perkins: Paul Fix.

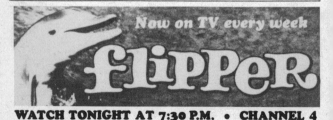

MONDAY, NOVEMBER 9, 1964

November 9, 1964 **Monday**
Evening

9:00 ❷ ❸ **LUCILLE BALL—Comedy**
Policewoman Lucy Carmichael is a kook-ie rookie, just the type needed to accompany Det. Bill Baker on a stakeout for Green Scarf Louie, the lovers' lane bandit. Viv: Vivian Vance. Bill Baker: Jack Kelly.

❹ **JONATHAN WINTERS**
[SPECIAL] [COLOR] First of Jonathan's eight comedy specials to be seen in this time spot. Details are in the Close-up on page A-48. (60 min.)

Andy Williams will not be seen tonight.

❺ **77 SUNSET STRIP—Mystery**
"The Grandma Caper." The Fenwick family are afraid that their eccentric Grandma might get into serious trouble. Frances Bavier, Jerome Cowan. (60 min.)

❼ ❽ **WENDY AND ME—Comedy**
Jeff invites the airline's president home to dinner, not knowing that Wendy has lent the apartment to Mr. Bundy, who wants to impress a visiting niece. Landlord: George Burns. Wendy: Connie Stevens. Jeff: Ron Harper. Bundy: J. Pat O'Malley. Danny: James Callahan.
Guest Cast
Willard Norton .... Bartlett Robinson
Sharon Bundy .... Colleen O'Sullivan

⓫ **DICK POWELL—Drama**
"Luxury Liner." Selena Royce looks forward to an ocean voyage with bitterness and whisky as her only shipmates. Rory Calhoun, Jan Sterling. (60 min.)

⓭ **ART OF FILM—Kauffmann**
"Serge Eisenstein." The work of the Russian film director is studied and excerpts from "Potemkin," "Ten Days that Shook the World" and "Ivan the Terrible" are shown. Stanley Kauffmann is host.

Repeated Tuesday at 10:30 P.M.
9:30 ❷ ❸ **MANY HAPPY RETURNS**
Walter tries to help out former fraternity brother Wesley Fosdick, a monumentally unsuccessful author by promoting his works in Krockmeyer's book department. Walter: John McGiver. Joe: Mickey Manners. Lynn: Elena Verdugo. Sharp: Russell Collins. Wesley Fosdick: Willard Waterman.

❼ ❽ **BING CROSBY—Comedy**
"The Dominant Male." Joyce's boy friend Don maintains that "the hand that rocks

the cradle rules the world," and he challenges Bing to prove otherwise. Bing sings "How Deep Is the Ocean" and Don joins him in "Hallelujah, I Just Love Her So." Ellie: Beverly Garland. Joyce: Carol Faylen. Don Turner: Gary Crosby.

❾ **MARSHAL DILLON—Western**
Little Peter Scooper is caught stealing potatoes from the Galloway farm. Jeanette Nolan, David Kent.

⓭ **CHALLENGE—Science**
"Breeder in the Desert." Host Norman Ross visits Idaho's National Reactor Testing Station, where a reactor is turning some of its own uranium fuel into another fuel—plutonium.

Repeated Tuesday at 4:30 P.M.
10:00 ❷ **SLATTERY'S PEOPLE**
"Question—Is Laura the Name of the Game?" When her husband is passed over as the party choice to fill a Congressional vacancy, Laura Tamiris goads him into finding an issue to gain some attention. Script by Norman Katkov. Slattery: Richard Crenna. Metcalf: Tol Avery. Radcliff: Edward Asner. (60 min.)
Guest Cast
Harry Tamiris .... Jack Warden
Laura Tamiris .... Georgann Johnson
Emmett Logan .... James Griffith
Gert .... Joyce Meadows
Bryan Chalmers .... Lew Brown

Cleveland Amory reviews this series in next week's TV GUIDE.

❸ **RICHARD DIAMOND—Mystery**
"Double Trouble." Diplomat Arnold Bascomb hires Diamond to locate his fiancée. David Janssen, Chris Waterfield.

❹ **ALFRED HITCHCOCK**
"See the Monkey Dance," by Lewis Davidson. A young London broker, heading for an illicit weekend with his married sweetheart, runs into a gentleman who has strikingly similar plans. (60 min.)
Cast
Stranger .... Efrem Zimbalist Jr.
George .... Roddy McDowall
Wife .... Patricia Medina

❺ **BREAKING POINT—Drama**
"The Bull Roarer." A construction worker is appalled as he watches his brother beat up a man who accosted them. Paul Knopf: Lou Antonio. Murray Knopf: Ralph Meeker. (60 min.)

December 28, 1964 **Monday**

9:00 4 ANDY WILLIAMS—Variety

Andy Williams, Milton Berle, Lawrence Welk

COLOR Andy's guests are band leader Lawrence Welk, who brings along his accordion, and comedian Milton Berle.

Uncle Miltie suggests that Andy should devote his musical talents to operetta, so the trio takes a fling at "Somewhere in Canada," a musical sketch in which a Mountie officer (Berle) tries to dissuade a trooper (Williams) from marrying the daughter of an Indian chief (Welk). Osmond Brothers, Nick Castle dancers, Good Time singers, Dave Grusin orchestra. (60 min.)

Highlights

"Sweet and Low" ....Williams, Welk
"Goodbye Charlie," "I'll Never Stop
  Loving You," "Hey, Look Me Over,"
  "Time After Time," "May Each Day"
....Williams
"Laura Lee" ....Williams, Osmonds
"Mr. Sandman," "I Wouldn't Trade the
  Silver in My Mother's Hair" ....Osmonds
Operetta medley, "Office Party Polka"
....All

9:30 7 8 CHRISTMAS DRAMA

SPECIAL The Dickens classic, "A Christmas Carol," was updated by Rod Serling for this first in a series of specials relating to activities of the United Nations. The 90-minute drama, the first TV effort by award-winning motion-picture director Joseph Mankiewicz, will be presented without commercial interruptions.

Daniel Grudge's 22-year-old son Marley was killed in action on Christmas Eve in 1944, a loss from which the industrial tycoon has never recovered. An insulated and embittered man, Grudge has nothing but scorn for any American involvement in international affairs. But tonight he will be visited by three strangers.

Cast

Daniel Grudge ....Sterling Hayden
Fred ....Ben Gazzara
Imperial Me ....Peter Sellers
Wave ....Eva Marie Saint

'CAROL FOR ANOTHER CHRISTMAS'

The Ghost of Christmas Past (Steve Lawrence) takes Daniel Grudge (Sterling Hayden) back through time to a World War I troopship.

Ghost of Christmas Past ..Steve Lawrence
Ghost of Christmas Present ....Pat Hingle
Ghost of Christmas Future ...Robert Shaw
Doctor ....James Shigeta
Ruby ....Barbara Ann Teer
Charles ....Percy Rodriguez
Mother ....Britt Ekland
Soldier ....Gordon Spencer

Turn on the laughs!

F Troop

9:00 ⑦ ⑧   NEW SHOW   abc

Petticoat Junction

9:30 pm. New season! Bea Benaderet, Edgar Buchanan and those lovely girls run the best hotel in television. In color.

CBS◉2

Tuesday

Evening

Richard Chamberlain. Gillesple: Raymond Massey. Nurse Lawton: Lee Kurty.
Guest Cast

Dr. Maxwell Becker	James Mason
Chris Becker	Margaret Leighton
Charles Shannon	Burt Brinckerhofi
Dr. Secaras	John Lodge
Garcia	Rodolfo Hoyos

⑤ **ROGUES—Drama**
"House of Cards." Tony has the task of avenging dear old Uncle Bertie, who was cheated out of his entire retirement fund at London's most elegant gambling house. Major Hamilton: Patric Knowles. Linda Tennant: Jessica Walter. Tony: Gig Young. Margaret: Gladys Cooper. Timmy: Robert Coote. Inspector Briscoe: John Williams. Alec: David Niven. Mike Weatherby: John Orchard. Head Croupier: Gil Stuart. Uncle Bertie: Donald Foster. (60 min.)

⑦ ⑧ **McHALE'S NAVY**
"McHale's Navy" begins its fourth season with a transfer from the Pacific to Southern Italy, and the sailors receive quite a welcome: Binghamton is shot at by the Germans and the local mayor prepares to fleece the crew. McHale: Ernest Borgnine. Binghamton: Joe Flynn. Parker: Tim Conway. Mayor: Jay Novello. Fuji: Yoshio Yoda. Carpenter: Bob Hastings. Gruber: Carl Ballantine. Christy: Gary Vinson. Tinker: Billy Sands. Virgil: Edson Stroll. Willy: John Wright.

⑬ **TO BE ANNOUNCED**

9:00 ④ **MOVIE—Drama**
COLOR Tuesday Night at the Movies: William Holden and Grace Kelly in "The Bridges at Toko-Ri," first in a series of TV movie debuts. For details, see the Close-up on page A-65. (Two hours)

⑦ ⑧ **F TROOP—Comedy**
DEBUT "Scourge of the West." After bumbling Private Parmenter inadvertently leads a charge against the Confederates, he is rewarded with a promotion—to captain—and given command of Fort Courage. O'Rourke: Forrest Tucker. Parmenter: Ken Berry. Corporal Agarn: Larry Storch. Wrangler Jane: Melody Patterson. Roaring Chicken: Edward Everett Horton. Wild Eagle: Frank de Kova. Hannibal Dobbs: James Hampton.
Guest Cast

Colonel Malcolm	Alan Hewitt

September 18, 1965 **Saturday**

8:30 **4** **GET SMART—Comedy**

DEBUT This half-hour spoof of the current cloak-and-dagger trend stars Don Adams as Maxwell Smart, CONTROL's Secret Agent 86. Aiding the inept Smart (who needs all the help he can get) are Agent 99 (portrayed by tiger-cat Barbara Feldon), the Chief (Edward Platt) and Fang, an undercover canine.

Tonight: The evil Mr. Big, leader of KAOS, has kidnaped Prof. Dante and his Inthermo, an invention capable of melting anything in its path. Mr. Big is willing to return both the professor and his machine—for 100 million dollars. No one has that kind of money, so Smart is assigned to rescue Dante, wreck KAOS and get Mr. Big.

Howard Morris directed this episode, written by comics Mel Brooks and Buck Henry, who developed the series.

'MR. BIG'

Barbara Feldon and Don Adams

Guest Cast

Mr. Big	Michael Dunn
Prof. Hugo Dante	Vito Scotti
Zelinka	Janine Gray
Garth	Kelton Garwood

9:00 **4** **MOVIE—Western**

'GUNFIGHT AT THE O.K. CORRAL'

COLOR On October 26, 1881, one of the most famous gun battles in Western history took place at the O.K. Corral in Tombstone, Ariz. When the smoke cleared from the brief but bloody battle, two more names were added to the pages of American folklore—Wyatt Earp and Doc Holliday.

This 1957 movie, directed by John Sturges, traces the events that led up to the encounter.

Novelist Leon Uris wrote the screenplay. Dimitri Tiomkin's score features Frankie Laine singing the title song. Oscar nominations went to the sound recording and Warren Low's film editing.

Cast

Wyatt Earp	Burt Lancaster
Doc Holliday	Kirk Douglas
Laura Denbow	Rhonda Fleming
Kate Fisher	Jo Van Fleet
Ringo	John Ireland
Cotton Wilson	Frank Faylen
Ike Clanton	Lyle Bettger
Bat Masterson	Kenneth Tobey
Charles Bassett	Earl Holliman
Billy Clanton	Dennis Hopper

Wednesday September 22, 1965

Evening

7 (8) GIDGET—Comedy

COLOR Gidget is dateless for the class luau, until brother-in-law John decides to escort her—in order to do research for a psychology paper on teen-agers. Gidget: Sally Field. John: Peter Deuel. Russ: Don Porter. Anne: Betty Conner.

Guest Cast

Siddo	Mike Nader
Pokey	Heather North
Treasure	Beverly Adams
Randy	Rickie Sorensen

Cast as a surfer, series star Sally Field had to cope with a board of education. See the article on page 24.

11 MOVIE—Comedy

"The Most Wanted Man." (French; 1961) A gun moll mistakes an innocent bungler for a wanted public enemy. Fernandel, Zsa Zsa Gabor. (90 min.)

13 FRENCH CHEF—Cooking

"Lest We Forget Broccoli and Cauliflower." Julia Child works with the cabbage family.

9:00 2 (3) GREEN ACRES—Comedy

COLOR Lisa gets her first look at the farm and, as far as she's concerned, one look is enough. Oliver: Eddie Albert. Lisa: Eva Gabor. Eb: Tom Lester. Haney: Pat Buttram. Joe Carson: Edgar Buchanan. Sam Drucker: Frank Cady. Floyd: Rufe Davis. Charley: Smiley Burnette. Kate: Bea Benaderet.

Guest Cast

Bennett	Lyle Talbot
Mrs. Bennett	Iris Adrian

4 BOB HOPE—Comedy

COLOR "The Crime." A pair of illicit lovers have been murdered, and prosecuting attorney Abe Perez is going all out to get a conviction against the dead man's wife, a wealthy socialite who used to date Perez. Bob Hope is the host. (60 min.)

Cast

Abe Perez	Jack Lord
Sarah Rodman	Dana Wynter
DA Hightower	Pat O'Brien
Mary	Sheree North
Winfield	Oliver McGowan
Mikel Hawley	Berkeley Harris
Frank Busch	Walter Woolf King
Fran Perez	Karen Steele
Selena Parker	Virginia Baker
Johnny Ridzik	Michael Pataki

5 MOVIE—Drama

"Force of Arms." (1951) During World War II, an American soldier and a WAC lieutenant discover romance in Italy. (Two hours, 5 min.)

Cast

Peterson	William Holden
Eleanor	Nancy Olson
Major Blackford	Frank Lovejoy
McFee	Gene Evans
Klein	Dick Wesson
Sheridan	Paul Picerni
Major Waldron	Katherine Warren

7 (8) BIG VALLEY—Western

COLOR Heath has trouble with some of the hired hands, who don't like the idea of taking orders from the new member of the family. Heath: Lee Majors. Nick: Peter Breck. Victoria: Barbara Stanwyck. Jarrod: Richard Long. Audra: Linda Evans. (60 min.)

Guest Cast

Wallant	Andrew Duggan
Barrett	John Milford
Lillard	Calvin Brown
McColl	Douglas Kennedy
Spock	Walker Edmiston
Cota	Allen Jaffe

9 DANGER IS MY BUSINESS

COLOR "Alligator Wrestler." A Seminole Indian living on an island off Miami supplements his income by wrestling alligators for the entertainment of tourists. Bobby Tiger. Lt. Col. John D. Craig is host and narrator.

13 CREATIVE PERSON—Schuman

This film portrait of composer William

Friday October 8, 1965

Evening

Hugo ....Sandy Kenyon
Foxx ....Harlan Warde

④ CAMP RUNAMUCK—Comedy

[COLOR] "Cow." Spiffy, who finds camp food "depressing," decides it might be a good idea to buy his own fresh beef—on the hoof. Written and directed by David Swift. Spiffy: David Ketchum. Wivenhoe: Arch Johnson. Doc: Leonard Stone. Mahala May: Alice Nunn. Pruett: Dave Madden. Caprice: Nina Wayne.

Guest Cast

Glefky ....Frederic Downs
Malden ....Mike Wagner
Mrs. Jackson ....Maidie Norman

⑤ OUTER LIMITS

"Counterweight." Six ordinary people board a spaceship for a simulated flight to another planet. Michael Constantine, Jacqueline Scott. (60 min.)

⑦ FLINTSTONES—Cartoon

[COLOR] Fred thinks it's pretty funny when Barney is called for jury duty, but his smile fades when he's summoned for service on the same jury.

⑨ MOVIE—Western

[COLOR] Million Dollar Movie: "Vera Cruz." See Monday 7:30 P.M. Ch. 9

⑪ LLOYD THAXTON—Variety

Gene Chandler sings "Just Be True" and "Good Times." (60 min.)

⑬ ERIC HOFFER—Comment

Eric Hoffer discusses his concept of the "New Age." James Day is the host.

8:00 ④ HANK—Comedy

[COLOR] Track coach Weiss thinks Indian Sam Lightfoot is another Jim Thorpe, but the light-footed lad who outran the track team is actually Hank in a new disguise. Hank: Dick Kallman. Weiss: Dabbs Greer. Doris: Linda Foster. Franny: Kelly Jean Peters. Ethel Weiss: Sheila Bromley.

⑦ TAMMY—Comedy

[COLOR] Uncle Lucius, low on gambling funds, fells a tree across the road to the Brent plantation—and plans to raise cash by moving the tree for passing motorists. Tammy: Debbie Watson. Lucius: Frank McGrath. Brent: Donald Woods. Grandpa: Denver Pyle. Florabelle Ellen Corby. Walter: Philip Ober.

⑬ CINDERELLA—Ballet

[SPECIAL] Margot Fonteyn stars in the Royal Ballet Company's TV adaptation of Serget Prokofiev's ballet "Cinderella," with choreography by Frederick Ashton. The ballet was directed by Mark Stuart and produced by Granada TV in association with London's Royal Opera House Covent Garden. (Two hours)

Cast

Cinderella ....Margot Fonteyn
Prince ....Michael Somes
Fairy Godmother ....Annette Page
Ugly Sisters
................Gerd Larsen, Rosemary Lindsey

8:30 ② ③ HOGAN'S HEROES

[COLOR] "The Late Inspector General." Hogan and his men have to delay their plan to blow up a munitions train when Inspector General von Platzen makes an unexpected visit to Stalag 13. Hogan: Bob Crane. Klink: Werner Klemperer. Schultz: John Banner. Newkirk: Richard Dawson. Carter: Larry Hovis. Kinchloe: Ivan Dixon. Helga: Cynthia Lynn General von Platzen: John Dehner.

Comic Bill "José Jiménez" Dana offers

WOR-TV
TONIGHT AT 9:30
BEN GAZZARA
CHUCK CONNERS
ARREST & TRIAL

Sunday October 17, 1965

Evening

plagued ranchers feud over the services of a charlatan rainmaker named Fenimore Bleek. Big John: Jim Davis. Bleek: Denver Pyle. Nate Freed: Roy Engel.

⑪ SECRET LIFE OF ADOLF HITLER—Documentary

[SPECIAL] Westbrook Van Voorhis narrates this documentary on the public and private life of Adolf Hitler. Newsreel footage documents the growth of the Fuehrer's public image, and home movies made by Eva Braun and others show Hitler at Berchtesgaden. In interviews, members of Hitler's personal staff describe his early days, his relationship with Eva Braun and their suicide during the siege of Berlin. (60 min.)

7:30 ② ③ MY FAVORITE MARTIAN

[COLOR] Martin feeds his brainpower pills to Tim. Martin: Ray Walston. Tim: Bill Bixby.

Guest Cast

Merrick	Lee Bergere
Nelson	Harry Holcombe

④ WALT DISNEY'S WORLD

[COLOR] "The Flight of the White Stallions," first of two parts. As Allied bombers attack Vienna toward the end of World War II, German colonel Alois Podhajsky tries to evacuate the famed Lipizzan stallions from the Spanish Riding School. (60 min.)

Cast

Col. Alois Podhajsky	Robert Taylor
Verena Podhajsky	Lilli Palmer
General Pellheim	Curt Jurgens
Otto	Eddie Albert
Countess Arco-Valley	Brigitte Horney

⑤ I SEARCH FOR ADVENTURE

'The Eighteen Captains." Host Jack Douglas interviews John Moore, one of the 18 California university students who sailed to Tahiti.

⑧ RIPCORD—Adventure

[COLOR] "Infiltration." A clever dope peddler brings narcotics into the U.S. by having a plane drop the cargo into a border state. Ted: Larry Pennell. Jim: Ken Curtis.

⑨ MOVIE—Adventure

[COLOR] Big Preview: "The Witch's Curse." (Italian; 1960) NY TV Debut. The legendary Maciste enters Hell to remove the curse a sorceress has cast

over a small town in Scotland. Kirk Morris, Hélène Chanel. (Two hours)

8:00 ② ③ ED SULLIVAN—Variety

[COLOR] In Hollywood, Ed's scheduled guests are Sid Caesar; actor Sean Connery; the singing McGuire Sisters; singer Pat Boone; the rock 'n' rolling Animals; comics Guy Marks and Totie Fields; and the Fiji Military Band. Ray Bloch orchestra. (Live; 60 min.)

⑤ DAKOTAS—Western

"Mutiny at Fort Mercy." Del and Vance return a prisoner to the Army base only to be held hostage when the prisoners revolt. Chad Everett, Michael Green. (60 min.)

⑦ ⑧ FBI—Drama

[COLOR] "The Insolents." When a man is found shot to death aboard a luxury liner, the ship's captain arrests the victim's millionaire stepson on a charge of murder. Erskine: Efrem Zimbalist Jr. Rhodes: Stephen Brooks. Ward: Philip Abbott. Barbara: Lynn Loring. (60 min.)

Guest Cast

Elizabeth	Joan Marshall
Roger York	James Ward
Mrs. Creighton	Eileen Heckart
Durant	Douglas Henderson
Captain Tillman	Ben Wright
Hewitt Pierce	Charles Robinson

⑪ MOVIE—Drama

[COLOR] "Mohawk." (1956) Frontier life in the Mohawk Valley finds three lovely lasses vying for the attention of the same man. Scott Brady, Rita Gam, Lori Nelson, Allison Hayes. (90 min.)

8:30 ④ BRANDED—Western

[COLOR] "Seward's Folly." Two unsavory characters plot to steal the maps McCord is making after his survey of the rich, recently purchased territory of Alaska. McCord: Chuck Connors.

Guest Cast

Leslie Gregg	Coleen Gray
Rufus Pitkin	J. Pat O'Malley
William Henry Seward	Ian Wolfe
Sobel	Charles Maxwell
Grimes	Robert Hoy
Millie	Lulu Porter

Why does series star Chuck Connors shun the social life of Hollywood and New York? See next week's TV GUIDE

October 19, 1965 **Tuesday**
Evening

naped and held for ransom by Ma Gufler and her unscrupulous mountain brood. Rowdy: Clint Eastwood. Jed: John Ireland. Wishbone: Paul Brinegar. Simon: Raymond St. Jacques. (60 min.)

Guest Cast

Ma Gufler	Mercedes McCambridge
Max Gufler	Robert Blake
Billie Lou Gufler	Sharon Farrell
Jesse Gufler	Warren Oates
Cousin Will	Hal Baylor
Skinner	Robert Beecher

④ MY MOTHER, THE CAR

COLOR "I'm Through Being a Nice Guy." In a desperate effort to get the car away from Dave, Captain Manzini resorts to Operation Unscrupulous: replacing the car with an exact replica. Dave: Jerry Van Dyke. Manzini: Avery Schreiber. Mother's Voice: Ann Sothern. Barbara: Maggie Pierce.

Guest Cast

Inge	Barbara Bain
Jennay	Fernando Roca
Pitt	Arthur E. Gould-Porter
Gentleman	Harmon L. Stevens

⑤ ROUTE 66—Drama

"The Quick and the Dead." In Riverside, Cal., for the Grand Prix stock-car races, Tod and Buz find themselves also involved in a complex marital mixup. Katherine: Susan Kohner. Tod: Martin Milner. Buz: George Maharis. (60 min.)

⑦ COMBAT!—Drama

"Evasion." In an ancient castle that the Germans are using as a POW camp, Hanley's fellow prisoners provide him with a fake Albanian uniform for his escape attempt. Script by Esther and Bob Mitchell. Hanley: Rick Jason. (60 min.)

Guest Cast

Major Thorne	Lloyd Bochner
Kopke	Jacques Aubuchon
Julie	Monique Lemaire
Major Ramsey	John Lodge

⑧ MARCH OF TIME

SPECIAL "Seven Days in the Life of the President." First in a series of "March of Time" TV documentaries. William Conrad narrates. For details, see the Close-up on page A-80. (60 min.)

⑨ MOVIE—Drama

Million Dollar Movie: "Crisis." See Monday 7:30 P.M. Ch. 9.

⑪ LLOYD THAXTON—Variety

⑬ BOOK BEAT—Interview

8:00 ④ PLEASE DON'T EAT THE DAISIES—Comedy

COLOR "Look Who's Talking." Joan buys baby clothes for a friend's shower, but the boys start spreading the word that it's their mother who is expecting. Joan: Patricia Crowley. Jim: Mark Miller. Kyle: Kim Tyler. Joel: Brian Nash. Trevor: Jeff Fithian. Tracy: Joe Fithian.

Guest Cast

Herb Thornton	Harry Hickox
Marge Thornton	Shirley Mitchell
Rob Miller	Roy Stuart

Want 9 feet of suds in the house when you wash your sheep dog? See page 10.

⑬ NEGRO PEOPLE—History

"The Negro and the South." Fact and fallacy about the "Southern way of life" are explored through interviews with Mississippi residents, white and Negro. Films show children being taught Southern history in segregated schools.

Repeated Wednesday at 4 P.M.

November 6, 1965 **Saturday**

Evening

is inside. Sandy: Luke Halpin. Porter: Brian Kelly. Ulla: Ulla Stromstedt.

Guest Cast

Pilot	Ron Hayes
Copilot	Courtney Brown
Manuel	Gilbert Frye
Paco	Warren Day

7 8 SHINDIG—Music

Scheduled guests: Jackie Wilson ("I Believe I'll Love On"); the Rolling Stones ("Good Times," "Have Mercy"); Billy Joe Royal ("I Knew You When"); The Strangeloves ("I Want Candy"); Tony and the Bandits ("It's a Bit of All Right"); and Fontella Brass ("Rescue Me"). Host: Jimmy O'Neill.

11 MOVIE—Science Fiction

"Four Sided Triangle." (English; 1953) Two young scientists have created an instrument which can duplicate anything in the world. Barbara Payton, James Hayter, Stephen Murray. (90 min.)

8:00 4 I DREAM OF JEANNIE

Jeannie feels that Tony is taking her for granted, so she decides to stop using her magic powers and start acting like a normal female. Jeannie: Barbara Eden. Tony: Larry Hagman. Roger: Bill Daily. Dr. Bellows: Hayden Rorke.

Guest Cast

Armand	Steven Geray
Sadelia	Tania Lemani
Sam	Del Moore
Pierre	Jacques Roux
Driver	Bobby Johnson
First Woman	Yvonne White
Second Woman	Jewell Lain

5 WRESTLING—Washington

7 8 KING FAMILY—Music

In a salute to Veterans' Day, Robert

Clarke reads a letter to President Johnson from a Marine in Vietnam. Musical highlights include "The Battle Hymn of the Republic" (All); a medley of George M. Cohan songs (Alvino Rey, Cousins); "Yesterday" (Sisters); "Walk like a Sailor" (Marilyn, Girl Cousins); "Something Wonderful" (Alyce); and, in the Cousins' Top 20 spot: "1-2-3," "A Lover's Concerto," "Everyone's Gone to the Moon," "Positively 4th Street" and "Not the Lovin' Kind." Mitchell Ayres conducts the orchestra.

8:30 2 3 TRIALS OF O'BRIEN

"The Trouble with Archie." Archie Vasoulian has been embezzling funds from the company he runs with his partner Ben. When things get too hot for him, he hires an arsonist to burn the books. Script by George Bellak. O'Brien: Peter Falk. Katie: Joanna Barnes. Great McGonigle: David Burns. Miss G.: Elaine Stritch. Garrison: Dolph Sweet. (60 min.)

Guest Cast

Ben Moravian	Theodore Bikel
Fay Roberti	Alice Ghostley
Harry	Simon Oakland
Archie Vasoulian	Lou Jacobi
Brody	Bernie West
Malcolm	Bruce Hyde

4 GET SMART—Comedy

[COLOR] In an attempt to discredit Smart's testimony against KAOS, the international hoods set out to confuse the overworked agent. Smart: Don Adams. Chief: Edward Platt.

Guest Cast

Cowboy	Simon Oakland
Blake	Phillip Pine
Mrs. Dawson	Iris Adrian
Dr. Fish	Howard Caine

Continued on next page

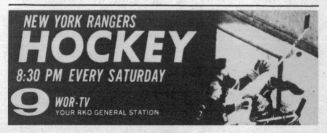

April 23, 1966 **Saturday**

Afternoon

At press time, these horses were outstanding among the eligibles. Exhibitionist (trained by Hirsch Jacobs) won the Santa Catalina Stakes. Amberoid (Roger Lawrin) finished second in the Fountain of Youth Stakes. Stupendous (Eddie Neloy) was second in the Louisiana Derby. Indulto (Max Hirsch) and Throne Room (L. S. Barrera) look strong for this race. Compiled by The Morning Telegraph

⑧ COMMUNITY SALUTE

5:00 ③ **HORSE RACE**—Aqueduct
See 4:30 P.M. Ch. 5 for details of this race taped earlier today.

④ **BOWLING CHAMPIONS**

⑤ **LAWMAN**—Western
Lily Merrill comes to Laramie and prepares to open a saloon. But Lily runs up against Troop, who thinks she is aiding an outlaw. Troop: John Russell. McKay: Peter Brown. Lily: Peggie Castle.

⑦ ⑧ **WIDE WORLD OF SPORTS**
Scheduled: highlights from "Wide World of Sports'" last five years. See the Close-up below. (90 min.)

⑪ **RAMAR**—Adventure
"Tree of Death." The reappearance of a missing timepiece provides Ramar with a clue to an unsolved jungle mystery. Jon Hall, Ray Montgomery.

5:30 ③ **BRAD DAVIS**—Music

④ **SAM SNEAD**—Golf
COLOR Sam demonstrates the uses of the short irons—the eight and nine irons and the wedge—in solving the problems posed by approach shots. Also: tips for the tall and the portly golfer.

⑤ **TRUE**—Jack Webb
Game warden Ernie Swift doesn't believe in special privileges for anyone, so when he catches mobster Frank MacErlane fishing out of season, he immediately serves him with a summons. Swift: James Best. MacErlane: David McLean. Claire: Shirley Ballard.

⑨ **MOVIE**—Drama
Million Dollar Movie: "Conspirator." See 11 A.M. Ch. 9 for details.

⑪ **COLOR** **ROCKY AND HIS FRIENDS**—Cartoons

TV CLOSE-UP GUIDE 5:00 ⑦ ⑧ **WIDE WORLD OF SPORTS**

For its fifth anniversary show, "Wide World" presents film clips from its first five years of sports coverage.

Scheduled sports and competitors include . . . **Track:** Russia's Valery Brumel setting a world high-jump record in the 1961 U.S.-Russian Track Meet; Bob Hayes setting the 9.1 world record in the 100-yard dash; and Jim Beattie running the world's first sub-four-minute indoor mile (3:58.9). **Golf:** Arnold Palmer's win in the 1962 British Open. **Auto Racing:** Jim Clark taking the 1965 Indianapolis 500 in the rear-engined Lotus Ford. **Tennis:** Chuck McKinley winning the men's singles at Wimbledon, England. **Figure Skating:** Peggy Fleming's victory in the 1966 U.S. National Championship.

Also: some of the more exotic sports, including Australian Rules Football, a firemen's competition, figure-eight auto racing, mountain climbing (on the Eiffel Tower), dog-sled racing, barrel jumping, a jeep derby, rattlesnake hunting, an auto-demolition derby and high diving from 100 feet—into 11 feet of water. Jim McKay is the host. (90 min.)

Valery Brumel

Arnold Palmer

Auto racing

Tuesday September 13, 1966

Evening

season. Clarence isn't himself after sustaining a head injury in a jeep accident. The usually mild-mannered lion growls at everyone and takes a vicious swipe at Paula. Tracy: Marshall Thompson. Paula: Cheryl Miller. Jack: Yale Summers. Mike: Hari Rhodes. (60 min.)

4 GIRL FROM U.N.C.L.E.

DEBUT COLOR "The Dog-Gone Affair." April Dancer delivers an unusual package to Mark Slate, who's on a Greek island where THRUSH is testing a devilish new drug. The package is a dachshund named Puzti—whose fleas carry the drug's only known antidote. Script by Tony Barrett. April: Stefanie Powers. Mark: Noel Harrison. Waverly: Leo G. Carroll. (60 min.)

Guest Cast

Apollo Zakinthios .... Kurt Kasznar
Tuesday Hajadakis ......Luciana Paluzzi
Antoine Fromage ....Marcel Hillaire
Patras ....Jan Arvan

Hostess ....Beth Brickell

Same organization—different agent. Read the series' profile on page 39

5 ROUTE 66—Drama

"Eleven, the Hard Way." The citizens of Broken Knee, Nev., have an off-beat plan to finance a tourist-trade project. They hand their life savings to Sam Keep, the town's leading gambler, and send him off to Reno. Keep: Walter Matthau. Francis Oliver: Edward Andrews. Tod: Martin Milner. Buz: George Maharis. Monty Knight: Guy Raymond. Dora Knight: Debbie Megowan. (60 min.)

7 8 COMBAT!—Drama

COLOR "The Gun." Saunders and his men begin their fifth year of duty. The GI's plan to use a captured 75-millimeter gun to knock out a German bunker, but there's one hitch: They must drag the piece over a mile of rough, enemy-infested terrain. Saunders: Vic Morrow. Kirby: Jack Hogan. Littlejohn: Dick Peabody. Caje: Pierre Jalbert. (60 min.)

Saturday September 17, 1966

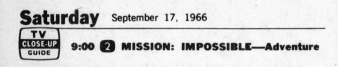

TV CLOSE-UP GUIDE 9:00 **2** MISSION: IMPOSSIBLE—Adventure

DEBUT **COLOR** Undercover agent Dan Briggs and his Impossible Missions Force are handed a deadly assignment that takes them to a Caribbean island.

Gen. Rio Dominquez, dictator of Santa Costa, has tipped the balance of power by obtaining two nuclear warheads. Briggs's assignment: remove the weapons, which are stashed—under constant surveillance—in the dictator's impenetrable vault.

Briggs's assault team includes Rollin Hand, master of disguise; Terry Targo, safecracker and demolition expert; electronics wizard Barnard Collier; Willy, a champion weightlifter; and a shapely diversion named Cinnamon Carter.

Bernard Kowalski directed from a script by series creator Bruce Geller. Briggs: Steven Hill. Collier: Greg Morris. Dominquez and Hand: Martin Landau. Cinnamon: Barbara Bain. Willy: Peter Lupus. (60 min.)

Guest Cast

Terry Targo	Wally Cox
Alisio	Harry Davis
Day Vault Clerk	Patrick Campbell

Steven Hill and Wally Cox

TV CLOSE-UP GUIDE 9:30 **7** **8** HOLLYWOOD PALACE—Variety

Bing Crosby

Jane Marsh

George Burns

COLOR Host Bing Crosby, ushering the "Palace" into its fourth season, presents comedians Sid Caesar and George Burns, and soprano Jane Marsh, who won first place at the Tchaikovsky competition in Moscow last July. Miss Marsh makes her TV network debut singing "Mi chiamano Mimi," from Puccini's "La Bohème."

Also on the bill: the Mamas and the Papas, folk-rock group; singer-dancer Lola Falana; French comic magician Mac Ronay; and the Rhodins, aerialists.

Bing portrays a star visiting suburbia in a comedy sketch featuring Sid,

Joyce Jameson and Mickey Deems.

Director: Grey Lockwood. Choreographer: Tom Hansen. Mitchell Ayres conducts the orchestra. (60 min.)

Highlights

"Strike Up the Band"	Bing, Dancers
"Pack Up Your Sins and Go to the Devil"	George
"You're Nobody till Somebody Loves You"	Bing, George
"Dancing Bear," "Dancing in the Streets"	Mamas and Papas
"Promise Her Anything"	Lola

Tuesday September 20, 1966

Evening

Cook ....Harold Fong
Dan Manners ....Norman Bartold
⓭ LINCOLN CENTER—NYC
[SPECIAL] New York's Lincoln Center for the Performing Arts celebrates its third anniversary with a three-part performance. 1. "Far Rockaway" focuses on a man who seeks punishment for an unpunishable crime. 2. "The Act," a ballet choreographed by Anna Sokolow. 3. "The Hero" is a one-act satirical opera by Marc Bucci. (60 min.)
㉛ SCIENCE SEMINAR
㊼ BOXING—Newark
9:00 **④ MOVIE—Comedy**
[COLOR] "My Six Loves." (1962) Musical-comedy star Janice Courtney retreats to her home in Connecticut for a rest—only to discover a flock of children living on her estate. Directed by Gower Champion. (Two hours)

Cast

Janice Courtney ....Debbie Reynolds
The Rev. James Larkin .....Cliff Robertson
Martin Bliss ....David Janssen
Ethel Swenson ....Eileen Heckart
Sheriff ....Jim Backus
Diana Soper ....Pippa Scott
Kingsley Cross ....Hans Conried
⑦ ⑧ PRUITTS OF SOUTH-AMPTON—Comedy
[COLOR] The Pruitts feel the pinch when they get the repair bill for their Rolls—so Phyllis heads for a pawnshop with some of the household items. Phyllis: Phyllis Diller. Uncle Ned: Reginald Gardiner. Sturgis: Grady Sutton.

Guest Cast

Desk Sergeant ....James Flavin
Henry Evans ....Jonathan Hole
Pawnbroker ....Eddie Quillan
⑨ MOVIE—Mystery
"The Ringer." (English; 1951) Edgar Wallace's tale about a master criminal

whose specialty is disguising himself and thus eluding the police. Herbert Lom, Mai Zetterling, Greta Gynt. (90 min.)
9:30 **② ③ PETTICOAT JUNCTION**
[COLOR] Crop duster Steve Elliott gets a good aerial view of the swimming Bradley girls—just before his plane crashes near Shady Rest. Steve: Mike Minor. Kate: Bea Benaderet. Uncle Joe: Edgar Buchanan. Betty Jo: Linda Kaye. Bobbi Jo: Lori Saunders. Billie Jo: Meredith MacRae. Major Corbett: Ed Deemer. Doc Stuart: George Chandler.
⑤ ALFRED HITCHCOCK—Drama
Detective Reardon vows to avenge the death of his son Phil, a rookie cop who was abducted and murdered by a gang of hoods. Reardon: Victor Jory. Philip Reardon: Peter Brown. (60 min.)
⑦ ⑧ LOVE ON A ROOFTOP
[COLOR] Dave and Julie are frantically trying to scrape some extra money: Dave's miserly boss has invited himself to dinner—and they haven't enough chairs in the apartment. Julie: Judy Carne. Dave: Peter Deuel. Stan: Rich Little. Carol: Barbara Bostock. Jim: Sandy Kenyon.

Guest Cast

Bert Bennington ....Charles Lane
Mrs. Bennington ....Hope Summers
⓭ STRUGGLE FOR PEACE
[SPECIAL] See Mon. 7:30 P.M. Ch. 13.
㉛ AMERICANS AT WORK
㊼ SPORTS QUIZ—Miranda
9:45 **㉛ NEWS**
10:00 **② ③ CBS REPORTS**
[SPECIAL] [COLOR] "The Poisoned Air," an examination of the growing problem of man-made air pollution, which costs the Nation an estimated $11 billion each year. CBS correspondent Daniel Schorr reports on the effects of air pollution on health, property, agriculture—and even the evolutionary process. Cam-

Wednesday October 19, 1966

Evening

engineer and the artistic ability of the architect, is discussed by Neal Mitchell, director of the Structures Workshop at Harvard. John Fitch is the host.

31 DESIGN DIMENSIONS

47 MARCELA—Serial

7:25 **47 NEWS—José Lanza**

7:30 **2 3 LOST IN SPACE**

COLOR "The Prisoners of Space." The Galaxy Tribunal of Justice notifies the space travelers that all—except Dr. Smith—will be tried for crimes committed in space. John: Guy Williams. Maureen: June Lockhart. Don: Mark Goddard. Smith: Jonathan Harris. Will: Billy Mumy. Judy: Marta Kristen. Penny: Angela Cartwright. (60 min.)

4 VIRGINIAN—Western

COLOR "The Challenge." Trampas, half conscious and badly injured, staggers to a near-by farm—where his loss of memory makes him a prime suspect in a recent murder and stage robbery. Trampas: Doug McClure. Virginian: James Drury. Grainger: Charles Bickford. Stacy: Don Quine. (90 min.)

Guest Cast

Ben Crayton	Dan Duryea
Jim Tyson	Don Galloway
Bobby Crayton	Michael Burns
Sarah Crayton	Barbara Anderson
Sheriff Milt Hayle	Ed Peck
Sam Fuller	Bing Russell
Hank Logan	Hal Bekar
Walt Sturgess	Grant Woods
Dr. Manning	Lew Brown
Marshal Coons	Clyde Howdy

5 COLOR TRUTH OR CONSEQUENCES—Quiz

7 8 BATMAN—Adventure

COLOR "An Egg Grows in Gotham" begins a two-part story. A new super-criminal, Egghead, has hatched a plan to get legal control of Gotham City. Batman: Adam West. Robin: Burt Ward. Gordon: Neil Hamilton. Aunt Harriet: Madge Blake. Alfred: Alan Napier.

Guest Cast

Egghead	Vincent Price
Chicken	Edward Everett Horton
Miss Bacon	Gail Hire
Foo Lung	Ben Welden
Benedict	Gene Dynarski
Tim Tyler	Steve Dunne
Pete Savage	Albert Carrier

9 MOVIE—Adventure

COLOR Million Dollar Movie: "The Far Horizons." See Mon. 7:30 P.M. Ch. 9.

11 HONEYMOONERS—Comedy

Ralph gets into a tangle with a wiseacre. Ralph: Jackie Gleason.

13 STRUGGLE FOR PEACE

What type of war might be fought by the U.S. and the USSR? This question is the starting point for an examination of the powers of NATO and the Warsaw Treaty Organization.

31 ON THE JOB—Fire Dept.

47 CARLOTA Y MAXIMILIANO

8:00 **5 UNTOUCHABLES—Drama**

Ness is not only beating the bootleggers —now he has joined them. He's supplying speakeasies with booze in an attempt to put mobster Jake Guzik out of business. Guzik: Nehemiah Persoff. Ness: Robert Stack. Moran: Harry Morgan. (60 min.)

7 8 MONROES—Western

COLOR "Ordeal by Hope." Little Twin is bitten by a rock chuck—an animal that Major Mapoy's men suspect is carrying rabies. Clayt: Michael Anderson Jr. Kathy: Barbara Hershey. Little Twin: Kevin Schultz. Big Twin: Keith Schultz. Amy: Tammy Locke. Jim: Ron Soble. Mapoy: Liam Sullivan. Sleeve: Ben Johnson. Ruel: Jim Westmoreland. (60 min.)

Guest Cast

Ferris	Edward Faulkner
Doctor	John Bryant
Corporal	Jack Williams

11 PATTY DUKE—Comedy

The girls want Cathy's father to write an autobiography—to show his ex-boss what a good man he lost. Patty and Cathy: Patty Duke. Martin and Kenneth William Schallert.

13 PLAY OF THE WEEK—Drama

"Burning Bright," by John Steinbeck. Circus acrobats Joe and Mordeen Saul are married and deeply in love, but are unable to have the child they both want. Out of love for her husband Mordeen conceives the child of Victor, the third member of the troupe. Myron McCormick, Colleen Dewhurst, Donald Madden, Crahan Denton. (Two hours)

31 JOURNEY THROUGH ARTS

"Graphic Designer."

47 WRESTLING—Newark

Thursday November 10, 1966

Evening

Robertson. Alice: Diane Brewster. Mr. Bolton: Richard Devon.

🄻 🄼 BEWITCHED—Comedy
[COLOR] When the lights go out in 12 Eastern states, the question is whether something went wrong with the generator at Niagara Falls or whether the whole blackout was the result of a goofed-up witch chant by bumbling Aunt Clara. Samantha: Elizabeth Montgomery. Darrin: Dick York. Aunt Clara: Marion Lorne. Abner: George Tobias. Gladys: Sandra Gould. Larry: David White.

Guest Cast
Ocky ....Reginald Owen
MacElroy ....Arte Julian

🄼 HAWAIIAN EYE—Mystery
Lopaka wonders how some obviously inept jewel thieves could have outsmarted guard Glen Thompson. Thompson: Dick Davalos. Adele Wafren: Betty Bruce. Lopaka: Robert Conrad. (60 min.)

🄼 WEDNESDAY REVIEW

🄼 TRIBUNA DEL PUEBLO

🄼 MYRTA SILVA—Variety

9:30 🄼 HERO—Comedy
[COLOR] Take one TV Western hero, one jealous wife, one new leading lady who just happens to be the hero's old girl friend, and one pair of handcuffs. Leave out the key to the handcuffs—and you've got trouble. Sam: Richard Mulligan. Ruth: Mariette Hartley. Dewey: Marc London. Angie Winters: Charlene Holt. Mike: Quinn Redeker.

🄼 OUTER LIMITS
Chino Rivera has one alternative to spending his life in prison—he can volunteer for the "inhabitant exchange" program being conducted with the planet Chromo. Rivera: Henry Silva. Julia Harrison: Diana Sands. (60 min.)

🄼 🄼 THAT GIRL—Comedy
[COLOR] Broadway star Sandy Stafford is delighted when her best friend Ann becomes her understudy. Then Sandy starts

having accidents—and second thoughts about her once best friend. Ann: Marlo Thomas. George: George Carlin. Judy Bonnie Scott. Sandy Stafford: Sally Kellerman. Jim: Robert Sampson.

Who is Marlo Thomas? Marlo wants to know too. See next week's TV GUIDE.

🄼 [COLOR] NFL GAME OF THE WEEK—Football

🄼 SPANISH NEWSREEL

9:45 🄼 NEWS

10:00 🄼 DEAN MARTIN—Variety
[COLOR] Guests: comedians Sid Caesar and Phyllis Diller, singer Diahann Carroll and the dancing Step Brothers, Ken Lane, Les Brown orchestra. (60 min.)

Highlights
"Falling in Love with Love," "Am I Blue?" "What Did I Have?"
....Diahann
"A Million and One" ....Dean
"A Hundred Years from Today"
..................................Diahann, Dean
"What Can I Say After I Say I'm Sorry?" ....Dean, Ken

🄼 🄼 HAWK—Drama
[COLOR] Stool pigeon Frankie Gellen, believing his life is in danger, has gone into hiding. Hawk's assignment: Find Gellen before the assassins reach him. Hawk: Burt Reynolds. Sam Crown: John Marley. (60 min.)

Guest Cast
Frankie Gellen ....Lou Antonio
Tommy ....Bernard Kates
Miro ....Carlos Montalban
Scaher ....Robert Burr
Anthony ....Michael Baselon
Chi Chi ....William Hickey

🄼 [COLOR] JETS HUDDLE

🄼 NAKED CITY—Drama
Dance hall hostess Ruby Redd reports that her co-worker has been missing three days. Vinnie: Dennis Hopper. Ruby Hilda Brawner. (60 min.)

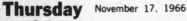

Thursday November 17, 1966
Evening

Bonaface ....Eduardo Ciannelli
Cesare Ferretti .... Titos Vandis

(4) DANIEL BOONE—Adventure
[COLOR] After giving Shawnee warrior Red Sky a rifle, Daniel faces a difficult task: He must dissuade Red Sky from his belief that the gun is magical—and more powerful than the law. Daniel: Fess Parker. Mingo: Ed Ames. Israel: Darby Hinton. (60 min.)

Guest Cast

Red Sky ....Michael Ansara
Jake Manning ....Robert Wilke

(5) [COLOR] TRUTH OR CONSEQUENCES—Game

(7) (8) BATMAN—Adventure
[COLOR] "The Joker's Proverbs," conclusion. The Joker threatens Gotham with a device that can stop, reverse or speed up time. Batman: Adam West. Robin: Burt Ward. Alfred/Egbert: Alan Napier.

Guest Cast

Joker ....Cesar Romero
Cornelia ....Kathy Kersh
Latch ....Louis Quinn
Bolt ....Larry Anthony

(9) MOVIE—Western
[COLOR] Million Dollar Movie: "The Unforgiven." See Mon. 7:30 P.M. Ch. 9.

(11) HONEYMOONERS—Comedy
Ralph appears in an amateur theatrical production. Ralph: Jackie Gleason.

(13) FRENCH CHEF—Cooking
Julia Child shows how to prepare quiches, fancy light custards.

(31) ON THE JOB—Fire Dept.

(47) MAXIMILIANO Y CARLOTA

8:00 (3) MR. ROBERTS—Comedy
[COLOR] The Reluctant's new chief warrant officer is getting everything he wants from the captain. Roberts: Roger Smith. CWO Al Briggs: Jack Carter.

(5) ALFRED HITCHCOCK—Drama
A lodge brother has been given only a short time to live, so the Grand Knight thinks now's the time to sell him a plot in the lodge's newly acquired cemetery. Clarence Weems: Russell Collins.

(7) (8) [COLOR] F TROOP—Comedy
[COLOR] Captain Parmenter's aggressive, husband-hunting sister breezes into the fort, where she sets her sights on bugler Dobbs. O'Rourke: Forrest Tucker.

Agarn: Larry Storch. Parmenter: Ken Berry. Dobbs: James Hampton. Wild Eagle: Frank de Kova. Crazy Cat: Don Diamond. Daphne: Patty Regan.

(11) MUNSTERS—Comedy
Eddie invites a schoolmate to spend the weekend with his family. Eddie: Butch Patrick. Lily: Yvonne DeCarlo.

(13) FOOD BUYER'S QUIZ
[SPECIAL] This audience-participation quiz is designed to help you learn more about food, for which the average American family spends about 20 percent of its weekly income. The home audience has the opportunity to answer questions concerning wise shopping. Celebrity panelists: Betty Furness, comic George Kirby, psychologist Joyce Brothers and saxophonist Boots Randolph. (60 min.)

(31) LATIN AMERICA—Education
"Francisco Pizarro and the Subjugation of the Incas." Prof. Ansel lectures.

(47) REVISTA DEL HOGAR

8:30 (2) (4) MY THREE SONS
[COLOR] The everyday routine in the Douglas household falls by the wayside when the family gets wrapped up in Robbie's successful new business venture—baking and selling birthday cakes. Robbie: Don Grady. Steve: Fred MacMurray. Charley: William Demarest. Chip: Stanley Livingston. Ernie: Barry Livingston. Welch: John Howard.

(4) STAR TREK—Adventure
[COLOR] In the first of a two-part story, Spock inexplicably abducts crippled Captain Pike, the Enterprise's former commander, locks the spaceship on a course for the only forbidden planet in the galaxy—and then turns himself in for court-martial. Spock: Leonard Nimoy. Kirk: William Shatner. McCoy: DeForest Kelley. (60 min.)

Guest Cast

Captain Pike ....Jeffrey Hunter
Vina ....Susan Oliver
Commander Mendez ....Malachi Throne

(5) BRANDED—Western
[COLOR] President Grant asks McCord to infiltrate a group of men rumored to be plotting against Grant's life. Chuck Connors, William Bryant.

(7) (8) [COLOR] DATING GAME

(11) HONEY WEST—Mystery
Honey, hired to find a missing girl,

Friday November 18, 1966

Evening

6:40 **⑧** COLOR **WEATHER—Francis**

6:45 **⑧** **NEWS—Peter Jennings**

㉛ **NEWS AND WEATHER**

7:00 **②** COLOR **NEWS—Cronkite**

③ **DEATH VALLEY DAYS**
COLOR A food-loving miner offers a chef partnership in a mine—in return for his culinary services. Paul Brinegar.

④ COLOR **NEWS—Chet Huntley, David Brinkley**

⑤ **McHALE'S NAVY—Comedy**
McHale and his crew are jailed during a visit to a French-owned island. McHale: Ernest Borgnine. Big Frenchy: George Kennedy. Binghamton: Joe Flynn.

⑦ **MOVIE—Drama**
"Sweet Smell of Success." (1957) Broadway columnist J. J. Hunsecker decides to use his influence to terminate his sister's romance with a musician. Burt Lancaster, Tony Curtis. (Two hours)

⑨ COLOR **SUB-MARINER**

⑪ **ZORRO—Adventure**
Ricardo del Amo plays so many pranks that no one believes his story of over-hearing a payroll-robbery plot. Diego: Guy Williams. Ricardo: Richard Anderson.

⑬ **MASTER CLASS—Segovia**
Segovia supervises performances of a gigue by Sylvius Leopold Weiss, an allemande by John Dowland, and "Zambra Granadina," by Isaac Albéniz.

㉛ **FILM FEATURE**

7:25 **㊼** **NEWS—José Lanza**

7:30 **② ③** **WILD WILD WEST**
COLOR The mad Dr. Loveless again tries to establish a Utopia of evil. This time, he promises a life without toil to a tribe of starving Indians, provided they obey his orders—and kill James West. West: Robert Conrad. Artemus: Ross Martin. (60 min.)

Guest Cast

Dr. Miguelito Loveless ....Michael Dunn
Bright Star ....Anthony Caruso
Old Chief ....Paul Fix

④ **TARZAN—Adventure**
COLOR Tarzan races to recover a serum stolen by a native chief. The drug is the only hope for the fever-wracked Jai, who has been bitten by a jaguar and is near death. Tarzan: Ron Ely. Jai: Manuel Padilla Jr. (60 min.)

Guest Cast

Dr. Haru ....Nobu McCarthy
Bwanichi ....Joel Fluellen
Totoni ....Chuck Wood

⑤ COLOR **TRUTH OR CONSE-QUENCES—Quiz**

⑦ **GREEN HORNET—Adventure**
COLOR The Green Hornet and Kato attempt to smash a vicious Chinatown protection racket. Hornet: Van Williams. Kato: Bruce Lee.

Guest Cast

Low Sing ....Mako
Wing Ho ....Allen Jung
Duke Slate ....Tom Drake

⑨ **MOVIE—Western**
COLOR Million Dollar Movie: "The Unforgiven." See Mon. 7:30 P.M. Ch. 9.

⑪ **HONEYMOONERS—Comedy**
Alice and Trixie try to share their husband's interests. Alice: Audrey Meadows. Trixie: Joyce Randolph.

⑬ **ART OF FILM—Discussion**
Stanley Kauffmann reviews new films.

㉛ **BROOKLYN COLLEGE**
"American Poets," Part 1. Edgar Allan Poe is discussed.

㊼ **MAXIMILIANO Y CARLOTA**

8:00 **⑤** **MOVIE—Drama**
"Double Indemnity." (1944) Walter Neff runs into Phyllis Dietrichson when he goes to sell some insurance. He also runs into her little scheme to kill her husband for insurance money. Fred MacMurray, Barbara Stanwyck. (Two hours)

⑦ **TIME TUNNEL—Adventure**
COLOR Tony and Doug find themselves in Paris during the French Revolution's Reign of Terror. Anxious to escape the death-ridden city, the time travelers agree to aid the imprisoned Marie Antoinette and her son in exchange for safe-conduct passes. Tony: James Darren. Doug: Robert Colbert. Kirk/Querque: Whit Bissell. Swain: John Zaremba. Ann: Lee Meriwether. (60 min.)

Guest Cast

Shopkeeper ....David Opatoshu
Marie Antoinette ....Monique Lemaire
Simon ....Louis Mercier

⑪ **PATTY DUKE—Comedy**
A doctor has traced Patty's allergy to its source—Cathy. Patty and Cathy: Patty Duke. Ted: Skip Hinnant.

Monday December 5, 1966

Evening

8:30 🄺 **LUCILLE BALL—Comedy**

COLOR Lucy seeks the help of a psychiatrist. The befuddled redhead swears that banker Mooney keeps changing from a man into a monkey . . . then back to Mooney . . . then back to a monkey! Mooney: Gale Gordon.

Guest Cast

Bob Bailey .... Hal March
Dr. Parker .... Lew Parker
Monkey .... Janos Prohaska

🄸 **ROGER MILLER—Variety**

COLOR Roger's guests are French actor-singer Charles Aznavour and the singing-dancing Doodletown Pipers.

Highlights

"Happy Anniversary," "You've Let
  Yourself Go," "Reste" ...... Charles
"Don't Rain on My Parade" ..... Pipers
"Home," "The Moon Is High, and So
  Am I," "Chug-A-Lug" .... Roger

🄶 🄷 **RAT PATROL—Drama**

COLOR A French traitor endangers Troy, his men and their attempt to help a French general escape from North Africa. Troy: Christopher George. Moffitt: Gary Raymond. Hitchcock: Lawrence Casey. Pettigrew: Justin Tarr.

Guest Cast

Mathias .... Emile Genest
Violette .... Monique Lemaire

See the article on page 14 for a background story on this series.

Cleveland Amory reviews this series in next week's TV GUIDE.

⑪ **DR. KILDARE—Drama**

Death by suffocation was the coroner's ruling when Kildare's friends Gordon and Ellen lost their son. Ellen: Joan Freeman. Henry Harris: Ed Begley. Gordon: Dick Davalos. (60 min.)

⑬ **ART OF FILM—Discussion**

Stanley Kauffmann's guest is actress Angela Lansbury, currently appearing in "Mame."

Repeated Friday at 7:30 P.M.

㉛ **EYE ON THE UNIVERSE**

Three types of stars producing varying amounts of light are described.

㊼ **EVA AND RAUL—Variety**

9:00 🄺 🄰 **ANDY GRIFFITH—Comedy**

COLOR Opie tries to make amends after ruining Aunt Bee's hybrid rose, which she was certain would finally top

Clara's entry in the annual flower show. Aunt Bee: Francis Bavier. Opie: Ronny Howard. Clara: Hope Summers. Simmons: Richard Collier. Tillie: Maxine Seman.

🄸 **ROAD WEST—Western**

COLOR Susan Douglass, wife of an Indian named Red Eagle, plans to raise her child in the white man's world, despite her relatives' bigotry, and threats from Red Eagle—who demands the return of his son. Ben: Barry Sullivan. Tim: Andrew Prine. Midge: Brenda Scott. Chance: Glenn Corbett. (60 min.)

Guest Cast

Susan Douglass .... Barbara Anderson
Red Eagle .... Donnelly Rhodes
Oliver .... Tom Drake
Mrs. Oliver .... Phyllis Hill

🄴 **MOVIE—Comedy**

"Julia Misbehaves." (1948) Attempting to get her daughter happily married, an English music-hall actress becomes involved in a number of escapades. Greer Garson, Elizabeth Taylor. (Two hours)

🄶 🄷 **FELONY SQUAD—Drama**

COLOR "The Killer Instinct." Sam plays a hunch to prove that the tough-talking owner of a professional football team committed murder, and framed a washed-up grid star to take the rap. Sam: Howard Duff. Jim: Dennis Cole.

Guest Cast

Roy Madden .... William Smithers
Bull Bradovich .... Edward Asner

⑬ **NY TELEVISION THEATER**

"No Why" is given to the small boy who is made to apologize in this play about the destruction of innocence by John Whiting. Elinor: Teresa Hughes. Jacob: Joseph Lamberta. Max: Tom Ligon. A second play, not announced at press time, will also be presented. (60 min.)

㉛ **FOCUS ON BOOKS—Discussion**

See Sunday 7:30 P.M. Ch. 31.

㊼ **MYRTA SILVA—Variety**

9:30 🄺 🄰 **FAMILY AFFAIR—Comedy**

COLOR Jody, feeling a little insecure, deliberately misbehaves to find out if Uncle Bill loves him enough to punish him. Bill: Brian Keith. Jody: Johnnie Whitaker. French: Sebastian Cabot. Buffy: Anissa Jones. Cissy: Kathy Garver.

🄶 🄷 **PEYTON PLACE—Serial**

COLOR Rachel tells where she found

Sunday
Afternoon
January 15, 1967

TV CLOSE-UP GUIDE

4:00 **2** **3** **4** AFL-NFL CHAMPIONSHIP

SPECIAL **COLOR** Who's No. 1? The Green Bay Packers and the Kansas City Chiefs settle the issue in the first Super Bowl championship at Los Angeles. The game also marks the first meeting between AFL and NFL teams.

The Packers, 34-27 victors over the Dallas Cowboys in the NFL title clash, combine a balanced attack with a solid clutch defense. Coach Vince Lombardi's philosophy: simplicity in formations, and perfection in execution. The club moves behind the quarterbacking of Bart Starr,

the NFL's MVP and leading passer. On the ground, Jim Taylor and Elijah Pitts grind out the yardage via the air, Starr favors Carroll Dale and Boyd Dowler.

The Chiefs, coached by Hank Stram, coasted to a 31-7 win over the Buffalo Bills in the AFL title game. During the season, they put a league-leading 448 points on the scoreboard behind Len Dawson's aerials to Otis Taylor and Chris Burford, and the rushing of Curtis McClinton, Bert Coan and rookie Mike Garrett. (Live)

CHIEFS

10 Beathard	...QB	58 Rice	....DT
14 Ply	...........DB	60 Reynolds	......G
15 Mercer	.......K	61 Biodrowski	...G
16 Dawson	...QB	64 Merz	...........G
17 Smith	......DB	65 Gilliam	C
18 Thomas, E.	...DB	66 Frazier	C
20 Hunt	........DB	69 Headrick	...LB
21 Garrett	.......B	71 Budde	......G
22 Mitchell	...DB	72 DiMidio	......T
23 Coan	.........B	73 Hill	...........T
24 Williamson	..DB	75 Mays	.......DE
25 Pitts	........E	77 Tyrer	.......T
32 McClinton	..FB	78 Bell	..........LB
35 Stover	.....LB	80 Carolan	.......E
42 Robinson	...DB	84 Arbanas	......E
44 Wilson	......FB	85 Hurston	...DE
45 Thomas, G.	...B	86 Buchanan	...T
52 Abell	.......LB	87 Brown	......DE
55 Holub	......LB	88 Burford	.......E
56 Corey	........LB	89 Taylor	........FL

PACKERS

5 Hornung	....B	63 Thurston	......G
12 Bratkowski	..QB	64 Kramer	.......G
15 Starr	.....QB	66 Nitschke	...LB
21 Jeter	......DB	68 Gillingham	...G
22 Pitts	........B	72 Wright	.........T
24 Wood	......DB	73 Weatherwax	...T
26 Adderley	...DB	74 Jordan	........T
27 Mack	.......FL	75 Gregg	......G-T
31 Taylor	......FB	76 Skoronski	....T
33 Grabowski	..FB	77 Kostelnik	......T
34 Chandler	....K	78 Brown, B.	...DE
37 Vandersea	..LB	80 Long	...........FL
40 Brown, T.	...DB	81 Fleming	......E
43 Hart	........DB	82 Aldridge	...DE
44 Anderson, D.	..B	84 Dale	.........FL
45 Hathcock	...B	85 McGee	.......E
50 Curry	C	86 Dowler	......E
56 Crutcher	...LB	87 Davis	........DE
57 Bowman	C	88 Anderson, B.	..E
60 Caffey	......LB	89 Robinson	...LB

August 29, 1967 **Tuesday**

Evening

cinity. Mickey Rooney, Robert Strauss, Bill Goodwin, Hal March. (90 min.)

11:35 ④⑦ NEWS—Arturo Rodriguez

12:00 ⑨ MOVIE—Biography

[COLOR] Time approximate. "The Eddie Cantor Story." (1954) The life of the famous entertainer with the banjo eyes, from the time he was a young boy mixed up with a gang of toughs on New York City's East Side. Eddie Cantor: Keefe Brasselle. Ida: Marilyn Erskine. Grandma Esther: Aline MacMahon. Harry Harris: Arthur Franz. Rocky: Gerald Mohr. Gus Edwards: Hal March. Will Rogers: Will Rogers Jr. (Two hours)

12:45 ⑤ NEWS

1:00 ④ [COLOR] NEWS—Bob Teague

⑦ ⑧ **NEWS**

1:05 ⑦ MOVIE—Mystery

Best of Broadway: "Poison Ivy." (French; 1953) In Tangiers, an FBI agent learns of a plot to hijack a shipment of gold. Eddie Constantine, Dominique Wilms, Howard Vernon. (One hour, 45 min.)

1:10 ③ NEWS AND WEATHER

1:15 ④ MOVIE—Drama

"Sunday Dinner for a Soldier." (1944) Nostalgic World War II story about an impoverished family who scrimp so they can invite a soldier to dinner. Tessa: Anne Baxter. Eric Moore: John Hodiak. Grandfather: Charles Winninger. Agatha: Anne Revere. (One hour, 25 min.)

1:35 ② NEWS

1:40 ② MOVIE—Musical

Time approximate. Late Late Show: "Sound Off." (1952) A conceited night-club entertainer is drafted. Mickey Rooney, Anne James, John Archer, Sammy White, Wally Cassell, Arthur Space, Pat Williams. (One hour, 40 min.)

2:00 ⑨ NEWS AND WEATHER

3:20 ② MOVIE—Drama

Time approximate. "Up Periscope" (1959) During World War II, a recent graduate of the underwater Demolition School is assigned to a secret mission. James Garner, Edmond O'Brien, Andra Martin, Alan Hale, Frank Gifford, Edward Byrnes. (Two hours, 20 min.)

TV CLOSE-UP GUIDE

10:00 ⑦ ⑧ FUGITIVE—Drama

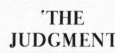

'THE JUDGMENT'

The Fugitive   The One-armed Man

[COLOR] Who murdered Helen Kimble? Was it the fugitive, the mysterious one-armed man—or someone else?

Tonight, viewers learn whodunit as the conclusion of this two-part drama brings to an end the series' four-year nighttime run.

The elusive Kimble returns to the small Indiana town where it all began. There, a flashback takes us to the moment in time when Helen Kimble faced her murderer.

Don Medford directed from a script by George Eckstein and Michael Zagor. Kimble: David Janssen. Gerard: Barry Morse. Johnson: Bill Raisch. (60 min.)

Guest Cast

Chandler	J. D. Cannon
Jean Carlisle	Diane Baker
Donna Taft	Jacqueline Scott
Len Taft	Richard Anderson
Betsy Chandler	Louise Latham
Helen Kimble	Diane Brewster

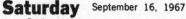

Saturday September 16, 1967

Evening

husbands die abruptly—leaving her millions. Screenplay by Betty Comden and Adolph Green. (Two hours, 15 min.)

Cast

Louisa	Shirley MacLaine
Larry Flint	Paul Newman
Rod Anderson	Robert Mitchum
Leonard Crawley	Dean Martin
Jerry Benson	Gene Kelly
Dr. Steffanson	Bob Cummings
Edgar Hopper	Dick Van Dyke
Painter	Reginald Gardiner
Mrs. Foster	Margaret Dumont

31 **BIG PICTURE—Army**

"The New First Team," a report on the First Cavalry Division (Airmobile). Since 1965, the combat group has used helicopters to fight in the jungles of Vietnam.

47 **MOVIE—Drama**

"Prisioneros de la Tierra." (Argentine; 1950) In Spanish with Francisco Petrone, Angel Magana and Elisa Galvez. (90 min.)

9:30 **2** **3** **PETTICOAT JUNCTION**

[COLOR] Kate's mettle is tested on two fronts: Billie Jo gives an eyebrow-raising night-club audition, and Betty Jo makes a stomach-wrenching debut as a cook. Kate: Bea Benaderet. Billie Jo: Meredith MacRae. Betty Jo: Linda Kaye. Steve: Mike Minor. Bobbi Jo: Lori Saunders. Barney: Herb Vigran.

7 **8** **IRON HORSE—Western**

[COLOR] "Iron Horse" begins its second season (new day and time) with a venture in horse racing. It all comes about when three con artists sucker Ben into buying a share of Diablo, the speediest thoroughbred in the West. Ben: Dale Robertson. Dave: Gary Collins. Barnabas: Bob Random. (60 min.)

Guest Cast

Applegate	Strother Martin
Fitzpatrick	Lloyd Gough
Reese	Martin Brooks
Luft	Harry Raybould
Mayor	Forrest Lewis
Jessup	Jim Nolan
Turkey	Kay E. Kuter

31 **FILM FEATURE**

Mike Connors

MANNIX—Mystery

10:00 **2** **3**
The Name Is Mannix

[DEBUT] [COLOR] This series focuses on the adventures of Joe Mannix, a tough, independent private-eye who works for Intertect, an ultra-modern, computerized detective agency.

Tonight, Mannix tries to rescue the kidnaped daughter of former rackets boss Sam Dubrio. Mannix bets that he can find the girl and return her safely, if he can persuade her father to risk the $500,000 ransom—plus another $500,000 to bribe the kidnapers' pick-up man.

Script by executive producer Bruce Geller, the Emmy-winning producer of "Mission: Impossible." Directed by Leonard J. Horn. Mike Connors stars as Mannix. Joe Campanella plays his boss, Lew Wickersham. (60 min.)

Guest Cast

Sam Dubrio	Lloyd Nolan
Louise Dubrio	Kim Hunter
Eddie	John Colicos
Angela	Barbara Anderson

Tuesday October 10, 1967

Evening

to sell his prize possession—a fishing boat. His plans to win it back are complicated by a persistent girl and a hard-bargaining opportunist. Songs include "Return to Sender." (Two hours)

Cast

Ross Carpenter	Elvis Presley
Robin Gantner	Stella Stevens
Laurel Dodge	Laurel Goodwin
Wesley Johnson	Jeremy Slate
Chen Yung	Guy Lee
Kin Yung	Benson Fong
Mme. Yung	Beulah Quo

⓫ PERRY MASON—Mystery
"The Lonely Eloper." Merle Telford is planning to elope on her 21st birthday. Merle: Jana Taylor. Danny Pierce: Jack Ging. Mason: Raymond Burr. Della: Barbara Hale. (60 min.)

⓭ ACTORS COMPANY—Drama
In the third rehearsal of "The Winter's Tale," the company works on the blocking of one of the most difficult scenes in Act IV. Directed by Barry Boys. Players include Stacy Keach, Michael Kermoyan, Earle Hyman and Colgate Salsbury. (60 min.)

9:30 ❷ ⑶ GOOD MORNING WORLD
[COLOR] While hosting a telethon, the disc jockeys learn a well-kept secret about their stuffy boss: Hutton was once a vaudeville headliner. Billy De Wolfe (Hutton) does a routine with impressions of a chorus line, a snake charmer and a monster. Dave: Joby Baker. Larry: Ronnie Schell. Linda: Julie Parrish.

Guest Cast

Jackie Sullivan	Paul Gilbert
Ernie	A. G. Vitanza
Congressman Zukor	Remo Pisani

❼ ⑻ N.Y.P.D.—Drama
[COLOR] "Money Man." Detectives Corso and Ward go undercover as laborers to investigate a series of construction-site accidents. Script by Edward Adler. Corso: Frank Converse. Ward: Robert Hooks. Haines: Jack Warden.

Guest Cast

Uncle Cheech	Vincent Gardenia
Joey	Charles Grodin
Kiddy Keil	Ralph Dunn
Head Tough	John Ryan

㉛ FILM SHORT

9:45 ㉛ NEWS—Herbert Boland

10:00 ❷ ⑶ CBS NEWS SPECIAL
[SPECIAL] [COLOR] Scheduled: "Barry Goldwater's Arizona." For details, see the Close-up on page A-57. (60 min.)
May be pre-empted by a late-news show.

❺ [COLOR] NEWS—Bill Jorgensen

❼ ⑻ HOLLYWOOD PALACE
[COLOR] Milton Berle hosts an all-comedy show with guests Kaye Ballard of "The Mothers-in-Law," Joe Besser, Irving Benson, Prof. Irwin Corey and the Bottoms Up troupe. (60 min.)
If the NABET strike prevents the taping of this program, it will be replaced by a "Palace" rerun hosted by Phil Silvers.

❾ OUTRAGEOUS OPINIONS
[COLOR] Actor Ossie Davis and comedian-playwright Woody Allen, who talks about his childhood, guest. (60 min.)

⓫ PAT BOONE—Variety
[COLOR] Guests include Sammy Davis Jr., singer Vikki Carr, Hollywood columnist Army Archerd and comedian Jan Murray. (90 min.)

⓭ NEWSFRONT—Mitchell Krauss

㉛ REPORT TO THE PHYSICIAN

㊼ SPANISH DRAMA—Serial

10:25 ㊼ WEATHER—José I. Lanza

10:30 ❺ [COLOR] ALAN BURKE

㉛ AMERICAN PRINTMAKERS
Harold Altman shows his traditional landscapes and figures.

㊼ NEWS—Arturo Rodriguez

10:35 ㊼ PUMAREJO—Variety

11:00 ❷ [COLOR] NEWS—Tom Dunn
⑶ [COLOR] NEWS AND SPORTS
❹ [COLOR] NEWS—Jim Hartz
❼ [COLOR] NEWS—Bill Beutel
⑻ [COLOR] NEWS—Bob Norman
❾ MOVIE—Drama
"Conspirator." (1949) A young girl weds a man, not knowing he is a Communist. Filmed in England. Robert Taylor, Elizabeth Taylor, Harold Warrender. (One hour, 45 min.)

⓭ NET JOURNAL—Report
See Mon. 9 P.M. Ch. 13. (60 min.)

11:10 ❹ [COLOR] WEATHER—Frank Field
❹ [COLOR] WEATHER—Antoine

11:15 ❹ [COLOR] NEWS—Jim Hartz
❺ [COLOR] WOODY WOODBURY

Wednesday November 8, 1967

Evening

es' courtship features flashback scenes of Oliver and Lisa's first meeting in wartime Hungary, and later encounters in Paris and New York. Oliver: Eddie Albert. Lisa: Eva Gabor. Sam: Frank Cady. Kimball: Alvy Moore.

④ ㉚ BOB HOPE—Variety
[SPECIAL] [COLOR] Bob's guests include singer Bobbie Gentry. See the Close-up on page A-64. (60 min.)

"Kraft Music Hall" will not be seen.

⑦ ⑧ MOVIE—Drama
[COLOR] "Where Love Has Gone" (1964), an adaptation of Harold Robbins' novel about a domineering mother who seizes every opportunity to dictate her daughter's life. (Two hours)

Cast
Valerie Hayden ....Susan Hayward
Mrs. Gerald Hayden ....Bette Davis
Luke Miller ....Michael Connors
Dani ....Joey Heatherton
Marian Spicer ....Jane Greer
Sam Corwin ....DeForest Kelley
Gordon Harris ....George Macready
Dr. Sally Jennings ....Anne Seymour
Judge Murphy ....Willis Bouchey

⑪ PERRY MASON—Mystery
"The Reluctant Model." After millionaire Otto Olney buys a masterpiece, art dealer Colin Durant starts a rumor that the painting is a forgery. Olney: John Larkin. Durant: John Dall. Grace Olney: Joanna Moore. (60 min.)

⑬ CREATIVE PERSON—Arden
A profile of the British playwright John Arden, a "new Elizabethan" dramatist who believes in "relating one's work to the sort of life one leads outside it." Films show Arden working on a book at his home in Kirby Moorside, Yorkshire, and rehearsing a play he wrote for amateur groups.

9:30 ② ③ HE & SHE—Comedy
[COLOR] Paula immerses herself in the problems besetting Dick, who is fresh out of ideas for his comic strip—and facing a plagiarism suit from a rival cartoonist. Paula: Paula Prentiss. Dick: Richard Benjamin. Jerry Sargent: Hal Buckley. Mrs. O'Connor: Jane Dulo.

⑬ U.S.A.—Art
Cameras examine the work of Dutch-born artist Willem de Kooning, a leader of the

abstract - expressionist movement. De Kooning discusses his paintings as films (taken over the past 10 years) show the evolution of his slashing style.

㉛ ITALIAN PANORAMA
9:45 ㉛ NEWS—Herbert Boland
10:00 ② ③ DUNDEE AND THE CULHANE—Western
[COLOR] British character actor George Coulouris makes a rare TV appearance in "The 3:10 to a Lynching Brief." A poker game aboard a westbound train ends in murder. Dundee is railroaded into defending the chief suspect, an Irish friend of Culhane's. Dundee: John Mills. Culhane: Sean Garrison. (60 min.)

Guest Cast
Jeremiah Scrubbs ....George Coulouris
Dugan ....Larry Perkins
Conductor ....Dub Taylor
Henry Taylor ....Lonny Chapman
Jesus Padilla ....Joaquin Martinez

④ ㉚ RUN FOR YOUR LIFE
[COLOR] In the tiny country of Andorra, Paul and his friend Ramon da Vega risk a prison sentence—and their hides—to kidnap the long-lost daughter of an American millionaire. Paul: Ben Gazzara. (60 min.)

Guest Cast
Ramon da Vega ....Fernando Lamas
Susanne ....Letitia Roman
Andrew Dawson ....Edward Andrews
Eduardo Alonzo ....Joe De Santis
Ignacio Romero ....Frank Puglia

⑤ [COLOR] NEWS—Bill Jorgensen
⑪ PAT BOONE
[COLOR] Guests are Dorothy Lamour, Ronnie Schell of "Good Morning World," Israeli singer Ron Eliron and comic Marty Ingels. (90 min.)

⑬ NEWSFRONT—Mitchell Krauss
㉛ INTERNATIONAL INTERVIEW
㊼ SPANISH DRAMA—Serial
10:25 ㊼ WEATHER—José I. Lanza
10:30 ⑤ [COLOR] ALAN BURKE
㉛ SURVEY OF THE ARTS
"Contemporary Italian Poetry."
㊼ NEWS—Arturo Rodriguez
10:35 ㊼ PUMAREJO—Variety
10:55 ⑨ [COLOR] OUTDOOR WORLD
11:00 ② [COLOR] NEWS—Tom Dunn

Thursday December 21, 1967

Evening

⑬ FRENCH CHEF—Cooking
"Hot Turkey Ballottine." Julia Childs shows how to bone a turkey.

㉛ CONTINENTAL COMMENT

㊼ COOKING SCHOOL

7:30 ② ③ CIMARRON STRIP
[COLOR] Bo Woodard hires on as Marshal Crown's deputy, with plans to use his badge as a shield for murder. The targets of his revenge: former cohorts who left him for dead after a payroll robbery. Crown: Stuart Whitman. MacGregor: Percy Herbert. (90 min.)

Guest Cast

Bo Woodard	J. D. Cannon
Tate	Lyle Bettger
Zena	Marj Dusay
Rapp	Larry Pennell
Benji	Anthony James

④ ⑳ DANIEL BOONE—Adventure
[COLOR] An unusually large amount of fan mail praising guest star Neville Brand prompted the repeat showing of this story about a deranged outcast who seizes Israel. Daniel tries desperately to find the pair before a pursuing mob of settlers frighten the outcast into harming the boy. Daniel: Fess Parker. Israel: Darby Hinton. Rebecca: Patricia Blair. Tanner: Neville Brand. (Rerun; 60 min.)

⑤ [COLOR] TRUTH OR CONSEQUENCES—Game

⑦ ⑧ BATMAN—Adventure
[COLOR] Vincent Price and Anne Baxter make return appearances as Egghead and Olga, who thunder into Gotham City with their plundering Cossack legions. Batman: Adam West. Robin: Burt Ward. Batgirl: Yvonne Craig. Commissioner: Neil Hamilton. Alfred: Alan Napier.

⑬ COMMUNICATIONS AND EDUCATION—Lecture
"Growing Up in America." Prof. Siepmann discusses the American educational process with Dr. Edgar Friedenberg, sociologist and educator at the University of Buffalo.

㉛ ON THE JOB—Fire Dept.

㊼ BATTLE OF KNOWLEDGE

7.55 ㊼ SPORTS—Fausto Miranda

8:00 ⑤ HAZEL—Comedy
[COLOR] Golf pro Tony Lema visits the Baxters, and Hazel manages to lose his clubs. Hazel: Shirley Booth. Tony Lema:

Himself. George: Don DeFore. Dorothy Whitney Blake. Policeman: Hal Baylor. Mrs. "Baby" Gollard: Kathie Browne.

⑦ ⑧ FLYING NUN—Comedy
[COLOR] A white Christmas, courtesy of Sr. Bertrille, makes a homesick Norwegian nun very happy, but threatens San Juan's tourist trade and leaves the economy—and the weatherman—in shreds. Sr. Bertrille: Sally Field. Carlos: Alejandro Rey. Sr. Sixto: Shelley Morrison. Mother Superior: Madeleine Sherwood. Sr. Jacqueline: Marge Redmond. Weatherman: Woodrow Parfrey. Sr. Olaf: Celia Lovsky.

⑨ MOVIE—Drama
Million Dollar Movie: "Captain China." See Sun. 4 P.M. Ch. 9. (Two hours)

⑪ PASSWORD—Game
[COLOR] Guests are Sheila MacRae and George Grizzard. Host: Allen Ludden.

⑬ TONIGHT IN PERSON
In this first of two programs, the folksinging Trio Los Paraguayos offers a concert of Latin-American tunes. Selections include "La Bamba" and "Noche del Paraguay."

㉛ NAVY FILM FEATURE

㊼ MIGUELITO VALDES—Variety

8:30 ④ ⑳ IRONSIDE—Drama
[COLOR] "Girl in the Night." Sgt. Ed Brown is mugged after a date with a troubled night-club singer. Probing the girl's past, Ironside, Brown and company slowly unravel a tawdry skein of coercion, betrayal—and murder. Ironside: Raymond Burr. Brown: Don Galloway. Mark: Don Mitchell. Eve: Barbara Anderson. (60 min.)

Guest Cast

Elaine Moreau	Susan Saint James
Joe Varona	Donnelly Rhodes
Johnny Foster	Steve Carlson
Stulka	George Keymas

⑤ [COLOR] MERV GRIFFIN

⑦ ⑧ BEWITCHED—Comedy
[COLOR] Echoes of "A Christmas Carol" ring clear when a Scrooge-like client of Darrin's intrudes on Christmas Eve. Samantha: Elizabeth Montgomery. Darrin: Dick York. Larry: David White.

Guest Cast

Mortimer	Charles Lane
Santa Claus	Don Beddoe

TUESDAY, FEBRUARY 20, 1968

an English lawyer who is handling Tony's inheritance, a quaint old house in England. On an inspection trip, Tony, Jeannie and Roger face a harrowing night with some not-so-quaint old ghosts. Tony: Larry Hagman. Jeannie: Barbara Eden. Roger: Bill Daily.

Guest Cast

James Ashley .... Jack Carter
Willingham .... Ronald Long
Chauncey .... Leslie Randall

5 [COLOR] **TRUTH OR CONSEQUENCES—Game**

7 **8** **GARRISON'S GORILLAS**
[COLOR] "The Plot to Kill," conclusion of a drama about a scheme to assassinate Hitler. The success of the mission hinges on Garrison's ability to train a New York gangster to impersonate a German field marshal. Garrison: Ron Harper. Actor: Cesare Danova. Casino: Rudy Solari. Goniff: Christopher Cary. Chief: Brendon Boone. (60 min.)

Guest Cast

Frank Keeler .... Richard Kiley
Carla .... Faith Domergue

9 **OUTRAGEOUS OPINIONS**
[COLOR] Scheduled guest is TV personality Ed McMahon.

13 **COMMUNICATIONS AND EDUCATION—Lecture**
"Broadcasting: Giving the Public What It Wants." Prof. Siepmann presents his idea of an equitable program service.

31 **HUMAN RIGHTS FORUM**
See Sun. 6 P.M. Ch. 31.

47 **SPANISH DRAMA—Serial**

7:55 **47** **SPORTS—Fausto Miranda**

8:00 **4** **20** **JERRY LEWIS—Variety**
[COLOR] Tony Randall and singer Nancy Ames visit Jerry. Sketches: 1. Tony emcees a TV quiz show that transforms two lovers (Jerry and Nancy) into vicious, greedy contestants. 2. Jerry clerks in a store where the salesman is always wrong. 3. The Nutty Professor (Jerry) lectures on dancing history. Nick Castle dancers, George Wyle singers, Lou Brown orchestra. (60 min.)

Highlights

"Dear Hearts and Gentle People," "See

CLOSE-UP 9:00 **4** **20** MOVIE—Suspense

Prescription: Murder

[COLOR] Gene Barry and Peter Falk star in a suspense movie made especially for TV. Dr. Ray Flemming, a successful psychiatrist, thinks he can get away with murder. With a young actress as his accomplice, Flemming ends his 10-year marriage by killing his wife.
Flemming hadn't counted on Lieutenant Columbo, a rumpled, cigar-chewing detective who knows a little psychology himself. Playing a skillful game of cat-and-mouse, Columbo conducts a deceptively casual investigation that is calculated to catch Flemming off guard.
Script by Richard Levinson and William Link, who created TV's "Mannix." (Two hours)

Cast

Lieutenant Columbo .... Peter Falk
Dr. Ray Flemming .... Gene Barry
Carol Flemming .... Nina Foch
Joan Hudson .... Katherine Justice
Burt Gordon .... William Windom

Peter Falk

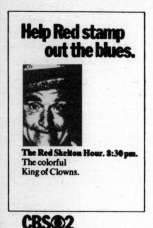
Tuesday
Evening

teed annual income. Through a state-county arrangement, the program provides funds to supplement salaries of low-income families.

31 HUMAN RIGHTS FORUM
47 COLOR MIGUELITO VALDES

8:00 **5 PAY CARDS!—Game**
COLOR Celebrity player: comedienne Joanne Worley. Host: Art James.

9 STEVE ALLEN—Variety
COLOR Guests include comic professor Irwin Corey, jazz vibraphonist Terry Gibbs, former football star Roosevelt Grier, comedian Ed Antak and Jayne Meadows. Sketch: Jayne portrays an overprotective secretary. (90 min.)

11 RUN FOR YOUR LIFE—Drama
COLOR Bryan goes behind the Iron Curtain as navigator to an American race driver. Ben Gazzara. (60 min.)

13 NET PLAYHOUSE—Biography
"Victoria Regina: Autumn." Prime Minister Disraeli clashes with the Queen in this third of a four-part adaptation of

CLOSE-UP

DORIS DAY
9:30 2 3

DEBUT COLOR Doris Day, a longtime TV holdout, makes her series debut. The formula involves an occasional song and lots of family-style comedy.

Young widow Doris Martin leaves the city to raise her two young sons on her father's ranch. The newcomers will be sharing the chores with housekeeper Aggie and a mildly inefficient hired hand.

Tonight: the Martin boys take their mom out to dinner as a surprise birthday treat. The real surprises begin when they arrive at the eatery—a rowdy roadhouse.

Doris Martin: Doris Day. Buck Webb: Denver Pyle. Aggie: Fran Ryan. Leroy B. Simpson: James Hampton. Billy: Philip Brown. Toby: Tod Starke. Nelson, the sheepdog seen on the show, was the household pet on "Please Don't Eat the Daisies."

Guest Cast

Restaurant Manager	Norman Alden
Waiter	Leonard Stone

Tuesday
Evening
September 24, 1968

CLOSE-UP
10:00 ❷ ③ 60 MINUTES

[SPECIAL] [COLOR] The first edition of a new magazine hits the TV stands—a topical review with an unlimited range of subjects.

The magazine format is as literal as possible: the sets are huge magazine pages; the cover (and cover story) change for each edition. ("60 Minutes" is scheduled as a biweekly journal.) On-the-air editors: Mike Wallace and Harry Reasoner.

A prominent regular feature is "Viewpoint," the by-lined opinions of guest columnists. Contributors include Cleveland Amory, Luigi Barzini, Art Buchwald, Godfrey Cambridge, Sen. Everett Dirksen, David Frost, Barry Goldwater, Norman Mailer, Marshall McLuhan, Malcolm Muggeridge, James Reston, Leo Rosten, Bayard Rustin, Ted Sorenson, Harriet Van Horne and Roy Wilkins.

Among tonight's features . . . exclusive footage of Vice President Hubert Humphrey winning the Democratic nomination; and films of Richard Nixon with his family and advisors, watching the former Vice President's nomination on TV in their hotel. Also: "Cops." An interview with Attorney General Ramsey Clark opens with this question: "Dick Gregory says 'Today's cop is yesterday's nigger.' Do you understand that?" What are the bitter grudges that cops and citizens hold against each other?

Excerpts from "Why Man Creates," a film by Saul Bass, rounds out the hour.

The show will also use previously telecast contributions. (Upcoming: a BBC report on Britain's race turmoil, and "A Black Man Is a Man," created by WBTV, Charlotte, N.C.)

Producers are Palmer Williams, Joe Wershba, Bill McClure and Jack Beck. (They and Hewitt go back to Ed Murrow's famed "See It Now.") Also: Andrew Rooney and John Sharnik.

CLOSE-UP
THAT'S LIFE
10:00 ❼ ⑧

[DEBUT] [COLOR] Robert Morse and E.J. Peaker star in this weekly Broadway-paced blend of comedy, music and romance.

Tonight we learn how Robert Dicksen met Gloria Quigley. The match was arranged by Cupid (guest star George Burns), who first had to show Gloria that marriage to a dream ideal (Rodney Wonderful, played by Tony Randall) would be dull, dull, dull.

Marvin Marx created the series' concept. Choreography: Tony Mordente. Music: Elliot Lawrence orchestra. The Rocking Turtles perform in a discotheque sequence. Robert: Robert Morse. Gloria: E.J. Peaker. Bubbles La Tour: Maureen Arthur. (60 min.)

Songs

"Pardon Me, Haven't We Met Before?" "Do I Hear a Waltz?"	Robert, E.J.
"The Glory of Love"	George Burns
"Baby, That's Life," "The Girl That I Marry"	Robert
"Eleanor," "Battle of the Bands"	Turtles

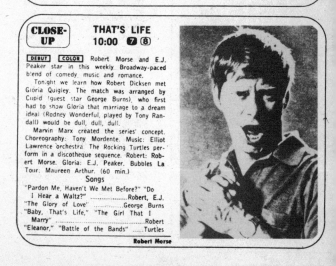

Robert Morse

Thursday   October 10, 1968

Evening

11 VOYAGE—Adventure
[COLOR] A fanatical UN official opposes Nelson's plan to put out a fire ranging through the radiation belt over the Southern Hemisphere. Nelson: Richard Basehart. Dr. August Weber: David J. Stewart. (60 min.)

13 FRENCH CHEF—Cooking
Julia Child shows how to roast a saddle of lamb.

How to cook Hungarian style. See the article on page 32.

31 FILM

6:45 31 NEWS

6:55 7 [COLOR] WEATHER—Antoine

7:00 2 [COLOR] NEWS—Walter Cronkite
3 MAN FROM U.N.C.L.E.
[COLOR] THRUSH plants a listening device in the apartment adjoining Waverly's office. Sandy Wyler: Judi West. Vincent Carver. Ray Danton. Solo: Robert Vaughn. Illya: David McCallum. Tiger Ed: Harvey Lembeck. (60 min.)

4 20 [COLOR] NEWS —Chet Huntley, David Brinkley
5 I LOVE LUCY—Comedy
En route to Rome, an Italian producer offers Lucy a role in a film about the wine industry. Lucy: Lucille Ball. Ricky: Desi Arnaz. Ethel: Vivian Vance.

7 [COLOR] NEWS—Frank Reynolds
8 [COLOR] TRUTH OR CONSE-QUENCES—Quiz
13 JAPAN SOCIETY
[COLOR] Beate Gordon shows Japanese flower arrangements.

31 HEALTH EDUCATION

7:30 2 BLONDIE—Comedy
[COLOR] Blondie attires herself in the modest fashion as she tries to unload the "outta sight" car she purchased from a hippie. Blondie: Patricia Harty. Dagwood: Will Hutchins. Alexander: Peter Robbins. Cookie: Pamelyn Ferdin. Dithers: Jim Backus. Tootsie: Bobbi Jordan. Cora Dithers: Henny Backus.

Guest Cast
Big Cat ....Corey Fischer
Doggie ....Lloyd Batista

4 20 [COLOR] DANIEL BOONE—Adventure
[COLOR] David Scott, a dandified painter, is foisted on Dan for two months'

training in the art of becoming a man. The apprenticeship takes on special significance when David is abducted by Longknife, a Shawnee seeking a replacement for the tribal artist. Daniel: Fess Parker. (60 min.)

Guest Cast
David Scott ....David Watson
Longknife ....Johnny Cardos

5 [COLOR] TRUTH OR CONSE-QUENCES—Quiz

7 8 UGLIEST GIRL IN TOWN
[COLOR] An appendicitis attack while in his "Timmy" guise lands Tim in a hospital, ricocheting between the men's and women's wards—the naked truth riding on every bounce. Tim: Peter Kastner. Gene: Gary Marshal. Julie: Patricia Brake. Courtney: Nicholas Parsons.

Guest Cast
Eunice ....Helen Ford
Hester ....Mai Bacon
Dr. Bledsoe ......Howard Marion Crawford

9 WHAT'S MY LINE?—Game
[COLOR] Panelists are Karen Machon, Bert Convey, Arlene Francis and Soupy Sales. Host is Wally Bruner.

11 RAT PATROL—Drama
[COLOR] Troy brings the German army down on his neck when he kidnaps a German field commander. Troy: Christopher George. Von Helmreich: Wolfgang Preiss. Shtengler: Max Slaten.

13 COMMUNICATIONS AND SOCIETY—Lecture
Prof. Siepmann continues his discussion on the history of communication.

31 ON THE JOB—Fire Department

47 [COLOR] MAID AT YOUR SERVICE—Comedy

8:00 2 3 HAWAII FIVE-O
[COLOR] A kidnaping—planned as a publicity stunt—becomes the real thing when the millionaire father of singer Bobby George offers a fat reward for his son's return. McGarrett: Jack Lord. Danny: James MacArthur. Chin Ho: Kam Fong. Kono: Zulu. (60 min.)

Guest Cast
Bobby George ....Sal Mineo
D. J. Georgiatti ....Harold J. Stone
Allen ....Ian Berger
Jerry ....Sam Melville

Monday October 14, 1968

Evening

④ ㉑ JEANNIE—Comedy
[COLOR] A bouncing baby boy named Abdullah (Jeannie's nephew) triggers a rash of wild misunderstandings when his genie aunt drops him in Tony and Roger's laps for a day. Tony: Larry Hagman. Roger: Bill Daily. Dr. Bellows: Hayden Rorke. Jeannie: Barbara Eden.

Guest Cast
Sally .... Shyrl Formberg
Frank .... Jack Riley

⑤ [COLOR] TRUTH OR CONSE-QUENCES—Quiz

⑦ ⑧ SEX EDUCATION SPECIAL
[SPECIAL] [COLOR] "How Life Begins." See the Close-up on page A-47 for details. (Rerun; 60 min.)

"The Avengers" will not be seen tonight.

⑨ WHAT'S MY LINE?—Game
[COLOR] Guests are Arlene Francis, Gawn Grainger, Pia Lindstrom and Soupy Sales. Host is Wally Bruner.

⑪ RAT PATROL—Drama
[COLOR] Troy wakes up in a field hospital unable to see. Troy: Christopher George. Dietrich: Hans Gudegast.

⑬ COMMUNICATIONS AND SOCIETY—Lecture
Prof. Siepmann discusses the history of radio from the 1890's to the 1920's with Erik Barnouw, professor of radio and television at Columbia.

㉛ ON THE JOB—Fire Department
㊼ [COLOR] LILIA LAZO—Variety

8:00 ④ ㉑ ROWAN AND MARTIN
[COLOR] Guest Bobby Darin sings "Mack the Knife" in Russian with Rozmenko (Arte Johnson), and chairs a poll-takers' convention; Mod, Mod World surveys all sorts of pollution; Ruth Buzzi answers lonely-hearts letters; and Rowan and Martin present the Discovery of the Week: the rocking Holy Modal Rounders ("Right String, Baby, but the Wrong Yo

Yo"). Judy Carne, Henry Gibson, Goldie Hawn, Dave Madden, Gary Owens, Alan Sues and Jo Anne Worley. (60 min.)

⑤ [COLOR] PAY CARDS!—Game
⑨ STEVE ALLEN—Variety
[COLOR] Tentatively scheduled guests: actors Ricardo Montalban and Genevieve Bujold, comedian Pat Harrington and jazz pianist Oscar Peterson. (90 min.)

⑪ RUN FOR YOUR LIFE—Drama
[COLOR] A sportscar firm is on the skids. Paul: Ben Gazzara. (60 min.)

⑬ ART AND MAN
[SPECIAL] "Chess Games with Marcel Duchamp," a profile of the French artist, written and produced by host Jean Marie Drot. A pioneer in the post-World War I Dada ("anti-art") movement, Duchamp is—according to critics as well as colleagues—still a powerful influence in art, especially among pop artists. Duchamp talks about his life and work as cameras view him visiting a friend in New York's Greenwich Village. (60 min.)

㉛ CONSULTATION—Medicine
"The Expensive Nurse."

8:30 ② HERE'S LUCY—Comedy
[COLOR] Two-time Oscar-winner Shelley Winters makes her debut in a situation comedy as a big star who's unfortunately getting bigger and bigger. Lucy is hired to somehow slam the brakes on the movie queen's compulsive eating. Lucy: Lucille Ball. Harry: Gale Gordon. Shelley Summers: Shelley Winters.

⑤ MERV GRIFFIN—Variety
[COLOR] Tentatively scheduled guests: Twiggy, actors Scott Jacoby and Genevieve, comedian Charlie Manna, and singers Monti Rock III and D'Aldo Romano. Arthur Treacher. (90 min.)

⑦ PEYTON PLACE—Serial
[COLOR] Young Lew-Miles arrives home from New York, giving the vague impres-

run for your life
8:00 PM MONDAY THRU FRIDAY
TELEVISION 11 COLOR
wpix tv

TUESDAY, OCTOBER 29, 1968

October 29, 1968 **Tuesday**
Evening

5 ⬛ **TRUTH OR CONSE-QUENCES—Quiz**

7 **8** **MOD SQUAD—Drama**
⬛**COLOR** "You Can't Tell the Players Without a Programmer." An attempted suicide brings the Mod Squad into a case involving a computerized blackmail racket. Pete: Michael Cole. Linc: Clarence Williams III. Julie: Peggy Lipton. Greer: Tige Andrews. (60 min.)

Guest Cast

Louise	Linda Marsh
Roy Tilson	Mark Goddard
Samantha	Julie Adams
Arthur Quinn	Byron Foulger
Morgan	Jerry Harper

Read the article about Peggy Lipton in next week's TV GUIDE.

9 **STEVE ALLEN—Variety**
⬛**COLOR** Guests include Doug McClure of "The Virginian," singer Damita Jo and comedian John Barbour. Sketch: Steve interviews a swashbuckler. (90 min.)

11 **RAT PATROL—Drama**
⬛**COLOR** A German Army unit traps the Rat Patrol at the ammo dump they planned to blow sky-high. Troy: Christopher George. Moffitt: Gary Raymond.

31 **HUMAN RIGHTS FORUM**

47 ⬛**COLOR** **MIGUELITO VALDES**

8:00 5 ⬛**COLOR** **PAY CARDS!—Game**

11 **RUN FOR YOUR LIFE—Drama**
⬛**COLOR** Paul returns to San Francisco to see his old friend Duke Smith, who is the favorite to win the middle-weight title. Ben Gazzara. (60 min.)

31 **REPORT TO THE DENTIST**

47 ⬛**COLOR** **PUMAREJO—Variety**

8:30 2 **3** **RED SKELTON**
⬛**COLOR** Guests: George Gobel and the Mills Brothers, who sing "Opus One" and "Shy Violet." Sketch: con man San Fernando Red goes to work on a pint-sized tycoon (George) who is too shy to propose to his tall sweetheart (Jan Davis). Silent Spot: Red plays twin brothers dealing with trick-or-treaters. Barbara Bostock, Beverly Powers, Bob Duggan, Fred Villani, Tom Hansen dancers, Alan Copeland singers, David Rose orchestra. (60 min.)

After the program: a five-minute Republican political message.

4 ⬛ **JULIA—Comedy**
⬛**COLOR** "Am I, Pardon the Expression, Blacklisted?" Julia fears she may be out of a job when there's a delay in granting her security clearance at the aerospace center. Julia: Diahann Carroll. Dr. Chegley: Lloyd Nolan. Corey: Marc Copage. Earl: Michael Link. Marie: Betty Beaird. Hannah: Lurene Tuttle.

5 **MERV GRIFFIN—Variety**
⬛**COLOR** Tentatively scheduled guests: Gov. John Connally (D-Texas), composer-guitarist Joao Gilberto, comedy writer Jack Douglas and his wife Reiko, comedienne Betty Walker. (90 min.)

7 **8** **IT TAKES A THIEF**
⬛**COLOR** "The Packager." Ambrose Billington, a jet-set bon vivant, gives Mundy a choice of pulling an espionage snatch (for half a million dollars) or losing his SIA cover (which would send him back to prison). Mundy opts for the money, but adds spice to the heist by absconding with Billington's girl friend. Mundy: Robert Wagner. (60 min.)

Continued on next page

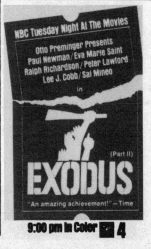

NBC Tuesday Night At The Movies

Otto Preminger Presents
Paul Newman/Eva Marie Saint
Ralph Richardson/Peter Lawford
Lee J. Cobb/Sal Mineo
in

(Part II)

EXODUS

"An amazing achievement!"—Time

9:00 pm in Color **4**

TUESDAY, DECEMBER 3, 1968

Tuesday December 3, 1968

Evening

④ ⑳ JULIA—Comedy

`COLOR` Groucho Marx makes a cameo appearance in this episode. Worry and wonder mark Julia's reaction to the attentions—and gifts—given to her by the plant's retiring messenger. Julia: Diahann Carroll. Dr. Chegley: Lloyd Nolan. Hannah: Lurene Tuttle. Eddie: Eddie Quillan. Corey: Marc Copage.

⑤ MERV GRIFFIN—Variety

`COLOR` Tentatively scheduled guests: actors Eli Wallach and Anne Jackson (Mrs. Wallach), musical-comedy performer Jack Cassidy, Hermione Gingold, comedienne Marge Greene and singer Frankie Scinta. Arthur Treacher. (90 min.)

⑦ ⑧ IT TAKES A THIEF

`COLOR` "The Galloping Skin Game." Ricardo Montalban returns as international fence Nick Grobo, Mundy's match in suave skulduggery. Now on opposite sides of the law, Mundy and Grobo savor the skilled professionalism that permits them to compete mercilessly for secret documents while ostensibly co-operating in the theft of the Kimberly diamond. (The authentic 60-karat Kimberly gem is seen in this episode.) Mundy: Robert Wagner. (60 min.)

Guest Cast

Nick Grobo .... Ricardo Montalban
Christina .... Martine Beswick

Continued on page A-61

CLOSE-UP

ELVIS
9:00 ④ ⑳

`SPECIAL` `COLOR` Surrounded by musicians and adoring fans, Elvis Presley headlines his first TV special.

The fans are up and screaming as Elvis rocks through a nostalgic medley of his hits: "Heartbreak Hotel," "Hound Dog," "All Shook Up," "Can't Help Falling in Love with You," "Jailhouse Rock" and "Love Me Tender." He also sings "Memories" and his seasonal hit "Blue Christmas."

The Blossoms vocal group and dancer-choreographer Claude Thompson open a gospel medley with "Sometimes I Feel like a Motherless Child," and Elvis joins them for "Where Could I Go but to the Lord?" "Up Above My Head" and "Saved!"

A rocking production number stars Elvis as a traveling musician (singing "Guitar Man") who leaves a dull job ("Nothingsville") for an amusement park ("Big Boss Man") and modest night-club success ("Trouble"). Finale: "If I Can Dream," written for Elvis by vocal arranger Earl Brown. (60 min.)

nine o'clock news MONDAY THRU FRIDAY CHANNEL 11

Sunday December 22, 1968

Evening

⑦ LAND OF THE GIANTS

COLOR The travelers encounter another earthling—a half-crazed astronaut whose captured spaceship could provide a chance for escape. Steve: Gary Conway. Kagan: Glenn Corbett. Mark: Don Matheson. Barry: Stefan Arngrim. Dan: Don Marshall. Valerie: Deanna Lund. Betty: Heather Young. Fitzhugh: Kurt Kasznar. Watchman: Don Gazzaniga. (60 min.)

⑧ CHRISTMAS SHOW—Ray Conniff

SPECIAL COLOR See Sat. 7:30 P.M. Ch. 5 for details. (60 min.)

⑨ MOVIE—Drama

COLOR "The Brave One." (1956) A Mexican boy loses his pet bull after the owner of the ranch is killed in an accident. Michel Ray, Rodolfo Hoyos, Joi Lansing. (Two hours)

⑪ 12 O'CLOCK HIGH—Drama

"POW," first of two parts. Savage and three crewmen are shot down and imprisoned in Staglag Luft 12. Richter: Alf Kjellin. (60 min.)

⑬ CRITIQUE

COLOR Scenes from "Huui Huui," a modern comedy of the absurd, are performed. Theodore Hoffman of Drama Review interviews playwright Anne Burr and Joseph Papp, director of the play and producer of the New York Shakespeare Festival. Critics include David Goldman (WCBS Radio, New York) and Clayton Riley (Manhattan Tribune). Host: John Daly. (60 min.)

㉛ COLOR BIG PICTURE—Army

㊼ MOVIE—Drama

"Si Yo Fuera Millonairo." (Mexican; 1958) In Spanish. Maria Felix. (90 min.)

7:30 ② ③ GENTLE BEN—Adventure

COLOR An act of honesty spells danger for Mark as he searches deep in the Everglades for a swamp tramp who overpaid for goods at the general store. Tom: Dennis Weaver. Mark: Clint Howard. Hobo Jim: Albert Salmi.

④ ⑳ WALT DISNEY'S WORLD

COLOR A 40th birthday party for Mickey Mouse. See the Close-up opposite. (60 min.)

For pictures of highlights from Mickey's career, see page 18.

⑤ MOVIE—Drama

"Going My Way." (1944) A young priest interested in athletics and music, is assigned as assistant to a grumbling old priest. Bing Crosby, Barry Fitzgerald, Rise Stevens. (Two hours)

㉛ FOCUS ON BOOKS—Interview

Eileen Riols interviews Millen Brand on his book "Savage Sleep."

8:00 ② ③ ED SULLIVAN

COLOR Tentatively scheduled guests: Mike Douglas; singers Patti Page, Lovelace Watkins, and the Vogues; comedians Flip Wilson, and Stiller and Meara; dancer Peter Gennaro; the Muppets puppets, and the Chung Trio, singer-instrumentalists. In a special segment, New York Jets quarterback Joe Namath visits with Ed. Ray Bloch conducts the orchestra (Live; 60 min.)

Highlights

"Christmas Song," "Winter Wonderland" .... Mike
"Till" .... Vogues
"Sleigh Ride" .... Peter

⑦ ⑧ FBI

COLOR A con man kills his wealthy bride when she learns that her handsome "Major" borrowed his identity from a Vietnam casualty list. As the FBI investigates, he's already courting another rich—and gullible prospect. Erskine: Efrem Zimbalist Jr. Colby: William-Reynolds. 'Vard: Philip Abbott. (60 min.)

Guest Cast

Daniel Sayres	Chad Everett
Margaret Caine	Kathleen Widdoes
Julia Caine	Carmen Mathews
Eugenia Sayres	Virginia Christine
Muir	Paul Smith

⑪ HONEYMOONERS—Comedy

1. The Nortons join the Kramdens for Christmas. 2. Ralph refuses to pay an increase in his rent. Jackie Gleason. (60 min.)

⑬ PBL—Report

COLOR "The Whole World is Watching" an investigation of TV news reporting. Examined: the heavily criticized TV coverage of the Democratic National Convention in Chicago; pressures that determine what news gets on the air; competition for news-program ratings; and the honesty of TV interviews with poli-

Sunday January 12, 1969

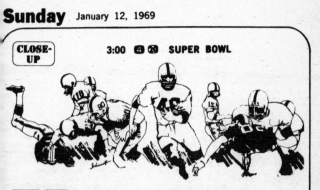

CLOSE-UP   3:00 **4** **29** **SUPER BOWL**

SPECIAL **COLOR** The AFL's New York Jets meet the NFL's Baltimore Colts in the third annual Super Bowl. Curt Gowdy, Kyle Rote and Al DeRogatis report from Miami. (Live)

Jets. A heart-stopping 27-23 win over Oakland brought superstatus and the long-awaited league title to New York. AFL Player of the Year Joe Namath passed less but won more in '68. The receivers are led by George Sauer (66 catches for second in the league). Don Maynard (all-time pro yardage leader) and tough Pete Lammons. The running game gets straight-ahead power from Matt Snell (sixth in the league) and fancy stepping from Emerson Boozer. Aggressive pass rushers Gerry Philbin and Verlon Biggs are standouts on the fine defensive unit.

Colts. Earl Morrall starred as the Cinderella quarterback, leading the league in passing and winning the MVP award as the Colts marched to the title. Top receivers are tough John Mackey and fleet Willie Richardson. Tom Matte, a tenacious, hard-bitten runner (10th in the league), leads the ground game. The defense is stubborn and stingier than Scrooge. Teamwork was the word as the defense turned in three regular-season shutouts, and then picked up another one in the NFL championship game, handing the Browns their first shutout in 143 games.

NEW YORK JETS

11 Turner, J. K-QB	50 McAdams	....LB
12 Namath	...QB	51 Baker	....LB
13 Maynard	...FL	52 Schmitt	C
15 Parilli	.QB	56 Crane	.LB-C
22 Hudson	...DB	60 Grantham	.LB
23 Rademacher DB	61 Talamini	.G
24 Sample	...DB	62 Atkinson	..LB
25 Richards DB-FL	63 Neidert	..LB
29 Turner, B.	...E	66 Rasmussen	...G
30 Smolinski	...B	67 Herman	...G
31 Mathis	...B	71 Walton	.......T
32 Boozer	...B	72 Rochester	.......T
33 Johnson	K-E	74 Richardson	..G
41 Snell	...B	75 Hill	...........T
42 Beverly	..DB	80 Elliott	..T-LB
43 Dockery	..DB	81 Philbin	..DE
45 Christy	..DB	83 Sauer	....E
46 Baird	..DB	85 Thompson	..DF
47 D'Amato	..DB	86 Biggs	.....DE
48 Gordon	.....DB	87 Lammons	....E

BALTIMORE COLTS

2 Brown	...B	53 Gaubatz	.LB
15 Morrall	..QB	55 Porter	..LB
16 Ward	...QB	61 Johnson	...G
19 Unitas	..QB	62 Ressler	....G
20 Logan	..DB	64 Williams, S. .LB
21 Volk	...DB	66 Shinnick	..LB
26 Pearson	..DB	71 Sullivan	....G
27 Perkins	.E	72 Vogel	.......T
28 Orr	...FL	73 Ball	.......T
32 Curtis	.LB	74 Smith, B.R. .T
34 Cole	...B	75 Williams, J. G
37 Austin	..DB	76 Miller	.......T
40 Boyd	..DB	78 Smith, B. ..DE
41 Matte	...B	79 Michaels .DE-K
43 Lyles	..DB	80 Cogdill	....E
47 Stukes	..DB	81 Braase	..DE
49 Lee	...K	84 Mitchell	....E
50 Curry	LB-C	85 Hilton	..DE
51 Grant	..LB	87 Richardson .FL
52 Szymanski	..C	88 Mackey	......E

JULY 17, 1969

sell. Jerry: Bernie Kopell. Ruth: Carolan Daniels. (Rerun)

⓫ PATTY DUKE—Comedy
Patty's devotion to Richard is shaken when the captain of the football team asks her for a date. Patty /Cathy: Patty Duke. Roddy: John Wolfe. Richard: Eddie Applegate. Cynthia: Shannon Gaughan. Louis: Robert Diamond.

⓭ WASHINGTON: WEEK IN REVIEW—News Analysis ©

㉛ MOZART SONATAS
Violinist Marvin Morgenstern and pianist Doris Konig Brooks play the Sonata in A and Six Variations on "J'ai Perdu Mon Amant."

㊶ JOSE FELICIANO—Music ©

㊼ MYRTA SILVA—Variety ©

8:30 ❹ ⓴ IRONSIDE ©
Ironside investigates the strange case of a widow tormented by freak accidents and an eerie voice threatening death. Is she really losing her mind, or is someone driving her to it? Ironside: Raymond Burr. Eve: Barbara Anderson. Ed: Don

Thursday
EVENING

Galloway. Mark: Don Mitchell. (Rerun; 60 min.)

Guest Cast
Karen .... Betsy Jones-Moreland
Avery Corman .... Ray Danton
Judith Corman ...... Victoria Shaw
Dr. Braven .... Fred Beir
Zuppas .... Nick Dennis
Howard Geary .... Alan Caillou
Alise Asher .... Issa Arnal
Israel Sanchez ...... Johnny Silver

❺ DAVID FROST—Variety ©

❼ ⑧ BEWITCHED—Comedy ©
Conclusion: Darrin and Samantha frantically try to change an Italian vamp back into a woman. The incorrigible Serena has literally made a monkey out of her for flirting with Darrin. Samantha/Serena: Elizabeth Montgomery. Clio: Nancy Kovack. Darrin: Dick York. Larry: David White. Scibetta: Cliff Norton. Policeman: Richard X. Slattery. (Rerun)

❾ HIGH ROAD TO DANGER ©
Steve Brodie rides with stuntman Harry Woolman. Daring motorcycle and car stunts are shown.

---

close up

GOLDDIGGERS ©
10:00 ❹ ⓴

Return: This saucy summer replacement for Dean Martin begins a nine-week run.

Headlining the musical-comedy hour: singers Lou Rawls and Gail Martin (Dean's daughter); comics Paul Lynde and Stanley Myron Handelman (both were featured last summer); and the Golddiggers, a 12-girl song-and-dance line. Music: Les Brown orchestra.

Newcomers include Tommy Tune, a limber singer-dancer; comic Albert Brooks (as a bad ventriloquist and hapless animal trainer); and singer Danny Lockin.

Highlights . . . **Lou:** "My Buddy," "St. Louis Blues." **Gail:** "He Loves Me." **Golddiggers:** "Opus One." **Tommy:** "Mountain Greenery," "Ain't She Sweet?" (with Danny). **Lou, Gail:** Old songs medley ("Barney Google," "Row, Row, Row," "Heart of My Heart"). Finale: songs from 1941. (60 min.)

Lou Rawls

TV GUIDE A-71

APOLLO 11

Details of the Manned Lunar Landing and TV Coverage

As this issue went to press, July 16 was still the launch date for the lunar-landing mission. If the launch went on or near schedule, two Americans will be on the moon this Sunday—well ahead of the 1970 target date envisioned by President Kennedy.

The flight plan for a perfect mission calls for the Apollo capsule to enter lunar orbit on Saturday, July 19, at about 1 P.M. (EDT). On Sunday afternoon, the landing operation begins, with touchdown set for approximately 4:20 P.M. Then, shortly after 2 A.M. Monday morning, *the* moment: astronaut Neil Armstrong, a civilian, leaves the "Eagle" lunar module (LEM) and becomes the first man to set foot on the moon. He is joined minutes later by Edwin Aldrin, while Michael Collins continues to circle the moon in the mother ship "Columbia." For complete details on the moon walk—and the astronauts' own TV coverage—see the story on page 6.

The lunar visit is set to end on Monday, just before 2 P.M., when the LEM lifts off from the surface. Ahead lie rendezvous with the mother ship, maneuvers to begin the trip home, and splashdown in the Pacific on Thursday. President Nixon is expected to be aboard the recovery ship.

TV Coverage

The Apollo mission is incredibly complex. It is also great adventure, perhaps man's greatest. For television, it presents a mammoth challenge—and opportunity—to exercise virtually every aspect of the medium.

Each network will provide mock-ups of the Apollo vehicles and equipment, simulations of the orbital mechanics of the flight, examinations of why we're going to the moon (including histories of manned space flight) and, to put the story in perspective, reports on the state of the earth at this historic juncture. Most of this coverage will be concentrated between Sunday morning and early Monday evening, when the networks plan at least 31 hours of continuous reporting. (Bulletins and progress reports will be provided throughout the mission as usual.) Some of the people you'll be seeing . . .

ABC: anchor men Frank Reynolds and Jules Bergman; astronaut Frank Borman; physicist Robert Jastrow, who has written a number of books about space; Rod Serling and a panel of science-fiction writers, including Isaac Asimov; and Steve Allen, with a piano essay on moon songs. Also: space travel as the movies have pictured it.

CBS: anchor man Walter Cronkite; former President Johnson, discussing his role in the U.S. space program; retired astronaut Walter Schirra; Arthur Clarke, author of "2001: A Space Odyssey"; and several entertainers, including Bob Hope, Orson Welles (who created a panic with his 1938 radio play about Martians invading the earth), Buster Crabbe (who played Buck Rogers and Flash Gordon in film serials), Keir Dullea of "2001" and Stan Freberg (with his space puppet Orville). Also: HAL 10,000, a special-effects device that uses nine projectors to simulate space travel.

NBC: anchor men Chet Huntley, David Brinkley and Frank McGee; astronauts Tom Stafford and Eugene Cernan; and Nobel Prize chemist Harold Urey, an expert on the formation of the solar system. Also: telescopes in Denver and Atlanta, with TV camera pickups, will attempt to spot the Apollo capsule as it circles the moon.

In addition to the telecast from the lunar surface, the flight plan lists color telecasts from the mother ship. The press-time schedule: Saturday at about 4 P.M., Sunday at 1:50 P.M. (just before the moon-surface telecast), Tuesday at 9 P.M. and Wednesday at 7 P.M.

The executive producers who engineered what should be TV's finest hour: James Kitchell, NBC; Walter Pfister, ABC; and Robert Wussler, CBS.

Monday
EVENING

JULY 28, 1969

7:00 ② NEWS—Walter Cronkite Ⓒ
③ MOVIE—Musical Ⓒ
"It Happened at the World's Fair." (1963) A bush pilot finds romance at the Seattle Fair. Elvis Presley, Joan O'Brien, Gary Lockwood. (2 hrs.)
④ ㉒ NEWS—Huntley/Brinkley Ⓒ
⑤ I LOVE LUCY—Comedy Ⓒ
Ricky's determined to teach Lucy the importance of punctuality—so he puts her on a rigid time schedule. Lucy: Lucille Ball.
⑦ NEWS—Reynolds/Smith Ⓒ
⑨ TRUTH OR CONSEQUENCES Ⓒ
⑪ HEY LANDLORD!—Comedy Ⓒ
A girl uses a sob story to get a free room in the brownstone. Melanie: Marilyn Mason.
⑬ JOHN LENNON AND YOKO ONO Special: See Sun. 7:30 P.M. Ch. 13.
㉛ FILM
㊶ ALMA DE MI ALMA—Serial
㊼ SIMPLEMENTE MARIA —Serial Ⓒ

7:30 ② GUNSMOKE Ⓒ
Doc Adams is marked for death when he saves a killer's life and at-

tends a woman whose baby is born dead. Outraged because the child was lost, the father vows to kill Doc —and so does the man Doc saved. Matt: James Arness. Doc: Milburn Stone. Kitty: Amanda Blake. Festus: Ken Curtis. Newly: Buck Taylor. Sam: Glenn Strange. (Rerun; 60 min.)

Guest Cast

Tom Butler . . . . Joe Don Baker
Garth . . . . Jack Lambert
Sara . . . . Eunice Christopher
④ ㉒ JEANNIE—Comedy Ⓒ
"Jeannie, the Governor's Wife." Jeannie's determination to make Tony governor of the state causes large headaches, beginning with a magical look-see into his political future. Jeannie: Barbara Eden. Tony: Larry Hagman. Roger: Bill Daily. Bellows: Hayden Rorke. Peterson: Barton MacLane. (Rerun)
⑤ TRUTH OR CONSEQUENCES Ⓒ
⑦ ⑧ AVENGERS—Adventure Ⓒ
Acclaimed British horror actor Christopher Lee appears in "Never, Never Say Die." Steed and Mrs.

close up

DOC Ⓒ
8:00 ④ ㉒

Forrest Tucker, formerly of "F Troop," plays the title role in this pilot for a projected comedy series (by "Green Acres" creator Jay Sommers).

In the sleepy little town of Stubbville (pop. 348), retiring Doc Fillmore awaits the arrival of his replacement before setting out to tour the world. But his replacement is young Orville Truebody, who's been upsetting Doc's plans ever since his singularly ill-timed birth 25 years earlier (recounted in flashback). As events develop, Orville proves he still has the knack.

Cast

Doc Fillmore . . . . Forrest Tucker
Dr. Orville Truebody . . . . Rick Lenz
Amy Fillmore . . . . Margaret Ann Peterson
Will: J. Pat O'Malley. Tillie: Mary Treen. Sheriff Bart: Guy Raymond. Luke: John Qualen. Conductor: Parley Baer. Watkins: Roland Winters. Mrs. Dobson: Norma Varden. Toby: Bob Steele.

Forrest Tucker

A-38 TV GUIDE

BILL COSBY ©
8:30 ④ ⑳

THE FATAL PHONE CALL

Debut: Emmy winner Cosby returns to weekly network TV in a half-hour comedy series set in a Los Angeles high school.

Cos plays Chet Kincaid, physical-education instructor, athletic coach for all sports and occasional substitute teacher. Series regulars include Lillian Randolph as his widowed mother, Lee Weaver as his brother Brian and De De Young as sister-in-law Verna.

Tonight: Chet interrupts his jogging to answer a ringing telephone in an outdoor booth. Better he should have kept running. The call gets him embroiled in a domestic squabble, and, seconds later, the police are out to arrest him.

Guest cast . . . Calvin: Victor Tayback. Mack: John Hawker. Policemen: Jay Powell, Craig Chudy, Jack R. Clinton. Detectives: Victor Millan, Robert Johns. Little Old Ladies: Edna O'Dell, Freda Jones.

Bonanza
9:00, In Color

Starring as the Cartwrights are (who else?) Lorne Greene, Dan Blocker and Michael Landon. It's the 11th big year for this adventure bonanza!

The Bold Ones
10:00, New Show, In Color

Exciting, contemporary dramas about now-style professionals in stories modern as today. Tonight: medics E. G. Marshall, John Saxon, David Hartman.

SEPTEMBER 21, 1969

WOODY ALLEN ©
9:00

WOODY'S FIRST SPECIAL

Special: A visit to the weird, wacky world of Woody Allen, underweight champion of the underdog. His guests: the Rev. Billy Graham (who fields questions from the audience with Woody), actress Candice Bergen and the 5th Dimension.

Sketches: in a comment on nudity in the theater, Woody and Candice rehearse for an off-Broadway play (Joe Silver is the director); a silent film features Woody as a poor soul wooing Candice, the gorgeous rich girl; "The Tutor" presents Rabbi Woody and dumb pupil Candice.

The 5th Dimension do "Workin' on a Groovy Thing," "Wedding Bell Blues" and "Aquarius/Let the Sunshine In."

Woody wrote the show, directed by Alan Handley. Robert Herget dancers, Elliot Lawrence orchestra. (60 min.)

MOVIE ©
9:00

OUR MAN FLINT

James Coburn graduated to star status as the swinging hero of this 1966 spoof on the James Bond genre. Aiding and abetting him: beautiful girls and slam-bang action, directed with tongue in cheek by Daniel Mann ("For Love of Ivy").

A mysterious organization called Galaxy plots to gain world control by harnessing the weather. To squelch the scheme, girl-happy superagent Derek Flint is selected by computer and dispatched to Europe. His target: Galaxy's electric weather-control plant. His only weapon: a sophisticated cigarette lighter that can eliminate an enemy agent 83 ways!

Cast . . . Flint: James Coburn. Cramden: Lee J. Cobb. Gila: Gila Golan. Malcolm Rodney: Edward Mulhare. Dr. Schneider: Benson Fong. Gina: Gianna Serra. Anna: Sigrid Valdis. Leslie: Shelby Grant. Sakito: Helen Funai. Dr. Krupov: Rhys Williams. Gruber: Michael St. Clair.

Gila Golan and James Coburn

★★★—Daily News.

"MIRAGE"

GREGORY PECK
& DIANE BAKER

Amnesia leads to a web of intrigue.

SUNDAY FILM FESTIVAL
11:30PM / WNBC-TV 4

Sunday
EVENING

BOLD ONES—Drama (C)
Burl Ives, Joseph Campanella and
James Farentino star in the lawyers'
segment of this rotating series. Tonight: "A Game of Chance." Junior
partner Neil Darrell (Farentino) goes
beyond the courtroom to expose a
rogue cop—Lt. William Anderson,
who gets a kick out of circumvent-
ing what he calls "Constitutional
technicalities." Walter Nichols: Burl
Ives. Brian Darrell: Joseph Campanella. (60 min.)

Guest Cast
Anderson	Steve Ihnat
Sgt. Joe Benedict	John Milford
Anne Carlin	Sandra Smith
Ralph Horton	John Ragin
Det. Mike Perkins	Mike Carr
Kevorkian	David Mauro
Milly	Leslie Perkins
Freddie	Paul Prokop
Valerie	Elissa Dulce
Steven Holt	Todd Martin
Judge	Bill Gordon

NEWS—George Scharmen (C)
GOLF TOURNAMENT
—Saratoga Springs, N.Y.
Special: Final-round action in the
New York PGA Golf tournament,
concluded yesterday at the Saratoga Springs Golf Course. Cameras
cover the final two holes. Press-
time entrants after approximately
$30,000 in prizes included Dave Marr,
Doug Ford and Tom Nieporte. The
par-72, 6900-yard layout features
fairways heavily lined with pines
and large, undulating greens. It is
well-trapped. If there's a tie, cameras
will cover a sudden-death play-off.
(90 min.)
[Pre-empts regular programming.]
NEWS—Marc Howard (C)
NET PLAYHOUSE—Drama
"The Tale of Genji," conclusion.
Genji's wife dies, leaving him to
ponder his many thoughtless flirta-
tions. Genji: Ichizo Itami. Murasaki:
Manami Fuji. Sannomiya: Mariko
Kaga. Kashiwagi: Choichiro Karawa-
zaki. (1 hr., 15 min.)
SILVIA Y ENRIQUE—Musica
10:30 **HELLUVA TOWN—Variety** (C)
NEW YORK CLOSE-UP (C)
MAGIC LANTERN—Movies
Film on the history of the cinema.

OCTOBER 17, 1969

Kerr Singers and comedian Don Sherman. Sandy Baron. (60 min.)

11 BEAT THE CLOCK—Game Ⓒ
Guest: Jack Cassidy.

13 NEW JERSEY SPEAKS
Two couples discuss interracial adoption. Guests: Mr. and Mrs. Ken Smith of Hackensack and Mr. and Mrs. Kenneth Knowlton of Plainfield.

31 BROOKLYN COLLEGE
A tour of the New York Aquarium with Brooklyn College students.

41 ESPEJISMO BRILLABA

8:00 2 3 GOOD GUYS—Comedy Ⓒ
A visit by a health inspector (guest star Vincent Price) leaves the diner on probation and the guys up tight. Another inspection is imminent, but when? Rufus: Bob Denver. Bert: Herb Edelman. Claudia: Joyce Van Patten. Bender: Jack Perkins. Middleton: Vincent Price.

5 TO TELL THE TRUTH—Game Ⓒ

7 8 BRADY BUNCH Ⓒ
With a new Brady lady at the helm, housekeeper Alice figures it's time for her departure. Dilemma: how to explain her decision without stepping on any toes, especially Carol's. Mike: Robert Reed. Carol: Florence Henderson. Alice: Ann B. Davis. Peter: Christopher Knight. Greg: Barry Williams. Cindy: Susan Olsen. Marcia: Maureen McCormick. Jan: Eve Plumb. Bobby: Mike Lookinland.

11 HE SAID! SHE SAID!—Game Ⓒ

13 BOOK BEAT—Interview Ⓒ
Sumner Locke Elliott discusses his critically acclaimed novel "Edens Lost." Focusing on a family at an Australian resort area, the book deals with family dynamics and interdependency and, says Elliott, "with what we find out too late about ourselves." Host: Bob Cromie. [Repeated Sun., Oct. 19 at 4:30 P.M.]

31 AMERICAN HISTORY
41 ROSENDO ROSELL Ⓒ
47 DANCE SPECIAL

8:30 2 3 HOGAN'S HEROES Ⓒ
A fueling pump for German convoys is being raised next to Barracks 12. That leaves Hogan with the ticklish task of blasting the thing, without blowing up the inmates. Hogan: Bob Crane. Klink: Werner Klemperer. Schultz: John Banner. Kinchloe:

Friday

EVENING

Ivan Dixon. Carter: Larry Hovis. Newkirk: Richard Dawson. LeBeau: Robert Clary. Louisa: Marianna Hill.

4 20 NAME OF THE GAME Ⓒ
"Chains of Command," a prison drama. Crime reporter Dan Farrell takes a job at a small state prison near the Mexican border. His assignment: prove that prison officials are growing rich by forcing the convicts to work on neighboring farms. Dan: Robert Stack. Joe: Ben Murphy. (90 min.)

Guest Cast
Hank . . . . .Pernell Roberts
Superintendent . . .Sidney Blackmer
Stella Fisher . . . . . .Dorothy Lamour
Zack Whitten . . . . . .Jay C. Flippen
Captain Oliver . . . . .Steve Ihnat
Judge O'Donnell . . . . .Paul Fix

5 DAVID FROST—Variety Ⓒ
Scheduled: actors William Redfield and Virna Lisi, comedian Pat Henry, and singers Brook Benton and Turley Richards. (90 min.)

7 8 MR. DEEDS—Comedy Ⓒ
The sight of a once-beautiful Vermont town that was ruined by a Deeds Enterprise sends the dragon slayer from Mandrake Falls on a crusade against air and water pollution. Deeds: Monte Markham. Tony: Pat Harrington. Masterson: Herbert Voland. Emily: Anna Hagan. Barney: Charles Seel. Burt: Stanley Farrar.

9 JOE NAMATH—Interview Ⓒ
New York Jets quarterback Joe Namath welcomes prominent sports figures and entertainers on this half-hour series. Co-host: sports writer Dick Schaap.

TV GUIDE A-85

MONDAY, NOVEMBER 10, 1969

NOVEMBER 10, 1969

MORNING

6:00 ③ SUNRISE SEMESTER (C)
Geology: natural environment.

6:10 ⑤ NEWS

6:15 ⑤ INFINITE HORIZONS—Religion

6:30 ② SUNRISE SEMESTER (C)
Geology: natural environment.
③ CONGRESSIONAL REPORT (C)
See Sat. 1 P.M. Ch. 3 for details.
④ EDUCATION EXCHANGE (C)
"A Bridge for Tomorrow." First in a
10-part series which examines the
problems of senior citizens. Host is
Robert L. Krit, Chicago Medical
School. Today: "Aging and the
Aging Process."

6:45 ⑤ MORNING REFLECTION (C)

7:00 ② ③ NEWS—Joseph Benti (C)
④ ⑳ TODAY (C)
Scheduled: celebrity profiler Rex
Reed; a report on U.S. farm problems; Edward Brecher, discussing
his non-fiction account "Sex Researchers." (Live and film; 2 hrs.)
⑦ NEWS—Tom Dunn (C)
⑧ MR. GOOBER—Children (C)

Monday
MORNING

7:05 ⑦ ED NELSON (C)

7:15 ⑤ GLENN SWENGROS (C)
⑪ NEWS—Marc Howard (C)

7:25 ⑪ NEWS AND WEATHER

7:30 ⑤ ALVIN—Children (C)
③ DAPHNE—Cartoons (C)
⑪ TV HIGH SCHOOL

8:00 ② ③ CAPTAIN KANGAROO (C)
Cartoon: Part 1 of "Andy and the
Lion." Animals: piglet, chameleon
lizard. (60 min.)
⑤ PRINCE PLANET—Children
⑦ CARTOONS (C)

8:25 ⑬ CLASSROOM—Education

8:30 ⑤ MARINE BOY—Children (C)
⑦ GIRL TALK (C)
Guests include travel authority Myra
Waldo. Host: Betsy Palmer.
⑪ KIMBA—Children (C)

9:00 ② ③ LEAVE IT TO BEAVER—Comedy
June thinks Wally and Beaver have
editorial talents, and she starts them
out by buying them a typewriter. Beaver: Jerry Mathers. Wally: Tony Dow.
June: Barbara Billingsley. Ward:
Hugh Beaumont.
③ HAP RICHARDS—Children (C)

---

close up

PREMIERE WEEK

SESAME STREET (C)
11:30 and 4:30 ⑬

Debut: "Sesame Street" opens its doors—and the ABCs and the
1-2-3s—to preschool scholars.

The series title, suggesting "open sesame," couldn't be more
appropriate. The two-year experimental project was produced by
the Children's Television Workshop and conceived, says executive
director Joan Ganz Cooney, "to reach all pre-school children, but
particularly the disadvantaged. We want to teach them a number of
basic skills and give them an increased awareness of themselves and the world."

Fast-paced daily fare, intended to make learning fun, features
a potpourri of techniques (all tested by researchers): psychedelic
cartoons; puppet goings-on, with Jim Henson's Muppets; songs,
ranging from musical comedy to rock; stories and film shorts; and
"commercials" that teach rather than sell.

Sesame Street is an anywhere-city street, and the people who
live there (and host the show) are anywhere people: teachers
Gordon and Bob (Matt Robinson and Bob McGrath); Gordon's
wife Susan (Loretta Long); and candy-store owner Mr. Hooper
(Will Lee). Celebrities, including Carol Burnett and James Earl
Jones, also appear from time to time. (60 min.)

TV GUIDE A-37

Monday
EVENING

NOVEMBER 17, 1969

demns the soap opera; and William Henry who has found some value to them.

41 LA TABERNA INDIA Ⓒ

9:00 **2** **3** MAYBERRY R.F.D. Ⓒcountry comic-musician Glenn Ash plays a farm hand in this episode. Sam takes a back seat when his son begins dogging the new hand's every step, showering him with hero worship. Sam: Ken Berry. Mike: Buddy Foster. Rudy: Glenn Ash. Flora: Dorothy Konrad. Emmett: Paul Hartman.

4 **20** MOVIE—Crime Drama ⒸRichard Widmark and Henry Fonda in "Madigan." See the Close-up on page A-47. (2 hrs.)

7 **8** THE SURVIVORS—Drama ⒸChapter 7: Philip learns of the bank's commitment to Santerra; Jeff has a falling-out with student radicals; Riakos arranges to meet Jeff; Duncan and Belle have a date. Tracy: Lana Turner. Duncan: George Hamilton. Philip: Kevin McCarthy. Baylor: Ralph Bellamy. Belle: Diana Muldaur. Jeff: Michael Vincent. Riakos: Rossano Brazzi. Sheila: Kathy Cannon. Tom: Robert Lipton. Jack: Zalman King. Art: Hal Frederick. Stafford: Foster Brooks. Mike: David Cassidy. (60 min.)

9 MOVIE—Drama"Battleground." (1949) This World War II drama centers on the 101st Airborne Division, "the Screaming Eagles," who defend Bastogne during the Battle of the Bulge. Van Johnson, John Hodiak, James Whitmore, Marshall Thompson, George Murphy, Ricardo Montalban, Denise Darcel, Douglas Foley, Jerome Courtland. (2 hrs.)

11 BEN CASEY—Drama"Cardinal Act of Mercy," first of two parts. Attorney Faith Parsons is suffering from a spinal tumor, but she seems to need morphine more frequently than Dr. Zorba thinks is necessary. Faith: Kim Stanley. Marthan: Glenda Farrell. (60 min.)

13 NET JOURNAL—Report"Guns Before Bread," a profile of the restless Philippines. German and Irish films highlight .. U.S. military presence: after 24 years, many Filipinos decry American military (and economic) influence; anti-U.S. demonstrations are shown. Political climate: elections were held November 11; at press time, the government headed by President Ferdinand Marcos (interviewed during the hour) seemed a sure winner. Economic life: hemp and copra production, the development of new strains of wheat and rice. Sociological problems: slums, rising crime, prostitution. (60 min.)

31 NEW YORK REPORT Ⓒ**41** MAS ALLA DE LA MUERTE Ⓒ9:30 **2** **3** DORIS DAY—Comedy ⒸDoris is ordered to guard two skinny French models whose wild eating binges could jeopardize a fashion scoop for the magazine. Nicholson: McLean Stevenson. Myrna: Rose Marie. Ron: Paul Smith.

Guest Cast

Montagne ....Johnny Haymer

Continued on page A-53

"Madigan"
NBC MONDAY NIGHT
AT THE MOVIES

Richard Widmark
Henry Fonda
Inger Stevens

A Manhattan detective puts his professional reputation on the line!

9:00PM IN COLOR ■ 4

Tuesday
EVENING

Cast

Rosie Lord ....Rosalind Russell
Daphne Shaw.........Sandra Dee
Oliver Stevenson ......Brian Aherne
Mildred Deever ...Audrey Meadows
David Weelwright ..James Farentino
Edith Shaw ....Vanessa Brown
Cabot Shaw ....Leslie Nielsen
Mae ....Margaret Hamilton
Patrick ....Reginald Owen
Nurse ....Juanita Moore
Mrs. Peters ....Virginia Grey

⑨ MOVIE—Science Fiction
"The Day the Earth Caught Fire."
(English; 1961) Nuclear tests shift
the earth's orbit and send it hurtling
toward the sun. Edward Judd, Janet
Munro, Leo McKern. (2 hrs)

⑪ BEN CASEY—Drama
Student nurse Michael Ann Bower-
sox is in a very embarrasing position—her father is picketing outside
the hospital. Luther Adler. (60 min.)

**⑬ NEW YORK TELEVISION
THEATER ©**
Special: Ten of America's most im-
portant young playwrights have writ-
ten a series of dramatic statements
on the pollution crisis in the world.
Nine actors perform the plays which
have a collective title—"Foul!" The
playwrights and titles of their plays:
Anne Burr "TSK"; John Guare,
"Kissing Sweet"; Israel Horovitz,
"Play for Trees"; Arthur Kopit, "Pro-
montory Point Revisited"; Terrence
McNally "Last Gasps"; Leonard
Malfi, "Puck! Puck! Puck!" Ronald
Ribman, "The Most Beautiful Fish";
Megan Terry, "One More Little Drin-
kie"; Jean-Claude Van Itallie, "Take

NOVEMBER 25, 1969

a Deep Breath"; and Lanford Wilson,
"Stoop". The performers: Philip
Bruns, Barbara Cason, Eileen Dietz,
Stanley Green, George S. Irving,
Charlotte Rae, Frances Sternhagen,
Sam Waterston and Arnold Wilkin-
son. (60 min.)
[Repeated Friday at 11 P.M. and
Sun., Nov. 30, at 8 and 11 P.M.]

㊋ MAS ALLA DE LA MUERTE ©
9:30 **② ③ GOVERNOR AND J.J. ©**
A man known for building monu-
ments to himself has offered a million dollars toward a new hospital
wing—providing the structure is
named after him and is dedicated
by the governor (Don Dailey). J.J.:
Julie Sommars. George: James Callahan. Gulley: Elliott Reid.

㊼ PUNAL Y LA CRUZ—Novela ©
10:00 **② ③ 60 MINUTES ©**
Special: Articles in this TV news
magazine are detailed in the Close-
up opposite.

⑤ NEWS—Bill Jorgensen ©
⑦ ⑧ MARCUS WELBY, M.D. ©
"Homecoming." Welby is drawn into
the turmoil of a family being torn
apart by LSD. After a year's absence,
a young dropout returns home suffering acute physical and emotional
kickbacks from tripping out. But the
terror of his illness cannot dull a
deep antagonism toward his father.
Welby: Robert Young. Kiley: James
Brolin. (60 min.)

Guest Cast

Max Behrman ...Nehemiah Persoff
Scott ....Robert Lipton
Madeline ....Peggy Cowles
Mary ....Jackie Lougherty

DECEMBER 3, 1969

(Petrovich). John Reardon (Shish-koff), David Lloyd (Skuratoff), Harry Danner (Alyosha), Frederick Jagel (Luka). Peter Herman Adler conducts. (90 min.)

㉛ ON THE JOB—Fire Department

㊶ UN GRITO EN LA OBSCURIDAD

8:00 ⑤ TO TELL THE TRUTH—Game ©
Guests: Kitty Carlisle, Peggy Cass, Bill Cullen and Durward Kirby.

⑦ EDDIE'S FATHER ©
Looks like wedding bells may sound for Tom's wacky secretary Tina and his way-out art director Norman. Admittedly, Norman was aghast when Tina first proposed, but then he started warming up to the idea . . . perhaps a motorcycle ceremony on Sunset Strip. Tom: Bill Bixby. Norman: James Komack. Tina: Kristina Holland. Eddie: Brandon Cruz. Mrs. Livingston: Miyoshi Umeki.

⑧ COLLEGE BASKETBALL
Fairfield vs. the University of Connecticut at the University of Connecticut. Dick Galliette and Tom Burns report. (Live)

⑪ HE SAID! SHE SAID!—Game ©

㉛ COMMUNICATIONS AND SOCIETY
Topic: "The Entertainers." Guest: Sylvester "Pat" Weaver, formerly of NBC. Charles Siepmann is host.

㊶ ALEGRIAS—Musica

㊼ TRES PATINES—Comedia

8:30 ② ③ BEVERLY HILLBILLIES ©
Shorty, the hillbillies' house guest, is getting a mad, mad rush from the girls at Drysdale's bank—who mistakenly believe the Ozark man has more millions than Jed Clampett. Jed: Buddy Ebsen. Granny: Irene Ryan. Jethro: Max Baer. Drysdale: Raymond Bailey.

Guest Cast
Shorty ....Shug Fisher
Gloria ....Bettina Brenna
Helen ....Danielle Mardi
Miss Leeds ....Judy McConnell
Miss Switzer ....Judy Jordan

⑤ DAVID FROST—Variety ©
Guests include Sen. George McGovern (D-S.D.), New York Jets wide receiver Don Maynard, Lou Rawls, producer Richard Brooks and actress Jean Simmons (Mrs. Brooks), and Jack Carter. (90 min.)

Wednesday
EVENING

⑦ ROOM 222 ©
"Clothes Make the Boy." Free-dress advocates clash with tradition-tied teachers and parents as Walt Whitman High considers a burning issue—the dress code. Script by Allan Burns, the series' newly-named producer. Pete: Lloyd Haynes. Kaufman: Michael Constantine. Liz: Denise Nicholas. Richie: Howard Rice. Jason: Heshimu.

Guest Cast
Charles Shaffer ......Kenneth Mars
Jerry ....Bud Cort
Wagner ....Ramon Bieri
Dragen ....Ivor Francis
Bonnie ....Jan Shutan
Abby Hundley ....Ta-Tanisha
Mick ....Frank Webb
Bellaver ....Arthur Adams

⑨ GAME GAME ©
Question: "How much of a gossip are you?" Celebrities: Richard Dawson, Rosie Grier and Sally Ann Howes.

⑪ FELONY SQUAD ©
Jim renews a romance with a former girl friend now implicated in a murder case. Jim: Dennis Cole. Linda Jo: Katherine Crawford. Sam: Howard Duff.

㉛ ALL ABOUT TV ©
"Violence in the Mass Media." (60 min.)

㊶ RISAS, SONRISAS—Variedad

㊼ SECUESTRO EN EL CIELO

9:00 ② ③ MEDICAL CENTER ©
Walter Pidgeon makes a rare TV appearance in "The Fallen Image." A cardiac condition is jeopardizing Ambassador Evarts's crucial negotia-

TV GUIDE A-67

DECEMBER 29, 1969

⑨ DIVORCE COURT—Drama ⓒ
A woman accuses her husband of desertion when he won't take her back after a trial separation. Pauline: Elaine Partnow.
⑪ HERE'S BARBARA ⓒ
⑬ WORLD PRESS ⓒ
㊶ NOTICIAS—Miguel Torres ⓒ
㊷ EL HIT DEL MOMENTO ⓒ
11:10 ④ WEATHER—Frank Field ⓒ
11:15 ④ NEWS—Jim Hartz ⓒ
④ WEATHER—Bob Jones ⓒ
11:20 ③ WEATHER ⓒ
⑧ SPORTS—Dick Galiette ⓒ
11:25 ④ SPORTS—Arnold Dean ⓒ
④ SPORTS—Kyle Rote ⓒ
11:30 ② ③ MERV GRIFFIN ⓒ
Merv continues his stay in Hollywood. (90 min.)
④ ⑳ JOHNNY CARSON ⓒ
Scheduled: New York City Mayor John V. Lindsay. (90 min.)
④ MOVIE—Drama ⓒ
"The Bad and the Beautiful." (1952) Film producer Jonathan Shields meets many people in his rise to the top—and makes enemies of most of them. Kirk Douglas. (2 hrs., 15 min.)
⑦ ⑧ DICK CAVETT ⓒ
Debut: Dick joins TV's late-night race with scheduled guests Woody Allen and British actor-playwright Robert Shaw. For further details, see the Close-up on page A-44. (90 min.)
⑨ MOVIE—Western
"Tall in the Saddle." (1944) When a cowboy turns up to start a new job, he learns that his employer has just been murdered. John Wayne, Ella Raines, Audrey Long. (2 hrs.)
⑪ PERRY MASON—Mystery
Steve Benton retains Mason to protect his brother, who has been forging checks to pay off gambling debts. Benton: Dick Foran. (60 min.)
㊶ CUERDAS Y GUITARRAS
㊷ NOTICIAS—Arturo Rodriguez
12:00 ⑪ FOCUS NEW JERSEY ⓒ
12:30 ⑪ PHIL DONAHUE ⓒ
Guest: Dr. Morris E. Chafetz, who discusses his methods of rehabilitating alcoholics.
1:00 ② NEWS ⓒ
③ MOVIE—Drama ⓒ
"The Bottom of the Bottle." (1956) An escaped convict, trying to get to

Monday
EVENING

his wife and three children in Mexico, goes to his brother for aid. Van Johnson, Joseph Cotten, Ruth Roman. (1 hr., 45 min.)
④ NEWS—Bob Teague ⓒ
⑦ MOVIE—Melodrama
"Voodoo Woman." (1957) A scientist experiments on a native girl, changing her into a monstrous creation. Marla English, Tom Conway, Touch Connors, Lance Fuller. (90 min.)
⑧ NEWS
⑪ HONEYMOONERS—Comedy
Ralph gets into a tangle with a wisacre. Ralph: Jackie Gleason. Norton: Art Carney.
1:10 ② MOVIE—Drama ⓒ
"Scarlet Angel." (1952) When a New Orleans widow dies, a saloon girl whom she befriended adopts both her identity and her infant son. Yvonne DeCarlo, Rock Hudson, Amanda Blake, Richard Denning, Whitfield Connor. (1 hr., 35 min.)
1:15 ④ MOVIE—Drama
"The Last Summer." (West German; 1960) A Scandinavian assassin falls in love with the daughter of the political leader he intends to murder. Hardy Kruger, Nadja Tiller, Lilo Pulver. (1 hr., 25 min.)
1:30 ⑨ JOE FRANKLIN ⓒ
⑪ NEWS ⓒ
1:45 ⑤ BIG BANDS ⓒ
Guests: Duke Ellington and his orchestra. Highlights: "Take the 'A' Train," "Rockin' Rhythm," "Step in Time," "Le Tigre," "Tuttie for Cootie" and "Jam with Sam."
2:30 ⑨ NEWS AND WEATHER
2:45 ② MOVIE—Drama ⓒ
"The Rose Bowl Story." (1952) Gridiron star Steve Davis falls in love with Denny Burke, one of the Rose Princesses in the Tournament of Roses. Marshall Thompson, Vera Miles, Natalie Wood, Richard Rober. (1 hr., 35 min.)
③ NEWS AND WEATHER ⓒ
4:20 ② MOVIE—Drama ⓒ
"Edge of Eternity." (1959) Three people have been murdered, and Arizona sheriff Les Martin tries to find the killer. Cornel Wilde, Victoria Shaw, Mickey Shaughnessy, Edgar Buchanan, Rian Garrick. (1 hr., 35 min.)

Tuesday
EVENING

⓭ NEW YORK TELEVISION THEATER—Drama Ⓒ
Special: "Father Uxbridge Wants to Marry." See the Close-up on page A-62. (60 min.)
[Repeated Sat., Feb. 21, at 7 P.M.]
㊵ PANDORAMA—Musica
㊼ PUMAREJO—Variedad Ⓒ
8:30 **② ③ RED SKELTON** Ⓒ
Guests: Barbara Feldon and the singing Lettermen. Sketch: Barbara is a lady scientist using Clem Kadiddlehopper as guinea pig in a chicken-feed experiment. Silent Spot: a man doing housework is undone by appliances and toys. Tom Hansen dancers, Jimmy Joyce singers. (60 min.)

Highlights
"Sweet Georgia Brown"
.....Barbara, Lettermen, Dancers
"Traces," "Memories" ...Lettermen

④ ⑳ JULIA—Comedy Ⓒ
Don Ameche directed and appears in this episode. Corey's accidental phone call gets Julia, the Waggedorns and Dr. Chegley involved in

FEBRUARY 17, 1970

the plight of a sick boy—3000 miles away. Julia: Diahann Carroll. Dr. Chegley: Lloyd Nolan. Corey: Marc Copage. Earl: Michael Link.
Guest Cast
Dr. Prestwick ....Don Ameche
Mrs. Appleton ....Janet Waldo
⑤ DAVID FROST—Interview Ⓒ
David talks with Jackie Gleason. See the Close-up on page A-62. (90 min.)
⑦ ⑧ MOVIE—Documentary Ⓒ
"The Journey of Robert F. Kennedy." See the Close-up opposite for details. (90 min.)
⑨ CANDID CAMERA—Comedy
Toddlers compose love songs; a sticky football dummy; surprising telephone calls.
⑪ HE SAID! SHE SAID!—Game Ⓒ
Guests are Jean-Pierre Aumont and his wife Marisa Pavan; Sonny Fox and his wife Gloria; Jack Cassidy and Shirley Jones; and Jessica Walter and Ross Broman. Joe Garagiola acts as the host.
㊵ AY! QUE FAMILIA—Comedia
9:00 **④ ⑳ MOVIE—Mystery** Ⓒ
"McCloud," a TV film with Dennis Weaver as a modern-day marshal. With a subpoenaed witness in tow, New Mexico lawman Sam McCloud travels to New York—and into trouble. His prisoner is kidnaped and the marshal gets entangled in a murder case involving Puerto Rican militants, a lady novelist, a Wall Street lawyer and a dead beauty queen. (2 hrs.)
Cast
McCloud ....Dennis Weaver
Whitman ....Craig Stevens
Clifford ....Mark Richman
Chris Coughlin ......Diana Muldaur
Broadhurst ....Terry Carter
Adrienne ....Julie Newmar
Peralta ....Mario Alcalde
[Turn to page 7 for an article about a scene in this film.]
⑨ MOVIE—Musical Ⓒ
"Because You're Mine." (1952) An opera star (Mario Lanza) gets drafted and falls for the sister (Doretta Morrow) of his sergeant (James Whitmore). Jones: Dean Miller. Francesca: Paula Corday. Mrs. Montville: Spring Byington. Patty: Jeff Donnell. Loring: Don Porter. (2 hrs.)

Saturday
EVENING

11 MOVIE—Melodrama
"Valley of the Zombies." (1946) An insane man believed long dead, returns bent on revenge. Robert Livingston, Adrian Booth. (60 min.)

31 FORSYTE SAGA—Drama
Chapter 19. Bicket learns of his wife's past; Fleur gives birth. Soames: Eric Porter. Bicket: Terry Scully. (Rerun; 60 min.)

47 MOVIE—Drama
"Marido Ambulante." (1938) Dubbed in Spanish. Lucille Ball, Joe Penner. (90 min.)

9:30 2 3 MARY TYLER MOORE ©
Debut: Rebounding from a broken romance, Mary Richards takes a job in a Minneapolis TV newsroom. For

Mike Connors as the private eye with a sharp bead on criminals. With Gail Fisher.

10 PM
CBS◉2

better or worse, her office mates will be a caustic boss who drinks (Edward Asner), a self-enamored news anchorman (Ted Knight) and an unflappable writer (Gavin MacLeod as Murray). Script by the series' producers, Emmy winners James L. Brooks and Allan Burns. Phyllis: Cloris Leachman. Rhoda: Valerie Harper. Phyllis's Daughter: Lisa Gerritsen ("My World and Welcome to It"). Guest Cast . . . Bill: Angus Duncan. Locksmith: Dave Morich. [Miss Moore's career is outlined in the article on page 34.]

7 8 ENGELBERT HUMPERDINCK ©
Guests: Liberace, singer Fay McKay and comedian Allen Drake. Jack Parnell orchestra. (Rerun; 60 min.)

Highlights

"Cabaret," "The Good Life,"
  "There's No You," "You've Lost
  That Lovin' Feelin'," "The Way
  It Used to Be," "Release
  Me" ....Engelbert
"Mixed Emotions," "There'll Be
  No New Tunes on This Old
  Piano," "Raindrops Keep Fallin'
  on My Head," love theme from
  "Romeo and Juliet" ....Liberace
[Last show of the series.]

10:00 2 3 MANNIX ©
Darren McGavin plays a psychotic killer bent on a senseless revenge against Mannix's former Army buddies. In "A Ticket to the Eclipse," the fourth-season opener, Mannix must convince the police to move fast before he too becomes a victim. Mannix: Mike Connors. Peggy: Gail Fisher. (60 min.)

Guest Cast
Mark ....Darren McGavin
Lieutenant Blaine ......Dane Clark
Toby ....Mark Stewart
Tommy ....Paul Mantee

5 NEWS—John Roland ©

SEPTEMBER 19, 1970

Monday
EVENING

berry when very turned-on Howard nags his cronies into forming a sensitivity group. Sam: Ken Berry. Howard: Jack Dodson. Millie: Arlene Golonka. Emmett: Paul Hartman. Goober: George Lindsey. Guest Cast ... Leader: Fred Sadoff. Jerry: Roy Applegate.

④ ㉒ MOVIE—Drama ©
"Boom!" (1968) kicks off a series of new Monday-night movies. On a Mediterranean island, a dying millionairess restlessly dictates her memoirs. Disrupting her final summer is an enigmatic poet—known as the "Angel of Death." Tennessee Williams' script is based on his play "The Milk Train Doesn't Stop Here Anymore." Filmed on Sardinia by Joseph Losey. (2 hrs., 15 min.)

Cast
Flora Goforth	Elizabeth Taylor
Chris Flanders	Richard Burton
Witch of Capri	Noel Coward
Blackie	Joanna Shimkus

⑦ ⑧ PRO FOOTBALL ©
Debut: This Monday-night football

NFL Monday Night Football
Jets vs. Browns

Premiere 9:00
ⓐⓑⓒ ⑦ ⑧

---

close up

PRO FOOTBALL ©
9:00 ⑦ ⑧

JETS VS. BROWNS AT CLEVELAND

New York Jets

7 Bell	....WR	32 Boozer	..RB	52 Schmitt	C	75 Hill	........T
11 T'rn'r, J. K-QB		33 Hicks	..CB-S	55 Ebersole	LB	77 Bayless	...G
12 Namath	..QB	34 White	....RB	56 Crane	▲LB-C	80 Elliott	....DT
13 Maynard	WR	35 Onkotz	LB-S	60 Grantham	LB	83 Sauer	...WR
18 Woodall	..QB	37 Nock	....RB	61 Finnie	..T-G	84 Lomas	...DE
20 O'Neal	K-WR	40 Battle	....S	62 Atkinson	LB	85 Thompson	DT
21 Tannen	CB-S	41 Snell	....RB	64 Perreault	..G	86 Briggs	..DE
22 Hudson	....S	43 Dockery	..CB	66 Rasmussen	G	87 Lammons	TE
28 Leonard	..CB	45 Thomas	CB-S	67 Herman	...G	88 Caster	..WR
29 Turner, B.	WR	51 Baker	....LB	70 Foley	....T	89 Stewart	...TE

Cleveland Browns

12 Cockroft	....K	29 Sumner	..CB	54 Hoaglin	C	77 Schafrath	..T
15 Phipps	...QB	34 Howell	....S	55 Reynolds	...LB	78 McKay	....T
16 Nelson	...QB	35 Scott	....RB	56 Matheson	DE	80 Jones, J.	..DE
20 Summers	CB	40 Barnes	....RB	64 Copeland	..G	81 Gregory	..DE
21 Brown, D.	..S	41 Jones, D.	WR	65 Demarie	..G	82 Houston	..LB
23 Stevens'n	CB	43 Hooker	..WR	66 Hickerson	G	83 Glass	....TE
24 Kellermann	S	44 Kelly	....RB	67 Langer	....G	85 Jones, H.	WR
25 Leigh	...RB	50 Garlington	LB	69 Shelly	....G	86 Collins	..WR
26 Morrison	RB	51 Lindsey	..LB	71 Johnson	..DT	88 Snidow	..DE
28 Davis	....CB	52 Andrews	...LB	72 Sherk	....DT	89 Morin	....TE

TV GUIDE A-61

OCTOBER 2, 1970

⓫ BROOKLYN COLLEGE
Disc jockey Rosko (Bill Mercer) is interviewed by Kenny Webb.

8:00 **⑤ TO TELL THE TRUTH** ⒞
⑦ ⑧ NANNY ⒞
"The Haunted House." The Professor is skeptical, Nanny is noncommittal, and a parapsychologist is a true believer as they investigate the children's report of ghostly manifestations in an abandoned house. Nanny: Juliet Mills. Everett: Richard Long. Hal: David Doremus. Butch: Trent Lehman. Prudence: Kim Richards. Guest Cast . . . Higgenbotham Botkin: Jack Albertson. Stinson: Joey Forman.

⑨ VIRGINIA GRAHAM ⒞
Guests: Karen Valentine ("Room 222") and Ricardo Montalban. Ellie Frankel Quartet. (60 min.)

⓭ BOOK BEAT—Interview ⒞
In "Stelmark: a Family Recollection," author Harry Mark Petrakis celebrates family life. Petrakis (who also wrote "A Dream of Kings") describes growing up in a colorful Greek-American community; and tells about his struggle to become a writer. Host: Bob Cromie.
[Repeated Sunday, Oct. 4, at 5:30 P.M.]

㉑ NET FESTIVAL
"Ellington on the Cote d 'Azur." On the French Riviera, Duke Ellington and Ella Fitzgerald cut loose in an all-Ellington concert, held in 1966 at the Antibes Jazz Festival. (60 min.)

㉛ FILM
㊶ MAURICIO GARCES ⒞
㊼ POLITO VEGA—Variedad ⒞

8:30 **② ③ HEADMASTER** ⒞
It takes the combined efforts of the headmaster, his wife and the athletic coach to prevent an honor student from cracking up. The boy is determined to match his brilliant older brother by scoring 100 on every test. Andy: Andy Griffith. Margaret: Claudette Nevins. Jerry: Jerry Van Dyke. Purdy: Parker Fennelly. Guest Cast . . . Billy: Mitch Vogel.

④ ⑳ NAME OF THE GAME ⒞
"Cynthia Is Alive and Living in Avalon" stars Robert Culp and Barbara Feldon, who turn in easy-going performances reminiscent of Culp's style in "I Spy." Culp plays a magazine reporter who poses as a thief to get an exclusive story on a kooky jet-setter turned protest organizer. (90 min.)

Guest Cast
Tyler . . . . .Robert Culp
Cynthia . . . . .Barbara Feldon
Les . . . . .Mickey Rooney
Pete . . . . .Tom Skerritt

⑤ DAVID FROST ⒞
Scheduled: Edward Bennett Williams, lawyer and associate of the late Vince Lombardi; the Carpenters; and film clips from Buster Keaton movies. (90 min.)

⑦ ⑧ PARTRIDGE FAMILY ⒞
Harry Morgan guest stars as Willie Larkin, an opportunistic motorist who turns a minor accident into a half-million dollar lawsuit. Song: "I'll Leave Myself a Little Time." Shirley: Shirley Jones. Keith: David Cassidy. Laurie: Susan Dey. Danny: Danny Bonaduce. Reuben: Dave Madden. Chris: Jeremy Gelbwaks. Tracy: Suzanne Crough.

The Partridge Family

New show 8:30
abc ⑦ ⑧

Friday
EVENING

OCTOBER 8, 1970

Thursday
EVENING

6:45 **47 NOTICIAS—Kevin Corrigan** ©
7:00 **2 CBS NEWS—Walter Cronkite** ©
3 TO ROME WITH LOVE ©
See Tuesday 9:30 P.M. Ch. 2.
4 20 NBC NEWS ©
5 I LOVE LUCY—Comedy
Ricky and Fred seem to do nothing
but watch the fights on TV.
7 ABC NEWS ©
8 TRUTH OR CONSEQUENCES ©
9 WHAT'S MY LINE?—Game ©
Panel: Bennett Cerf, Arlene Francis, Anita Gillette, Soupy Sales.
11 BEAT THE CLOCK—Game ©
Guest: Corbett Monica.
13 NEWSFRONT SPECIAL
21 PROFILE: LONG ISLAND
31 AROUND THE CLOCK ©
41 JUAN DE DIOS—Novela
47 SIMPLEMENTE MARIA—Novela
7:30 **2 3 FAMILY AFFAIR** ©
Comedienne Nancy Walker joins the
series as Emily Turner, the world's
sloppiest—and cheekiest—cleaning
woman. Bill hires her to help out
Mr. French, who is utterly appalled
at the idea. Bill: Brian Keith.
French: Sebastian Cabot.
4 20 FLIP WILSON ©
Scheduled: Perry Como, Denise
Nicholas ("Room 222"), comic Charlie Callas and singer-dancer Lola
Falana. Sketch: Geraldine Jones
(Flip) asks a psychiatrist (Perry)
about her boy friend's unwillingness
to get married. George Wyle orchestra. (60 min.)

Highlights
"I've Got You Under My Skin,"
"Hello, Young Lovers,"
"Impossible" ....Perry

5 TRUTH OR CONSEQUENCES ©
7 8 MATT LINCOLN ©
Exploring the problems of a young
drug addict . . . Matt tries group
therapy in an attempt to reach Nina
Conway, who uses drugs as a refuge
from loneliness and her neglectful
parents. Matt: Vince Edwards. Kevin: Michael Larrain. Jimmy: Felton
Perry. (60 min.)

Guest Cast
Nina Conway ..Belinda Montgomery
Doug Conway ....Jack Cassidy
Monica Conway ....Jeanne Bal
Georgie ....:Michael Bow
Nancy Cooper ...Dianne Turley
9 DIVORCE COURT—Drama ©
Charge: mental cruelty. Janice: Mal-
lory Jones. George: Jordan Rhodes.
11 STAR TREK—Adventure ©
McCoy falls in love with the high
priestess of a planet doomed to destruction. McCoy: DeForest Kelley.
Natra: Kate Woodville. Kirk: William
Shatner. (60 min.)
13 BOOK BEAT—Interview ©
John W. Gardner, chairman of the
National Urban Coalition, discusses
his book "The Recovery of Confidence." The work deals with the role
of dissent, increasing hostility to in-
stitutions, the loss of a sense of
community, the need for Governmen-
tal reforms, and "The hard, long, exciting task of building a new America." Host: Bob Cromie.
21 MAGGIE—Exercise
Maggie Lettvin show how to firm up
flabby inner thighs.
25 NET FESTIVAL—Profile
A profile of David Amram—composer

OCTOBER 21, 1970

UNITED NATIONS
Tapes of today's session of the UN General Assembly (if held) are shown on Ch. 31 from 9:45 to conclusion.

9:45 **㉑ UNEMPLOYMENT REPORT**
"Job Opportunities for Women."
㉛ FILM

10:00 **② ③ HAWAII FIVE-O** Ⓒ
Andrew Duggan is the heavy in "The Ransom." Duggan plays Obie O'-Brien, mastermind of a $250,000 kidnaping that backfires when Kono exchanges places with the victim, the young son of a millionaire. Question is, will the father pay that large a sum for Kono? McGarrett: Jack Lord. Dan: James MacArthur. Kono: Zulu. Chin Ho: Kam Fong. (60 min.)
Guest Cast
O'Brien .... Andrew Duggan
Nelson Blake .... Lloyd Gough
Earl .... Peter Bonerz

④ ⑳ FOUR IN ONE Ⓒ
"Our Man in Paris," the sixth and final "McCloud" episode, involves the New Mexico lawman in international intrigue. Chief Clifford is being held hostage by men who will kill him unless McCloud smuggles a million dollars into Paris on a police entry permit. McCloud's main problem is an armed escort he doesn't trust. McCloud: Dennis Weaver. Broadhurst: Terry Carter. Chief: J.D. Cannon. (60 min.)
Guest Cast
Anna Van Vliet .... Susan Strasberg
Mason .... Alfred Ryder

Wednesday
EVENING

Rissient .... John van Dreelen
Prideaux .... Marcel Hillaire
Polk .... Ken Scott
["San Francisco International," starring Lloyd Bridges, debuts in this time spot next week.]
⑤ NEWS—Bill Jorgensen Ⓒ
⑦ ⑧ DAN AUGUST Ⓒ
"In the Eyes of God" stars Bradford Dillman as a priest who recently left the church, but refuses to violate the secrets of the confessional. August is convinced that the multiple murderer he's after once made a confession to the priest, who will now be killed—despite the vow of silence he is so fiercely protecting. Dan: Burt Reynolds. Wilentz: Norman Fell. Rivera: Ned Romero. (60 min.)
Guest Cast
Matthew Costa .... Bradford Dillman
Wiley .... James Best
Edna Adams .... Donna Mills
[Read about Burt Reynolds in the article beginning on page 32.]
⑨ AVENGERS—Adventure Ⓒ
A top-level ministers' conference has become a nightmarish ordeal for key delegates. Mrs. Peel: Diana Rigg. Steed: Patrick Macnee. (60 min.)
⑪ NEWS—Lee Nelson Ⓒ
㉛ UNIVERSITY ROUNDTABLE
See Sunday 6:30 P.M. Ch. 31.
㊶ ELLAS—Novela
10:30 **⑬ FREE TIME** Ⓒ
Julius Lester is one of tonight's hosts. See the Close-up on page A-50 for details. (Live; 90 min.)
㉛ CASPER CITRON—Interview
See Monday 1 P.M. Ch. 31.

—Advertisement

OCTOBER 24, 1970

restore the appearance of his horribly disfigured daughter. Howard Vernon, Conrado San Martin. (90 min.)

⑬ NET PLAYHOUSE—Drama Ⓒ
"The Serpent," an unconventional re-creation of the story of Genesis. Members of the avant-garde Open Theatre perform in their street clothes on a bare stage. They writhe upon it in dance-like movements; their voices join in primitive chants. Playwright Jean-Claude Van Itallie: "the work is a ceremony . . . an exploration of ideas and images." (Rerun; 90 min.)

9:00 ② ③ ARNIE Ⓒ
Jack and Mabel Albertson, who are brother and sister, make a rare dramatic appearance together, playing Mr. and Mrs. Charlie Robinson. The episode deals with Arnie's reluctance to tell a healthy, vigorous 65-year-old that his retirement is now compulsory. Arnie: Herschel Bernardi. Lil: Sue Ane Langdon. Ogilvie: Herbert Voland. Majors: Roger Bowen. Fred: Alan Soule. Felicia: Elaine Shore.

④ ⑳ MOVIE—Adventure Ⓒ
"The Aquarians." A small East African nation sends a distress call to scientist Luis Delgado and his team of undersea explorers. Something is killing the fish in the area, but the pollutant is a total mystery. This made-for-TV movie was filmed at producer Ivan Tors' underwater studios in Miami and the Bahamas. (2 hrs.)

Cast
Luis Delgado ...Ricardo Montalban

Saturday
EVENING

Alfred Vreeland ....Jose Ferrer
Barbara Brand .....Kate Woodville
Bob Exeter ....Lawrence Casey
Jerry Hollis ....Tom Simcox
Official ....Leslie Nielsen
Ledring ....Chris Robinson

⑨ PRO HOCKEY Ⓒ
Rangers vs. Minnesota North Stars. See the Close-up opposite for details. (Live)

㉛ FORSYTE SAGA
Chapter 24. Jon and Fleur have an uneasy reunion; Michael becomes involved in clearing a slum. Soames: Eric Porter. Fleur: Susan Hampshire. Michael: Nicholas Pennell. Jon: Martin Jarvis. (Rerun; 60 min.)

㊶ LA CONSTITUCION—Novela

㊼ MOVIE—Drama
"Emboscada Mortal." (Mexican; 1964) In Spanish. Tony Aguilar, Dacia Gonzales. (90 min.)

9:30 ② ③ MARY TYLER MOORE Ⓒ
Comedienne Nancy Walker plays Rhoda Morgenstern's mother, charter member of the keep-'em-feeling-guilty school of child rearing, who has descended upon a bewildered Mary. Rhoda: Valerie Harper. Phyllis: Cloris Leachman. Grant: Edward Asner. Murray: Gavin MacLeod.

⑦ ⑧ MOST DEADLY GAME Ⓒ
"Gabrielle." Criminologist Jonathan Croft finds a former military intelligence buddy murdered. The only clue, a cryptic poem, sends Jonathan sifting through the ashes of his recent past to solve the mystery of $500,000 that disappeared during the dead man's last mission. Jonathan: George Maharis. Vanessa:

Thursday
EVENING

NOVEMBER 12, 1970

⓫ PERRY MASON—Mystery
A teen-age boy appears in Mason's office seeking legal advice. Nicky Renzi: Bobby Clark. Frank Anderson: James Anderson. Art Crowley: Elisha Cook. Mason: Raymond Burr. (60 min.)

㉑ GERMAN—Education
㉛ ONE TO ONE—Art Ⓒ
㊶ PASION GITANA—Novela

9:30 ❹ ⓴ NANCY Ⓒ
Seeking a quiet honeymoon far from Abby and the Secret Service, newlyweds Adam and Nancy Hudson carry out a plan to slip away to romantic Lake Bageneehawk. Adam: John Fink. Nancy: Renne Jarrett. Abby: Celeste Holm. Everett: Robert F. Simon. Guest Cast . . . Manager: Lew Brown. Bellboy: Dennis Fimple. Maid: Mimi Dillard.
[This episode was filmed on location at Lake Arrowhead, Cal.]

❼ ❽ ODD COUPLE Ⓒ
Felix creates a monster when he talks Oscar into posing for a men's cologne advertisement. Carried away by the glamour (and the girls), Oscar sets the scene for one of Felix's worst sinus attacks. Oscar: Jack Klugman. Felix: Tony Randall. Guest Cast . . . Rudy: Albert Brooks. Whitehill: Peter Brocco.
[A review of this series appears in next week's TV GUIDE.]

㉑ NEWS—George Morris
㉕ COMMUNITY REPORT
㉛ NEWS Ⓒ
㊸ NATACHA—Novela

UNITED NATIONS
Tapes of today's session of the UN General Assembly (if held) are shown on Ch. 31 from 9:45 P.M. to conclusion.

9:45 ㉑ COVER STORY
㉛ ITALIAN PANORAMA
10:00 ❹ ⓴ DEAN MARTIN Ⓒ
Scheduled: Zero Mostel, Tony Bennett and singer Gloria Loring. Comedy: Zero sells Dean a secret weapon and joins in a Shakespearian spoof. Ken Lane, Golddiggers, Les Brown orchestra. (60 min.)
Highlights
"If I Were a Rich Man" ....Zero

"Something," "Here, There and Everywhere" ....Tony
Broadway medley ....Dean, Tony
"Honey Wind Blows" ....Gloria
"Sing, You Sinners"
.............Tony, Golddiggers
"Is It True What They Say About Dixie?" "Make the World Go Away" ....Dean

❻ NEWS—Bill Jorgensen Ⓒ
❼ ❽ IMMORTAL Ⓒ
"The Queen's Gambit." A beautiful woman lures Ben out of the frying pan into the fire: from Fletcher's hands into those of Simon Brenner, a very wealthy young man whose estate lies deep in a seemingly inescapable jungle. Brenner has grandiose plans for Ben: a centuries-long partnership to better the world. Ben: Christopher George. Fletcher: Don Knight. (60 min.)
Guest Cast
Sigrid Bergen ......Lee Meriwether
Simon Brenner .....Nico Minardos
Dr. LiLorenzo ....Karl Swenson
[Why does Christopher George call himself "the lion in the jungle"? See next week's TV GUIDE.]

❾ AVENGERS—Adventure Ⓒ
Steed and Emma pursue two agents, who paid a huge sum for an alchemist's invisiblity formula. Steed: Patrick Macnee. (60 min.)

⓫ NEWS—Lee Nelson Ⓒ
⓭ SOUL! Ⓒ
Return: "Soul!" Begins its third season as Ossie Davis, Ruby Dee (Mrs. Davis) and gospel singer Marion Williams join in a tribute to American writer Langston Hughes (1902-1967). Ossie and Ruby read from Hughes' writings, which mirror his feelings about Harlem, the South, slum lords and drug addiction. Marion offers "Oh, Happy Days," "How I Got Over," "When Was Jesus Born?" and "Battle Hymn of the Republic." (Rerun; 60 min.)
[Repeated Sat., Nov. 14, at 7:30 P.M.]

㉛ SPORTS ALMANAC
㊶ ELLAS—Novela
10:30 ㊶ LUCECITA BENITEZ Ⓒ
㊸ EL ALMA NO TIENE COLOR Ⓒ
11:00 ❷ NEWS—Bob Young Ⓒ
❸ NEWS—Bill Hanson Ⓒ

JANUARY 12, 1971

Tuesday
EVENING

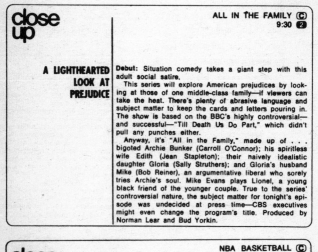

close up

ALL IN THE FAMILY Ⓒ
9:30 ②

**A LIGHTHEARTED
LOOK AT
PREJUDICE**

Debut: Situation comedy takes a giant step with this adult social satire.

This series will explore American prejudices by looking at those of one middle-class family—if viewers can take the heat. There's plenty of abrasive language and subject matter to keep the cards and letters pouring in. The show is based on the BBC's highly controversial—and successful—"Till Death Us Do Part," which didn't pull any punches either.

Anyway, it's "All in the Family," made up of . . . bigoted Archie Bunker (Carroll O'Connor); his spiritless wife Edith (Jean Stapleton); their naively idealistic daughter Gloria (Sally Struthers); and Gloria's husband Mike (Bob Reiner), an argumentative liberal who sorely tries Archie's soul. Mike Evans plays Lionel, a young black friend of the younger couple. True to the series' controversial nature, the subject matter for tonight's episode was undecided at press time—CBS executives might even change the program's title. Produced by Norman Lear and Bud Yorkin.

close up

NBA BASKETBALL Ⓒ
10:00 ⑦ ⑧

NBA ALL-STAR GAME

Special: The 21st All-Star game in San Diego, Cal. Chris Schenkel and Jack Twyman report the action. (Live)

At press time, sportswriters and broadcasters had chosen eight players per team. Six were to be added to each squad by NBA coaches.

Eastern Conference			
10 Frazier	Knicks	19 Reed	Knicks
11 Monroe	Bullets	25 Johnson	Bullets
12 White	Celtics	32 Cunningham	76ers
17 Havlicek	Celtics	41 Unseld	Bullets
Western Conference			
1 Robertson	Bucks	32 Lucas	Warriors
10 Love	Bulls	33 Alcindor	Bucks
13 Chamberlain	Lakers	42 Hawkins	Suns
21 Bing	Pistons	44 West	Lakers

TV GUIDE A-63

WEDNESDAY, SEPTEMBER 15, 1971

SEPTEMBER 15, 1971

Tate: Robert Hooks. Bronson: Michael Parks. (60 min.)

⓭ FIRING LINE Ⓒ
Host William F. Buckley plans to debate the leading contenders in the Presidential race. First up is Sen. Fred Harris (D-Okla.), who discusses his prospects and President Nixon's economic policies in an interview taped yesterday. (60 min.)

㉛ POLICE COMMISSIONER Ⓒ
㊶ LA LLORONA—Novela Ⓒ

9:30 **⓻ ⑧ SHIRLEY'S WORLD —Comedy** Ⓒ
Debut: Photojournalist Shirley Logan, (Shirley MacLaine) begins her globe-trotting adventures in London. "The Berkeley Club Caper" has her bent on infiltrating a veddy stuffy men's club to interview a VIP. Dennis Croft: John Gregson.

Guest Cast
Sir Harold ..... Charles Lloyd-Pack
Commissionaire .... Erik Chitty

㉑ NEWS
㉛ NASA FILM Ⓒ
㊶ MULATA DE CORDOBA Ⓒ

9:45 **㉑ SPANISH—Instruction**

10:00 **❷ ③ MANNIX** Ⓒ
Fast-paced action is the keynote of this private-eye series beginning its fifth season in this new time slot. Tonight: "Dark So Early, Dark So Long," a whodunit involving murder and the detective's ex-girl friend. Mannix: Mike Connors. Peggy: Gail Fisher. (60 min.)

Guest Cast
Leslie Fielding ... Rosemary Forsyth
Lt. Adam Tobias ...... Robert Reed
Glen Fielding ...... Guy Stockwell
Mrs. Oliver ... Josephine Hutchinson

❹ ⑳ NIGHT GALLERY —Drama Ⓒ
This former miniseries, now promoted to full-time status, starts a new season with four bizarre tales . . . 1. Clint Howard plays a young seer who suddenly refuses to continue making predictions. Wellman: Michael Constantine. Grandfather: William Hansen. 2. Joseph Campanella puts his teeth into a role as a Dracula type trying to hire a baby sitter (Sue Lyon). 3. "The Hand of Borgus Weems" is a hand with a murderous mind of its own. Lacland:

Wednesday
EVENING

George Maharis. Dr. Ravdon: Ray Milland. Dr. Innokenti: Patricia Donahue. Susan: Joan Huntington. Kazanzakis: Peter Mamakos. 4. Leslie Nielsen as the Phantom of the Opera, and Mary Ann Beck as his surprised—and surprising—victim. Host: Rod Serling. (60 min.)

❺ NEWS—Bill Jorgensen Ⓒ
⓻ ⑧ MAN AND THE CITY —Drama Ⓒ
Debut: Anthony Quinn's new series begins with "Hands of Love," a poignant drama about Mayor Thomas Jefferson Alcala's efforts to help a deaf couple adopt a normal boy. Bess Boyle's script offers a provocative insight into the problems of the handicapped in society. Alcala: Anthony Quinn. Andy Hays: Mike Farrell. Marian Crane: Mala Powers. (60 min.)

Guest Cast
Ellen Lewis .... June Lockhart
Ann Larrabee ...... Audree Norton
Richard Larrabee .... Lou Fant
Continued on next page

CBS❷2
MANNIX
NEW TIME, 10PM
MIKE CONNORS CONTINUES HIS ONE-MAN ASSAULT ON CRIME. GAIL FISHER IS HIS ONE-WOMAN BACKUP TEAM.

TV GUIDE A-75

SEPTEMBER 16, 1971

Ketcham: John Beck. Johnson: Paul
Hampton. (60 min.)

Guest Cast

Scully One ....John Quade
Scully Two ....Jesse Wayne
Salter ....John Harding
Mrs. Prine ....Florence Lake

⑦ ⑧ LONGSTREET—Drama ©
Debut: "The Way of the Intercepting
Fist" pits blind insurance investigator Michael Longstreet against a
hijack ring grabbing a million dollars a month in merchandise from
the New Orleans docks. Longstreet's
solution involves challenging a
dock-yard bully to a brawl. Bet on
Longstreet (James Franciscus). Nikki: Marlyn Mason. (60 min.)

Guest Cast

Li Tsung ....Bruce Lee
Sergeant Cory ....Lou Gossett
Bolte ....John Milford
Harv ....Johnny Haymer

㉛ POLICE COMMISSIONER ©
㊶ MULATA DE CORDOBA ©

9:30 **㉑ NEWS**
㊿ BOOK BEAT ©
See Tuesday 2 P.M. Ch. 31.
㊸ VIGOREAUX—Variedad ©

9:45 **㉑ SPANISH—Instruction**

10:00 **❹ ⑳ DEAN MARTIN** ©
Return: Art Carney, Petula Clark,
Liberace and actor Richard Castellano ("Lovers and Other Strangers")
help Dean open his seventh season.
Art and Liberace don hot pants in a
musical tribute to the short shorts,
and Art plays the President in a
sketch about national budgetary
problems. Tonight's situation-come-
dy spot centers on a morality conflict between mother (Kay Medford)
and daughter (Marian Mercer). (60
min.)

Highlights

"Non Dimenticar" ....Dean
"For Once in My Life," "We've
  Only Just Begun"; dream medley
  ....Dean, Petula
"On the Good Ship Lollipop"
  ....Ding-a-Lings
"Mean to Me," "Why Can't You
  Behave?" ....Dean, Ding-a-Lings

❺ NEWS—Bill Jorgensen ©
**⑦ ⑧ OWEN MARSHALL,
COUNSELOR AT LAW** ©
Debut: A man's good reputation is

Thursday
EVENING

jeopardized in "Legacy of Fear."
The drama follows attorney Owen
Marshall's efforts to help a friend
whose life could be wrecked by a
20-year-old incident that sent him
to prison. Marshall: Arthur Hill. Jess:
Lee Majors. Melissa: Christine Mat-
chett. (60 min.)

Guest Cast

Jerry Woods ......Glenn Corbett
Ivan Bock ....Dane Clark
Charlie Gianetta .....Pat Harrington
Marion Lerman ....Marian McCargo
[Patricia Morrow of "Peyton Place"
gave up acting to study law. See
next week's TV GUIDE.]

❾ BILLY GRAHAM CRUSADE ©
Special: See Saturday 11 P.M. Ch.
11 for details. (60 min.)

⑬ SOUL!—Music ©
Host Carla Thomas welcomes singers
Rufus Thomas (her father) and
Jimmy Scott; and Ida Lewis, former
editor of Essence. (60 min.)

㉛ FOCUS ON THE ARTS ©
See Wed. 2 P.M. Ch. 31. (60 min.)

㊶ MUY JOVEN PARA BESAR

TV GUIDE A-87

SATURDAY, SEPTEMBER 18, 1971

Saturday
EVENING

ries: the situations revolve around her role as Sandy Stockton, student teacher and star of TV commercials. Tonight, Sandy gets taken for a ride when she agrees to do pitches for a high-pressure used-car dealer. Alice: Valorie Armstrong. Kate: Kathleen Freeman. Guest Cast . . . John E. Appleseed: Tom Bosley. George: Matt Clark. Sue Ellen: Cynthia Hull.

④ ㉒ GOOD LIFE—Comedy Ⓒ
Debut: Tired of keeping up with the Joneses? Drop out in style with Albert and Jane Miller, who chuck their suburban woes to work as butler and cook for a millionaire. The dropouts are Larry Hagman and Donna Mills. Representing the establishment: David Wayne as their millionaire employer; Hermione Baddeley as his sister Grace; and Danny Goldman and his son Nick. Guest Cast . . . Chauffeur: Ben Wright. Manager: Alan Oppenheimer. Salesman: Dave Morick. Englishman: Gil Stuart. Cosgrove: Jack Riley.

❺ MOVIE—Science Fiction
"The Creeping Unknown." (English; 1955) A new rocket is sent hurtling into space with three men on board. When it crash-lands in England, there is only one man left on board. Brian Donlevy, Jack Warner, Margia Dean, Gordon Jackson. (90 min.)

❼ ⑧ MOVIE—Drama Ⓒ
Debut: "The Birdmen" (1971), a World War II escape drama, launches this TV-movie series. The story follows a daring attempt by Allied POWs to fly out of a top-security prison—in a glider constructed from bits of anything they can lay their hands on. David Kidd's script is based on a wartime incident. Directed by Philip Leacock ("The War Lover"). (90 min.)

Cast
Cook	Doug McClure
Crawford	Chuck Connors
Schiller	Richard Basehart
Brevik/Volda	Rene Auberjonois
Fitz	Tom Skerritt

DICK VAN DYKE Ⓒ
9:00 ❷ ③

SEEMS LIKE OLD TIMES

Debut: Dick Van Dyke and writer-director Carl Reiner strike again. (Last time out they produced the now-classic Van Dyke series of the early Sixties.)

Van Dyke plays TV talk-show host Dick Preston, whose work often leads to trouble. Like tonight, when a chimpanzee cons Dick into blowing smoke rings—and a no-smoking pledge.

Wife Jenny is played by Hope Lange, winner of two Emmys in "The Ghost and Mrs. Muir." Comic Marty Brill plays Dick's manager Bernie; Nancy Dussault is Bernie's wife Carol; Fannie Flagg plays Dick's sister Mike; and Angela Powell is the Prestons' daughter Annie.

Dick Van Dyke and Hope Lange

Friday
EVENING

41 **SYLVIA Y ENRIQUE**

47 **CHUCHO Y LISSETTE** Ⓒ

8:00 **2** **3** **CHICAGO TEDDY BEARS—Comedy** Ⓒ
Hans Conried has a small role as a far-out forger in a show about the latest battle of wits between Linc and his mobster cousin. Linc: Dean Jones. Nick: Art Metrano. Latzi: John Banner. Marvin: Marvin Kaplan.

4 **20** **D.A.—Crime Drama** Ⓒ
The crime is rape. The DA's challenge: proving that the victim can identify her hooded assailant beyond any doubt. Ryan: Robert Conrad. Staff: Harry Morgan. Katy: Julie Cobb. Ramirez: Ned Romero.

Guest Cast

Alice Conroe	Anne Whitfie'd
Judge Simmons	Victor Izay
Sgt. Ed Pettis	Lew Brown

5 **TRUTH OR CONSEQUENCES** Ⓒ

7 **BRADY BUNCH** Ⓒ
High school football may be rough on some parents, but it's rougher on kids like Greg, who's moping on the sidelines with a broken rib and the sting of defeat. Greg: Barry Williams. Mike: Robert Reed. Carol: Florence Henderson. Alice: Ann B. Davis. Bobby: Mike Lookinland. Guest Cast . . Coach: Bart LaRue. Linette: Elvira Roussel.

8 **SEVEN SEAS** Ⓒ
The Indian Ocean extends from the Persian Gulf, rich in oil and history, to Antarctica. (60 min.)

11 **PLEASE DON'T EAT THE DAISIES—Comedy** Ⓒ
Drawn into their neighbors' family argument, Joan and Jim soon find themselves fighting.

13 **ADVOCATES—Debate** Ⓒ
"Should Congress make strikers ineligible to receive public aid?" Proponents say public aid lengthens walkouts; opponents argue that cutting aid punishes strikers. Witnesses in favor of withholding aid include professors Thomas G.F. Christiansen and Armand Thieblot. Against: Rep. William D. Ford (D-Mich.) and steelworkers adviser John J. Sheehan. Advocates: Howard Miller and William Rusher. (60 min.)

21 **NET PLAYHOUSE**
"The Duel," an adaptation of Anton Chekhov's short story about the decaying love affair between a Russian civil servant (John Wood) and a married woman (Katharine Blake). (90 min.)

31 **UNIVERSITY BROADCAST LAB** Ⓒ
A rock concert featuring the sounds of "Blackjack."

47 **FIESTA CON VELDA** Ⓒ

8:30 **2** **3** **O'HARA, U.S. TREASURY —Crime Drama** Ⓒ
Air-cargo thefts spark an undercover investigation by special agent O'Hara (David Janssen). Exteriors filmed at Los Angeles International Airport. (60 min.)

Guest Cast

Karl Blake	Milton Selzer
Ben Hazzard	Stacy Harris
Dick Miles	Joe E. Tata
Harry Wilson	Leo Gordon
Pete McAdams	Richard Tate

4 **20** **CHRONOLOG** Ⓒ
Special: The November edition of NBC's newsmagazine. Press-time articles include . . . 1. Bob Rogers' report on war-ravaged East Pakistan. On screen: children scavenging in the rubble of shelled villages; West Pakistani troops on the march; and thousands of refugees pouring over the border into India. 2. On the lighter side: football fever in Omaha, where the Nebraska Cornhuskers are vying for the national championship. 3. "The Psychology of Imprisonment," a Stanford University class where students act out the roles of guards and prisoners. (2 hrs.) [Pre-empts the network movie.]

5 **DAVID FROST** Ⓒ
Scheduled guests in Hollywood: Andy Griffith, Lynn Redgrave, George Hamilton, comics Patchett and Tarses, and the Rev. Frank Stranges (who believes there are outer-space visitors living on earth). (90 min.)

7 **PARTRIDGE FAMILY** Ⓒ
The price of fame is costing teenage idol Keith more than he knows. Brother Danny is secretly selling Keith's possessions to buy mom a mink coat. Song: "Every Little Bit of You." Danny: Danny Bonaduce. Keith: David Cassidy. Shirley: Shirley Jones. Chris: Brian Forster.

DECEMBER 27, 1971

Court President: Clifford Parrish.
Prosecutor: David Kelb. (60 min.)

47 DESAFIANDO A LOS GENIOS C

8:30 5 DAVID FROST C
Scheduled: directors Vittorio de
Sica and Peter Bagdanovich; humorist Jean Shepherd, Cloris Leachman ("The Mary Tyler Moore Show")
actress-model Cybil Shepherd, and
singers Ella Mitchell and Helen
Reddy. Billy Taylor orchestra. (90
min.)

9 YEAR IN REVIEW C
31 NEW YORK REPORT C
47 MOVIE—Double Feature
1. "Mi Bella Genio." (Drama; 1951)
Dubbed in Spanish. Ray Middleton.
2. "Apartamento de Soltero." (Comedy; 1931) Dubbed in Spanish. Irene
Dunne, Lowell Sherman.
[Double feature: 4 hrs.]

9:00 2 3 HERE'S LUCY C
Comic calamities start brewing
when Harry gives Lucy a very unexpected $50 raise. Lucy: Lucille Ball.
Harry: Gale Gordon. Kim: Lucie

Monday
EVENING

Arnaz. Guest Cast . . . Dr. Cunningham: Parley Baer, Mary Jane: Mary
Jane Croft.

4 20 MOVIE—Drama C
Dark secrets and a crime of passion
. . . Part 2 of Thomas Hardy's "Far
from the Madding Crowd." Julie
Christie heads the cast in this 1967
film. For details, see the Close-up on
page A-16. (2 hrs.)

7 8 NORTH-SOUTH SHRINE GAME C
Special: The 26th annual North-South Shrine Game from Miami.
North: Iowa State coach Johnny
Majors has strong-armed QBs in
Northwestern's Maurie Diagneau (12)
and his own Dean Carlson (10). Wisconsin's Alan Thompson (22) leads
the ground game. South: Tennessee
coach Bill Battle will feel right at
home with his own star RB Curt
Watson (31) and All-America DB
Bobby Majors (47). Utah receiver
Fred Graves (86) caught 45 passes
this year. Frank Gifford and Don
Meredith report the action. (Live)

close up

SONNY & CHER COMEDY HOUR
C **10:00 2 3**

A SUMMER HIT IS BACK

Return: Sonny and Cher bring back their weekly
variety hour, opening the series with an operatic spoof of "All in the Family."

Robert Merrill is Archie Bunker, and Harvey
Korman plays both priest and rabbi, getting a
double dose of Bunker bigotry.

Cher vamps through blackouts as Nefertiti, Dietrich and Sadie Thompson; and sings "One Tin
Soldier," an antiwar song illustrated by a John
Wilson cartoon.

Glenn Ford and Carroll O'Connor make cameo
appearances. (60 min.)

Highlights
"Gypsies, Tramps and Thieves" . . . . Cher
"All I Ever Need Is You," "Love Grows," "Where
You Lead" . . . . Sonny, Cher

TV GUIDE A-39

Sunday
EVENING

❶ COMMENT! ©
Scheduled guests and topics; former ambassador Averell Harriman (recollections of Nikita Khrushev) and writer Gerald Tannebaum (life in the People's Republic of China).

❸ MOVIE—Western ©
"Man of the West." (1958) A tough tale of three people held captive by brutal outlaws in 1874 Texas. Gary Cooper, Julie London. (2 hrs.)

❼ MOVIE—Fantasy ©
"Journey to the Center of the Earth." (1959) Adaptation of Jules Verne's tale of a professor and his star pupil who explore a strange world. James Mason, Pat Boone. (2 hrs.)

❾ MOVIE—Drama ©
"Coogan's Bluff." (1968) Clint Eastwood as an Arizona deputy sheriff sent to New York. (2 hrs.)

❹ NOTICIAS—Frank Saldana ©

6:30   **❷ ❹ NBC NEWS—Utley ©**
❻ ELECTRIC COMPANY —Children ©
No. 81. Jimmy Boyd as J. Arthur Crank in a lesson on "tion."

FEBRUARY 20, 1972

❹ ALCALDIA DE MACHUCHAL
❺ PETER GUNN—Crime Drama
7:00   **❷ NEWS—Jim Lawrence ©**
❹ FACE THE STATE ©
❻ ❽ WILD KINGDOM ©
How Southwest African Bushmen hunt with primitive weapons.

❽ IT TAKES A THIEF ©
Mundy (Robert Wagner) adds spice to a billionaire's heist by beating it with his girl friend. (60 min.)

❻ ZOOM—Children ©
1. Films about the moon and a 12-year-old ballerina. 2. A Biblical production number called "Noah."
[Repeated Tuesday, 7:30 P.M. and February 26, 4:30 P.M.]

❹ SANTIAGO GREVI ©
❹ PAPA LO SABE TODO
7:30   **❷ ❹ MOVIE—Drama ©**
"Ben-Hur" concludes with the classic chariot race—an 11-minute sequence that took two months to shoot and cost $1,000,000. The 1959 Biblical epic won a record 11 Academy Awards, including best picture, actor (Charlton Heston), director

---

close up

PRESIDENT NIXON'S TRIP TO CHINA

President Nixon's historic visit to the People's Republic of China begins tonight, with his arrival at 10:30 P.M. (Monday, about noon in China).

At press time, the White House had not announced the exact times of the various telecasts of the President's activities, but extensive live and film network coverage is planned. Likely telecast times are regularly scheduled news periods, and early morning and late evening hours. (Background information on coverage of the trip on page A-1.) Announced programming:

ABC
Howard K. Smith anchors a half-hour preview of the trip on Saturday, 10:30-11 P.M. ABC covers the President's arrival on Sunday, 10:30 P.M., and plans early morning reports (with Frank Reynolds) and late evening telecasts (Smith anchors). China correspondents: Harry Reasoner, Howard Tuckner, Ted Koppel and Tom Jarriel.

CBS
Tonight: a report on changing attitudes toward the Chinese (6-7 P.M.), and a special edition of "60 Minutes" (9:30-10:30 P.M.) covering the President's arrival. Walter Cronkite, Bernard Kalb, Eric Sevareid and Dan Rather report from China.

NBC
Coverage is planned on "Today" and in the 11:30-12 mid. time slot through the week. Correspondents: Barbara Walters, Herbert Kaplow and John Rich.

PBS had not announced coverage at press time.

JULY 5, 1972

8:00 ❷ ③ MELBA MOORE AND CLIFTON DAVIS Ⓒ
Nancy Wilson and comic Marvin Braverman spice up the series' finale. Comedy highlight: Marvin playing a rookie cop in a missing-radio caper. Timmie Rogers, Ron Carey, Dick Libertini. (60 min.)

Highlights
"What a Little Moonlight Can Do," "Greatest Performance of My Life" ....Nancy
"For All We Know," "Your Feet's Too Big" ....Melba
"Fire and Rain" ....Clifton
"Looking Through the Window," "Here Comes the Sunrise" ....Melba, Clifton
[Last show of the series. Debuting July 19: comic David Steinberg's free-wheeling variety hour.]

❹ ⑳ ADAM-12 Ⓒ
An ecology crusader gives a manufacturer a taste of his own pollution in one of this week's cases. Malloy: Martin Milner. Reed: Kent McCord. MacDonald: William Boyett. (Repeat)

Guest Cast
Curtis Atherton ......Russ Conway
The Ferret ......Steve Franken
Larry Dent ......Jordan Rhodes
Doctor ....Marvin Miller

❺ TRUTH OR CONSEQUENCES Ⓒ

❼ ⑧ THE SUPER—Comedy Ⓒ
Super turns Cupid: Joe's trying to match his spaghetti-plate brother with a lox-and-bagel tenant. Joe: Richard S. Castellano. Frankie: Philip Mishkin. Francesca: Ardell Sheridan. Anthony: B. Kirby Jr. Joanne: Margaret E. Castellano. Guest Cast . . . Janice: Penny Marshall. Mrs. Stein: Janet Brandt.
[There's more to Richard Castellano than meets the eye. See the interview in next week's TV GUIDE.]

❾ BASEBALL Ⓒ
The San Diego Padres vs. the Mets at Shea Stadium. Lindsey Nelson, Ralph Kiner and Bob Murphy, sportscasters. (Live)

⓫ FATHER KNOWS BEST
Margaret's parents give her a small house they own, and she decides to rent it. Jim: Robert Young. Margaret: Jane Wyatt. Betty: Elinor Donahue. Bud: Billy Gray.

Wednesday
EVENING

⓭ A PUBLIC AFFAIR/ ELECTION '72 Ⓒ
A day in the life of a convention manager . . . profiling Richard Murphy as he prepares for next week's Democratic National Convention. In Miami Beach, cameras view Murphy working on security arrangements; setting up communication links for candidates, campaign workers and the media; and supervising hotel and transportation accommodations for delegates and their families.
[See next week's TV GUIDE for complete convention details.]

㉑ CONVERSATIONS IN THE PARK
The life and work of an artist are discussed in interviews with prominent Long Islanders. Special guest is composer Morton Gould. Host is Robert Payton, president, C.W. Post College.

㊶ ERNESTO ALONSO

㊼ CHUCHO Y LISSETTE Ⓒ

8:30 ❹ ⑳ McMILLAN & WIFE Ⓒ
The scent of perfume and a posh wake are the only clues in a string of jewel thefts masterminded by a criminal known as the Dutchman. Locations include Palm Springs, Cal. McMillan: Rock Hudson. Sally: Susan Saint James. (Repeat; 90 min.)

Cast
Freddie ....Claude Akins
Mayerling ......Edward Andrews
John T. Clarke ......Jon Cypher
Edmond Lake ......Clifford David
Paul Childs ......Richard Deacon
Iolanthe Simms ......Marj Dusay

❺ MERV GRIFFIN Ⓒ
Scheduled: All in the Family's" Sally Struthers, and seer Jeane Dixon. (90 min.)

❼ ⑧ CORNER BAR—Comedy Ⓒ
Six characters in search of relief: Harry and the gang are quarantined in Grant's Tomb — with a female version of Captain Bligh. Harry: Gabe Dell. Fred: J.J. Barry. Phil: Bill Fiore. Joe: Joe Keyes. Peter: Vincent Schiavelli. Meyer: Shimen Ruskin. Guest Cast . . . Henrietta: Betty Walker.

⓫ DRAGNET—Crime Drama Ⓒ
Friday and Gannon investigate the murder of a young widowed mother. Friday: Jack Webb. Gannon: Harry

SEPTEMBER 23, 1972

has a small but key role as the parents' attorney. DeSoto: Kevin Tighe. Gage: Randolph Mantooth. Brackett: Robert Fuller. Early: Bobby Troup. Dixie: Julie London. (60 min.)

Guest Cast

Mrs. Gentry	Anne Whitfield
Gentry	Roger Perry
Frankie	Christian Juttner
Jenny	Lori Busk
Sgt. Ed Pierce	William Bryant

⑤ MOVIE—Adventure
"The Heroes of Telemark." (1965) In Nazi-occupied Norway, Resistance fighters attempt to sabotage a top-secret research project. Kirk Douglas, Richard Harris, Ulla Jacobsson, Michael Redgrave. Arne: David Weston. Frick: Anton Diffring. Terboven: Eric Porter. Wilkinson: Mervyn Johns. Sigrid: Jennifer Hilary. Jensen: Roy Dotrice. (2 hrs.)

⑦ ⑧ ALIAS SMITH AND JONES
A shoot-out in Utah's Arches National Park climaxes "High Lonesome Country." Rod Cameron plays a bounty hunter tracking Curry and Heyes; Buddy Ebsen and Marie Windsor are the couple who got the boys into the mess. Curry: Ben Murphy. Heyes: Roger Davis. (60 min.)

Guest Cast

Phil Archer	Buddy Ebsen
Luke Billings	Rod Cameron
Helen Archer	Marie Windsor

⑨ MOVIE—Drama (BW)
"Pickup Alley." (English; 1957) A U.S. agent (Victor Mature) is sent to Europe to smash an international dope ring. Anita Ekberg, Trevor Howard. Salko: Alec Mango. (2 hrs.)

8:30 ② ③ BRIDGET LOVES BERNIE —Comedy
Sentimental journeys for the in-laws —complete with ethnic hangups. The Steinbergs have won an audience with the Pope; the Fitzgeralds have tickets to Israel. Problem is: how to switch tickets without causing offense? Sam: Harold J. Stone. Sophie: Bibi Osterwald. Walt: David Doyle. Amy: Audra Lindley. Bernie: David Birney. Bridget: Meredith Baxter. Moe: Ned Glass.
[Bridget and Bernie off-stage: meet the series' stars in next week's issue of TV GUIDE.]

Saturday
EVENING

⑪ HEE HAW
Patti Page ("Tennessee Waltz") and Charlie McCoy ("I'm So Lonesome I Could Cry") are the guests. Also: "Night Train to Memphis" (Buck), "Cold, Cold Heart" (Roy Clark), "Happiness Song" (Buddy Alan) and "Mama Tried" (Hagers). (60 min.)

⑬ VIOLENT UNIVERSE
Special: An absorbing geological survey examining the theory that shifting "plates" in the earth's crust reshape the continents. The global study includes films of Iceland, the Alps and Himalayas, Greenland and Iran. David Prowitt is host for the BBC-NET production, which was first telecast last February. (Repeat; 2 hrs.)

㉛ EVENING AT POPS
Guitar virtuosity . . . Charlie Byrd joins the Pops in a classical piece (Vivaldi's Concerto in D for Guitar and Orchestra) and swings with his trio to the bossa-nova beats of "El Gavilan," "Corcovado" and the haunting "Girl from Ipanema." Pops

TV GUIDE A-19

298

SEPTEMBER 24, 1972

Larry Corning and his buddy for the attempted murder of Larry's father. Judd: Carl Betz. Ben: Stephen Young. Lawrence Corning Sr.: Harold Gould. (60 min.)

⑬ TOY THAT GREW UP

㉛ JEAN SHEPHERD'S AMERICA
Flying is the subject. Shepherd is aloft in a Yankee Trainer to recall his dad's first plane trip (before airsick bags). Also: an exhibit of custom-built aircraft. (Repeat)

㊶ SANTIAGO GREVI

㊼ DETRAS DE LA FACHADA (BW)

7:30 ② ③ ANNA AND THE KING
—Comedy-Drama
Baby for sale. An old custom in Siam, but Anna's horrified—enough to buy the tot in defiance of the King (Yul Brynner). Anna: Samantha Eggar. Louis: Eric Shea. Prince: Brian Tochi.

④ ⑳ WORLD OF DISNEY
Movie slapstick with hoods vs. kids in the conclusion of "The Computer Wore Tennis Shoes." (1970) Sparking the comedy: a college student's information about an illegal gambling operation. (60 min.)

Cast

Dexter Riley	Kurt Russell
A.J. Arno	Cesar Romero
Dean Higgins	Joe Flynn
Prof. Quigley	William Schallert
Dean Collingsgood	Alan Hewitt
Chillie Walsh	Richard Bakalyan
Annie	Debbie Paine
Pete	Frank Webb

⑦ HALF THE GEORGE KIRBY COMEDY HOUR
Arte Johnson as a rabbi and George as a minister in a spoof on sermonettes. Songs: "Ballin' the Jack" and "Old-Time Religion".

㉛ ONE TO ONE—Art
An interview with Mary Stewart, assistant director for public education, Metropolitan Museum of Art.

㊼ CRIADA A LA ORDEN

8:00 ② ③ M*A*S*H—Comedy
How to deal with black marketeers, Korean War-style. Hawkeye: Alan Alda. Trapper John: Wayne Rogers. Henry: McLean Stevenson. Frank: Larry Linville. Hot Lips: Loretta Swit. Radar: Gary Burghoff. Guest Cast . . . Charlie Loo: Jack Soo.

⑤ LAWRENCE WELK

⑦ ⑧ FBI
"Eg of Desperation" centers on a man who arranges his own kidnaping to escape his wife and earn a quick $200,000 in getaway money. Erskine: Efrem Zimbalist Jr. (60 min.)

Guest Cast

Alan Graves	Michael Tolan
Dana Evans	Karen Carlson
Joan Graves	Jacqueline Scott
Lee Payne	Anthony Costello

⑪ STEUBEN DAY PARADE
Special: Taped highlights of yesterday's parade honoring General Friedrich Wilmer von Steuben. (60 min.)

⑬ EVENING AT POPS
Your Father's Mustache leads a sing-along to old standards: "Five Foot Two," "Ain't She Sweet?" "Yes Sir, That's My Baby" and "When the Saints Go Marching In." Pops selections include "Jalousie," the overture to "The Merry Wives of Windsor" and "Colonel Bogey March." (60 min.)

Double deceit

8:00
The FBI

A wealthy executive conceives an elaborate plot to deceive his wife and run away with his mistress. Starring Efrem Zimbalist, Jr.

close up

9:00 **CHINA** **7** **⑬**

"CHUNG KUO"—ANTONIONI'S "CHINA"

Special: A great moviemaker films "the ancient heart of the world."

Michelangelo Antonioni ("L'Avventura," "Blow-up")—a documentarian in his early career—toured China for five weeks this past year, amassing more than 100 hours of film. The result is a documentary that, says Antonioni "doesn't explain China, but observes it—an unknown human museum, a repertory of faces, gestures, habits."

High points of the journey . . . Peking, with scenes of strollers and a gymnastic exercise in Tien An Men Square, shoppers at a fruit market and workers in a cotton factory; the Ming tombs in a valley outside the capital; laborers on an agricultural commune in the southern province of Honan; restaurants, gardens and a Buddhist temple in the canal city of Suchow; Nanking school children in a relay race; and tea houses and junks in the port city of Shanghai.

Antonioni narrates the program, filmed for Italy's RAI-TV. (2 hrs.)

close up

AN AMERICAN FAMILY
9:00 **⑬**

A SLICE OF LIFE

Debut: Family drama that's compelling because it is real.

For seven months, producer Craig Gilbert filmed more than 300 hours in the life of the William C. Loud family. Because the Louds were a typical family group? No, says Gilbert, 'but we felt that if we stayed with them long enough, certain universals—about family behavior, attitudes and relationships—would surface." Margaret Mead, noting the series' impact in an article on page 21, writes that it offers "a new way in which people can learn to look at life, by seeing the real life of others."

There are 12 episodes. By way of introduction, the series opens with scenes from the last day's filming—at a New Year's Eve party in the Louds' California home. It is an affair mainly for the children—the Louds have separated. (Eight months after the filming was completed, the marriage had ended in divorce.) (60 min.)

TV GUIDE A-79

Sunday
EVENING

⑬ SKATING SPECTACULAR
Special: Olympic skating stars dominate this hour on ice. Highlights . . . '72 bronze medalist Janet Lynn demonstrates her freestyle; fellow Olympian Gordon McKellen Jr. dances to "MacArthur Park"; the Genesee (N.Y.) Figure Skating Club adds an international touch with a Viennese Waltz and a Dutch dance.
㊶ NOTICIAS—Saldana
㊼ TRIBUNA DEL PUEBLO
㊾ SILENT COMEDY FILM FESTIVAL (BW)
㊿ BOOK BEAT

6:30 ④ ⑳ NBC NEWS—Floyd Kalber
㉛ SKATING SPECTACULAR
Special: Popular melodies and lyrical skating dominate this hour on ice. Highlights: Olympic bronze medalist Janet Lynn demonstrates her freestyle skills; fellow Olympian Gordon McKellen Jr. dances to "MacArthur Park"; the Genesee (N.Y.) Figure Skating Club adds an international touch with a Viennese waltz and a Dutch dance treat. Filmed on location in Rochester, N.Y. (Repeat; 60 min.)
㊶ YO SE QUE NUNCA (BW)
㊼ CRIADA A LA ORDEN
㊿ WORLD PRESS

7:00 ② NEWS—Ralph Penza
③ FACE THE STATE
④ WILD KINGDOM
The sea otter—an endangered species. Scientists are shown capturing and tagging otters off the coast of California in an effort to save the animals from extinction. Also: family life, underwater and on the land.
⑦ OZZIE'S GIRLS
Ozzie's idea of harmless fun: telling Harriet he's just had a drink with the new and sexy next-door neighbor—but failing to mention she's 5 years old. The Nelsons portray themselves. Susie: Susan Sennett. Brenda: Brenda Sykes. Alice: Joie Guercio.
⑧ CONNECTICUT SCENE
Featured: a preview of this series' upcoming season.
⑪ AMERICA
Alistair Cooke's Peabody-winning history of the U.S.—"from the first of the Mohicans to the last of the hippies." Part 1, "New Found Land," focuses on early exploration, including Colum-

bus's voyage; French-Canadian trappers in the North; La Salle's trek across the Great Lakes. (60 min.)
⑬ ㊾ ㊿ ZOOM
The life style of a young Indian girl in Taos, N.M., is the subject of a film study. Also: how to make a terrarium; and a production number "Ubbi Dubbi Courtroom." (Repeat)
⑳ WILD KINGDOM
㊼ TRES MUCHACHAS

7:30 ② ③ PERRY MASON
Debut: Perry Mason is back, with Monte Markham in the shoes of Erle Stanley Gardner's famous courtroom lawyer. His first case: "The Horoscope Homicide," about a woman accused of murdering her astrologer husband. Sharon Acker plays Della Street. Perry's gal Friday; Harry Guardino is DA Hamilton Burger. Paul Drake: Albert Stratton. Tragg: Dane Clark. Gertie: Brett Somers. (60 min.)
Guest Cast
Nancy Addison . . Fionnuala Flanagan
Junius . . . . Peter Mark Richman

Continued on page A-41

Friday
EVENING

over alimony for his family. ("They're not supposed to be living in the manner to which they're accustomed.") Pat's dilemma: how to discipline five teen-age children with ideas of their own. (Repeat; 60 min.)
47 LA LOBA—Novela BW
49 MAN BUILDS, MAN DESTROYS
7:00 **2 CBS NEWS—Walter Cronkite**
3 THE WORLD AT WAR BW
4 20 NBC NEWS—John Chancellor
5 I LOVE LUCY—Comedy BW
Matters take a hairy turn when Lucy gives Ricky scalp treatments. Lucille Ball, Desi Arnaz.
7 ABC NEWS—Smith/Reasoner
8 TRUTH OR CONSEQUENCES
9 LUCILLE BALL—Comedy
Lucy gets plastered in a messy attempt to patch a hole in the kitchen ceiling. Viv: Vivian Vance. Mooney: Gale Gordon.
11 I DREAM OF JEANNIE—Comedy
A maniacal nutrition experiment makes guinea pigs of Tony and Roger. Tony: Larry Hagman. Porter: Paul Lynde. Roger: Bill Daily.
21 FOLK GUITAR—Instruction BW
Laura Weber reviews past lessons.
41 EL AMOR TIENE CARA DE MUJER BW
49 VINCE LOMBARDI
50 JAZZ SET
7:30 **2 DUSTY'S TRAIL**
4 POLICE SURGEON
A madman's false alarms endanger Dr. Simon Locke (Sam Groom). Joining in the hunt for the man is an old flame of Locke's (Skye Aubrey). Lieutenant Gordon: Larry Mann.
5 BEWITCHED—Comedy BW
Maurice Evans guest stars as Samantha's warlock-father. Endora: Agnes Moorehead. Samantha: Elizabeth Montgomery. Darrin: Dick York. Tate: David White.
7 LET'S MAKE A DEAL
8 DATING GAME
9 BOWLING FOR DOLLARS
11 MOVIE—Biography BW
"Pride of the Yankees." (1942) Stirring account of the life of Lou Gehrig (Gary Cooper), who rose to baseball heights. Teresa Wright, Walter Brennan, Babe Ruth, Dan Duryea, Virginia Gilmore, Ludwig Stossel, Elsa Janssen. Faithfully done

farewell speech is a moving climax to a fine film. (2 hrs., 30 min.)
13 41 WORLD PRESS
20 HUMAN DIMENSION
Topic: Christianity in Thailand.
21 EVENING AT POPS
"One of the most brilliant guitarists in the world," said Andres Segovia of Christopher Parkening. This young musician proves himself a worthy pupil of the great master in performing Albeniz' "Leyenda." (Repeat; 60 min.)
Boston Pops Selections
Medley from "South Pacific"
.... Richard Rodgers
"Malaguena" ...... Ernesto Lecuona
Overture to "Donna Diana"
.... Emil Reznicek

31 ON THE JOB
41 LOS POLIVOCES BW
47 COMO SER FELIZ EN EL MATRIMONIO
50 NEW JERSEY NEWS REPORT
8:00 **2 3 CALUCCI'S DEPT.—Comedy**
More bureaucratic buffoonery from the exasperating employees at Joe Calucci's state unemployment office. James Coco stars in this comedy series created by Renee Taylor and Joe Bologna ("Lovers and Other Strangers"). At press time, this week's episode had not been set. Gonzales: Jose Perez. Balukis: Candy Azzara. Fusco: Peggy Pope. Cosgrove: Jack Fletcher. Frohler: Bernard Wexler. Gordon: Rosetta Lenoire. Woods: Bill Lazarus.
4 20 SANFORD AND SON
Fred's ready to meet his maker. Thank Lamont, who's following an astrologist's advice and being nice to his dad. So nice, that the old man is sure he's dying. Fred: Redd Foxx. Lamont: Demond Wilson. Guest Cast . . . Aunt Esther: LaWanda Page.
5 THAT GIRL—Comedy
Unemployed Ann pressures Don to hire her as his secretary. Ann: Marlo Thomas. Don: Ted Bessell.
7 8 BRADY BUNCH
Joe Namath appears in this tale of a braggart whose bluff is called. Bobby says he knows the star QB—and now everyone wants to meet him. Bobby: Mike Lookinland. Cindy: Susan Olsen. Carol: Florence Henderson. Mike: Robert Reed. Alice: Ann B. Davis. Marcia: Maureen McCormick. Jan: Eve

Tuesday

EVENING

**31 MEN WHO MADE THE MOVIES
—Documentary**

"I'm an optimist," says director Frank Capra, "and comedy, to me, is victory. Victory over anything." Capra's comedies of the 30s and 40s blended social comment with large doses of idealism and sentiment. His typical hero, said one critic, was "a barefoot boy with brains." Now 76, Capra reflects on a career ranging from silent comedies with Harry Langdon to a string of box-office hits. Film highlights: "It Happened One Night" (1934), which swept five Oscars, including Best Director: "Mr. Smith Goes to Washington" (1939). (90 min.)

41 EL SHOW DE ROSITA

50 JERSEYFILE

9:00 4 20 BOB HOPE

Special: A spoof of TV private eyes highlights this tentatively scheduled hour with Bob Hope and guests Don Rickles, Redd Foxx, the Carpenters and Joey Heatherton. More comedy: a hillbillied Hope as football's hottest draft choice. Les Brown orchestra.

Songs

"We've Only Just Begun," "Top of the World" .... Carpenters
"It Amazes Me" .... Joey

11 BONANZA—Western

Two people come to Virginia City to take revenge on Joe (Michael Landon). Linda: Judi Meredith. Ben: Lorne Greene. Amos: Frank Overton. Hoss: Dan Blocker. (60 min.)

25 BORICUAS-PUERTO RICAN STUDIES (BW)

47 ESMERALDA (BW)

50 TO BE ANNOUNCED

9:30 2 3 HAWKINS—Crime Drama

William Windom stars as Hawkins' client, a neurotic man accused of murder. Helping the DA's case: the defendant believed that the victim murdered his son and had publicly vowed to take "A Life for a Life." Hawkins: James Stewart. (90 min.)

Guest Cast

Joe Hamilton	William Windom
Jeff Compton	John Ventantonio
Earl Coleman	James Hampton
Professor Hastings	Noam Pitlik

THE BLUE KNIGHT
10:00 4 20

Special: Four days in the life of a tough cop.

William Holden gives a gritty performance as Bumper Morgan, a 20-year man on the Los Angeles force. In his own words, he's a "helluva cop" whose beat has been his life. Now, facing retirement because he's "50 lousy years old, farsighted and just can't cut it anymore," Bumper is swept into the seamy hunt for a prostitute's killer.

This four-part adaptation of Joseph Wambaugh's novel continues in this time slot for the next three nights. Author Wambaugh discusses police work in a feature on page 13.

Lee Remick is featured as Cassie Walters, Bumper's fiancee. Sgt. Cruz Segovia: Joe Santos. Charlie Bronski: Sam Elliott. (60 min.)

Supporting Cast (Tuesday) . . . Marvin: David Moody. Yasser Hafiz: Jamie Farr. (Wednesday) . . . Grogan: Vick Tayback. Zoot: George Dicenzo. Hilliard: Raymond Guth. (Thursday) . . . Cites: Stanley Clay. Hughes: Kenneth Smedberg. (Friday) . . . Rudy: Ernest Esparza III. Ruthie: Gloria Leroy.

William Holden

Tuesday

EVENING

liam Wellman saw action firsthand in the French Foreign Legion and as a World War I flier. This portrait includes clips from among Wellman's best films . . . "Wings" (1927), Hollywood's first Oscar winner; "The Public Enemy" (1931), a vintage gangster classic. (60 min.)

41 ENTRE AMIGOS

50 JERSEYFILE

9:00 4 20 MAGICIAN—Drama
Joe Flynn stars as an amnesic ex-con who's being hunted by three wealthy businessmen. For reasons of their own, they want him to remember a robbery in which they were partners 30 years before. Tony: Bill Bixby. Max: Keene Curtis. (60 min.)

Guest Cast

Sam/George	Joe Flynn
Dunagan	George Murdock
Elizabeth Foster	Pamela Britton
Gordon	Hal Williams
Lubie	John Milford
Dr. Zabriskie	Yvonne Craig

11 BONANZA—Western
Townspeople are out to lynch a man

DECEMBER 11, 1973

acquitted of murder. John Degnan: Guy Stockwell. Will Griner: Walter Barnes. Jim: Ted Gehring. (60 min.)

21 CAROLING, CAROLING
Special: Christmas music by the Mormon Youth Symphony and Chorus . . . "O Thou That Tellest Good Tidings to Zion" and "For unto Us a Child Is Born," from Handel's "Messiah"; a choral rendition of "Twas the Night Before Christmas"; "Calypso Noel," a carol with a Caribbean beat; and Tchaikovsky's "Waltz of the Flowers." Host: Rex L. Cambell. (Repeat)

25 BORICUAS-PUERTO RICAN STUDIES BW

41 LA HIENA

47 EL HIJO DE ANGELA MARIA

50 TRIBUTE TO JIM CROCE

9:30 2 3 SHAFT—Crime Drama
"The Kidnaping" finds Shaft in a race with the clock to save a kidnaped suburbanite, while trying to evade local police who think he's a crook. Shaft: Richard Roundtree. Lieutenant Rossi: Ed Barth. (90 min.)

Guest Cast

Elliot Williamson	Paul Burke
Nancy Williamson	Karen Carlson
Matthew Potter	Nicolas Beauvy
Leo	Vic Brandt
Beck	Greg Mullavey
Hayden	Timothy Scott
Sheriff Bradley	Frank Marth

13 41 50 PERFORMANCE
A diverse repertoire of music from the Baltimore Chamber Players, featuring violinist Isador Saslov, pianist Lewis Shub, oboist Joseph Turner and Arthur Lewis on viola. Their selections include "Get Along Little Doggies."

21 BLACK EXPERIENCE
The role of blacks in the Civil War.

31 BERLIOZ'S REQUIEM—Music
Special: Hector Berlioz's celebrated Requiem "Grande Messe des Morts" is performed by the Music For Youth Symphony Orchestra and Concert Wind Ensemble. They are joined by a 150-voice chorus composed of students from the University of Wisconsin and Milton College. (90 min.)

10:00 4 20 POLICE STORY—Crime Drama
Martin Balsam stars as the "Man on a Rack," a veteran cop called before a ruthless internal affairs review board

Friday

EVENING

pleasantness at the Bellona Club."
Lord Peter Wimsey searches for the
last person to see the old soldier alive.
Wimsey: Ian Carmichael. Bunter: Derek Newark. (Repeat; 60 min.)
[After the drama: Early American
Windsor chairs are the subject of antiques expert John Kirk.]

**❸❶ INTERNATIONAL
PERFORMANCE**
Love motifs in music and dance . . .
1. Tchaikovsky's fiery and lyrical "Romeo and Juliet," a tone poem based
on Shakespeare's tragedy. 2. Dancers
Clair Motte and Milenko Banovitch,
seen earlier in "Firebird," have the
leads in a ballet set to Monteverdi's
17th-century dramatic cantata "The
Combat of Tancrede and Clorinda." (Repeat; 60 min.)

❹❶ ANA DEL AIRE
❹❼ PEREGRINA (BW)
❹❾ MALE MENOPAUSE—Report
**❺⓪ INTERNATIONAL
PERFORMANCE**

9:30 **❼ ⑧ TEXAS WHEELERS—Comedy**
Cagey Zack widens the credibility gap
when he offers a plethora of dubious
versions of how the family truck got
bashed in. Zack: Jack Elam. Truckie:
Gary Busey. Doobie: Mark Hamill. T.J.:
Tony Becker. Boo: Karen Oberdiear.
Sheriff: Noble Willingham. Bud: Dennis Burkley. Lyle: Bruce Kimball. Ray:
Sam Edwards.

㉑ FESTIVAL FILMS
The comic odyssey of a New Jersey
drum majorette who flees to New York
City unfolds in "Manhattan Melody," a
student film by Ken Weiderhorn.

10:00 **❹ ⓴ POLICE WOMAN—Crime
Drama**
William Windom stars as a modeling-agency owner who lures teen-age girls
into white slavery overseas. Cameo
roles: comedienne Judy Canova as an
upset mother and Olympic gymnast
Cathy Rigby as a police cadet. Pepper:
Angie Dickinson. Crowley: Earl Holliman. Styles: Ed Bernard. Royster:
Charles Dierkop. (60 min.)

Guest Cast
Ted Adrian ....William Windom
Debbie Sweet ......Kathleen Quinlan
Bonnie June ....Karen Lamm
Lieutenant Marsh ....Val Bisoglio
Don ....Harvey Jason

SEPTEMBER 20, 1974

Cora ....Jeane Byron
Rex ....Antonio Fargas
❺ NEWS—Bill Jorgensen
❼ ⑧ NIGHT STALKER—Drama
Gangsters are being savagely murdered by a former colleague they
thought they had killed and buried.
Kolchak: Darren McGavin. Vincenzo:
Simon Oakland. Monique: Carol Ann
Susi. (60 min.)

Guest Cast
Sposato ....Joseph Sirola
Francois Edmonds ....Earl Faison
Winwood ....Charles Aidman
Sweetstick ....Antonio Fargas
Mamalois ....Paulene Myers
[Postponed from an earlier date.]

⓫ NEWS—Joe Harper
⓭ PHANTOM INDIA
The caste system: how rigid a social
structure? Director Louis Malle's cameras focus on a village in Rajastan to
examine caste hierarchy on a community level. Each caste has its own wells
and economic activities. Also: religious and political aspects of the system. (60 min.)

㉑ HATHAYOGA—Exercise
Exercises: side push up, scale, side
leg split, hare head stand.
**❸❶ BLACK PERSPECTIVE ON THE
NEWS**
❹❶ HA LLEGADO UNA INTRUSA
❹❼ CELOS (BW)
❹❾ MASTERPIECE THEATRE
❺⓪ FESTIVAL FILMS
❻❼ NEWS—Gary Gunter

10:30 **❾ MOVIE—Musical**
"Viva Las Vegas" (1964) Sports-car
racers (Elvis Presley, Cesare Danova) compete for a night-club singer (Ann-Margret). Nicky Blair, William Demarest. Eye-filling musical
numbers and an exciting climactic
auto race. (2 hrs.)

⓫ THE TALKERS
John Bartholomew Tucker is the program host. (Live)
❸❶ CASPER CITRON
Kidney transplants and the treatment
of kidney disease are discussed.
❹❼ NOTICIAS—Iglesias/Valls
❺⓪ NEWS

11:00 **❷ NEWS—Smith/Marash**
❸ NEWS
❹ NEWS—Chuck Scarborough
❺ BEST OF GROUCHO (BW)

Wednesday

EVENING

sylvania at Franklin Field, Philadelphia. (2 hrs., 30 min.)
25 AMERICA—Documentary
31 CONSULTATION
Cosmetics and the enforcement of ingredient-listing are discussed.
41 ANA DEL AIRE
47 PEREGRINA ⓢ
48 RAGTIME
50 CANDIDATES—Report
57 MOVIE—Western ⓑⓦ
"The Man from Utah." (1934) John Wayne, Polly Ann Young. (60 min.)

9:30 **13** TIM WEISBERG—JAZZ ROCK
In concert: jazz-rock flutist Tim Weisberg and his group—organist Lynn Blessing, drummer Marty Foltz, guitarist Todd Robinson and bassist Doug Anderson. Selections include "Scrabble," "Tibetan Silver" and "Because of Rain," all written and arranged by Weisberg and Blessing.
25 EYE TO EYE—Art
31 FACING THE ISSUES
50 MOMENTS OF GLORY

10:00 **2** **3** MANHUNTER—Crime Drama
Danger awaits Dave when he brings his injured father to a small-town hospital. The place is invaded by fugitive bank robbers who want the local doctor to save their wounded cohort. Barrett: Ken Howard. Lizabeth: Hilary Thompson. (60 min.)

       Guest Cast
Bert Heideman ....Harry Guardino
Fred Lummit ....William Watson
Sonny Welch ....Bo Hopkins
Dr. Hurley ....William Schallert
Paul Tate ....Robert Hogan
Sheriff Grant ....Kelly Thordsen
4 **20** PETROCELLI—Drama
Petrocelli faces an agonizing decision when a murder suspect asks for his help. The man is accused of killing a consumer advocate who was one of the lawyer's closest boyhood friends. Petrocelli: Barry Newman. Maggie: Susan Howard. Pete Ritter: Albert Salmi. (60 min.)

       Guest Cast
Adam North ....William Shatner
Eleanor Warren .....Susan Oliver
Audrey North ....Lynn Carlin
Daley ....Dana Elcar
Dan Carter ....Morgan Paull
Robert Warren ......Glenn Corbett
Brannigan ....Harrison Ford

OCTOBER 2, 1974

5 NEWS—Bill Jorgensen
7 **8** GET CHRISTIE LOVE!—Crime Drama
In her investigation of a wino's murder, Christie's worst problem is her partner. She is working with a gruff homicide sergeant who doesn't think women belong on a police force. Christie: Teresa Graves. Reardon: Charles Cioffi. (60 min.)

       Guest Cast
Gus Marker ....Scott Brady
Lester Wheeler ....Quinn Redeker
Charlie Red ....Ken Tobey
Alex Dawson ....Dick O'Neill
Nick Varga ....Sid Haig
Mrs. Averdon ....Fritzie Burr
11 NEWS—Joe Harper
5 FESTIVAL FILMS
A comic look at the trials and tribulations of putting on an off-off-Broadway show is among this week's student films. Also on the program: "Greater Expectations," a satire.
11 WOMAN—Discussion
Return: This series designed "to go to the root of women's concerns" begins its second season with the first of two programs on women's sexual needs and capacities. Producer-moderator Sandra Elkin's guests include author Barbara Seaman.
41 HA LLEGADO UNA INTRUSA
47 CELOS ⓢ
50 CANDIDATES—Report
57 NEWS—Gary Gunter

10:30 **9** MOVIE—Drama ⓑⓦ
"I've Lived Before." (1956) An airline pilot (Jock Mahoney) is led to believe he is the reincarnation of a dead man. Ann Harding. Lois: Leigh Snowden. Hackett: Raymond Bailey. Bryant: John McIntire. Moderately interesting try at a baffling subject. (90 min.)
13 VIDEO VISIONARIES
"Some day, artists will work with capacitors, resistors and conductors as they work today with brushes, violins and more conventional tools," predicts video artist Nam June Paik. He demonstrates his own electronic versatility in an unusual symphony.
11 AMERICA—Documentary
In a profile of 17th-century colonial America, Alistair Cooke focuses on English-speaking settlements in the

Saturday

EVENING

resolution. Archie is developing a terrible complex about being the only one in the family who isn't earning any money. Conclusion of a four-part episode. Archie: Carroll O'Connor. Edith: Jean Stapleton. Guest Cast . . . Stretch: James Cromwell.

4 20 EMERGENCY!
Paramedics rescue a woman who has fallen into a lion's cage, and save a bleeding policeman from a sniper. Gage: Randolph Mantooth. DeSoto: Kevin Tighe. Brackett: Robert Fuller. Dixie: Julie London. Stanley: Michael Norell. Early: Bobby Troup. (60 min.)

Guest Cast

Caldwell	Robert Q. Lewis
Suzy	Lindsay Bloom
Mrs. Caldwell	Pamela Morris
Felix	John Wheeler

7 8 NEW LAND—Drama
A cholera epidemic threatens the community when an entire family is stricken with the contagious disease. While Anna and Molly risk infection to nurse the victims, Bo sets out on a two-day journey to the nearest doctor. Anna: Bonnie Bedelia. Bo: Kurt Russell. Christian: Scott Thomas. Molly: Gwen Arner. Guest Cast . . . Linka: Salome Jens. Arne: Don Dubbins. Gunnar: Lin McCarthy. Dr. Monroe: Robert Emhardt. Mrs. Monroe: Maxine Stuart. (60 min.)

9 MOVIE—Drama
"The Journey." (1959) Engrossing tale of people fleeing Budapest during the 1956 Hungarian revolt, with Yul Brynner as the Soviet officer who detains them. Deborah Kerr, Jason Robards Jr., Robert Morley. Rhinelander: E.G. Marshall. Eva: Anouk Aimee. Margie: Anne Jackson. Fine production, excellent acting. (2 hrs., 30 min.)

11 HEE HAW
Guests: Susan Raye ("Stop the World and Let Me Off"), and Danny Davis and the Nashville Brass ("Kaw-Liga," "St. Louis Blues"). Additional songs include "Amazing Love," "Chattanooga Shoeshine Boy," "Dear God," "Sleepy-Eyed John." (60 min.)

21 WALL STREET WEEK—Rukeyser
Investigating securities frauds is among the duties of Stanley Sporkin, director of enforcement for the SEC. Sporkin describes his work.

OCTOBER 5, 1974

31 WOMAN—Discussion
Return: This series designed "to go to the root of women's concerns" begins its second season with the first of two programs on women's sexual needs and capacities. Producer-moderator Sandra Elkin's guests include author Barbara Seaman.

47 MOVIE—Musical
"Esa Picara Pelirroja." (1972) In Spanish. Ethel Rojo, Ismael Merlo. (2 hrs.)

49 FAMILY THEATRE

50 INTERNATIONAL PERFORMANCE

8:30 2 3 FRIENDS AND LOVERS —Comedy
Robert and his violinist friend Fred are kept on their toes by Fred's visiting father, an affable violin virtuoso with a pronounced superiority complex. Robert: Paul Sand. Fred: Steve Landesberg. Guest Cast . . . Karl Meyerbach: Leon Askin. Thompson: Byron Webster. Sharon: Teri Garr. Engineer: Richard Reicheg.

5 MOVIE—Science Fiction (BW)
"Attack of the Crab Monsters." (1957) Mutants with brains provide the novelty in this low-budget Roger Corman thriller. Richard Garland, Pamela Duncan. Hank: Russell Johnson. Dr. Weigand: Leslie Bradley. Dr. Carson: Richard Cutting. Quinlan: Ed Nelson. Deveroux: Mel Welles. Ron: Beech Dickerson. (90 min.)

21 PARTIES AND THE ISSUES '74
How the U.S. will figure in world relationships is the topic discussed in Part 1 of a four-part program. Guests participating in the debate include Sens. Lloyd Bentsen (D-Texas) and William Brock (R-Tenn.). (60 min.)

31 JOURNEY TO JAPAN—Travel
The art of forming designs with gold powder on laquerware objects.

67 MOVIE—Drama
"Flame of the Islands." (1955) Yvonne De Carlo, Howard Duff. (2 hrs.)

9:00 2 3 MARY TYLER MOORE
Lou's new girl friend (Sheree North) is a nightclub singer who has fascinating stories to tell about her former marriages—stories which start to bother Lou after Ted and Murray rib him about falling for a woman with a "past." Lou: Edward Asner. Murray: Gavin MacLeod. Ted: Ted Knight. Sue

ABOUT THE EDITORS

STAN GOLDSTEIN is an award-winning advertising copywriter and designer residing in New York City. He grew up with the television constantly on, so this unique idea is a natural first venture for him in the world of books. He continues to explore the subject of TV, while actively working in two other areas which are of special interest to him, automobiles and science fiction.

FRED GOLDSTEIN was born, was schooled, and resides in New York City. At an early age he became a TV buff. His background is the advertising industry and this is his introduction into the book world. Future projects include both creating and producing books on various subject matters.

Bantam Book Catalog

Here's your up-to-the-minute listing of every book currently available from Bantam.

This easy-to-use catalog is divided into categories and contains over 1400 titles by your favorite authors.

So don't delay—take advantage of this special opportunity to increase your reading pleasure.

Just send us your name and address and 25¢ (to help defray postage and handling costs).